In the Nature of Cities

In the Nature of Cities engages with the long overdue task of re-inserting questions of nature and ecology into the urban debate. This path-breaking collection charts the terrain of urban political ecology, and untangles the economic, political, social and ecological processes that form contemporary urban landscapes.

Written by key political ecology scholars, the essays in this book attest that the re-entry of the ecological agenda into urban theory is vital, both in terms of understanding contemporary urbanization processes, and of engaging in a meaningful environmental politics.

The question of whose nature is, or becomes, urbanized, and the uneven power relations through which this socio-metabolic transformation takes place, are the central themes debated in this book. Foregrounding the socio-ecological activism that contests the dominant forms of urbanizing nature, the contributors endeavour to open up a research agenda and a political platform that sets pointers for democratizing the politics through which nature becomes urbanized and contemporary cities are produced as both enabling and disempowering dwelling spaces for humans and non-humans alike.

Nik Heynen is Assistant Professor in the Department of Geography at the University of Wisconsin-Milwaukee.

Maria Kaika is Lecturer in Urban Geography at the University of Oxford, School of Geography and the Environment, and Fellow of St. Edmund Hall, Oxford.

Erik Swyngedouw is Professor at the University of Oxford, School of Geography and the Environment, and Fellow of St. Peter's College, Oxford.

Questioning Cities

Edited by Gary Bridge, *University of Bristol, UK* and
Sophie Watson, *The Open University, UK*

The Questioning Cities series brings together an unusual mix of urban scholars under the title. Rather than taking a broadly economic approach, planning approach or more socio-cultural approach, it aims to include titles from a multi-disciplinary field of those interested in critical urban analysis. The series thus includes authors who draw on contemporary social, urban and critical theory to explore different aspects of the city. It is not therefore a series made up of books which are largely case studies of different cities and predominantly descriptive. It seeks instead to extend current debates, in most cases through excellent empirical work, and to develop sophisticated understandings of the city from a number of disciplines including geography, sociology, politics, planning, cultural studies, philosophy and literature. The series also aims to be thoroughly international where possible, to be innovative, to surprise, and to challenge received wisdom in urban studies. Overall it will encourage a multi-disciplinary and international dialogue always bearing in mind that simple description or empirical observation which is not located within a broader theoretical framework would not – for this series at least – be enough.

Published:

Global Metropolitan
John Rennie Short

Reason in the City of Difference
Gary Bridge

In the Nature of Cities
Urban political ecology and the politics of urban metabolism
Edited by Nik Heynen, Maria Kaika and Erik Swyngedouw

Ordinary Cities
Between modernity and development
Jenny Robinson

Forthcoming titles:

Cities and Race
America's new black ghettos
David Wilson

City Publics
The (dis)enchantment of urban encounters
Sophie Wilson

Small Cities
David Bell and Mark Jayne

Urban Space and Cityscapes
Christoph Lindner

In the Nature of Cities

Urban political ecology and
the politics of urban metabolism

Edited by Nik Heynen, Maria Kaika and Erik Swyngedouw

Routledge
Taylor & Francis Group

LONDON AND NEW YORK

First published 2006
by Routledge
2 Park Square, Milton Park, Abingdon, Oxon, OX14 4RN

Simultaneously published in the USA and Canada
by Routledge
270 Madison Ave, New York, NY 10016

Routledge is an imprint of the Taylor & Francis Group

© 2006 Nik Heynen, Maria Kaika and Eric Swyngedouw for selection
and editorial matter; the contributors for individual chapters

Typeset in Times New Roman by
Keystroke, Jacaranda Lodge, Wolverhampton

British Library Cataloguing in Publication Data
A catalogue record for this book is available from the British Library

Library of Congress Cataloguing in Publication Data
In the nature of cities : urban political ecology and the politics of urban
metabolism / edited by Nikolas C. Heynen, Maria Kaika and Erik Swyngedouw.
p. cm. – (Questioning cities series)
Includes bibliographical references and index.
1. Urban ecology. 2. Political ecology. I. Heynen, Nikolas C.
II. Kaika, Maria. III. Swyngedouw, E. (Erik) IV. Series.
HT241.U727 2005
307.76–dc22
2005012463

ISBN10: 0–415–36827–8 (hbk)
ISBN10: 0–415–36828–6 (pbk)

ISBN13: 9–78–0–415–36827–8 (hbk)
ISBN13: 9–78–0–415–36828–5 (pbk)

Contents

Illustrations

FIGURES

TABLES

Contributors

Julie-Anne Boudreau is Assistant Professor in the Department of Political Science at York University. Her research focuses on urban social movements and state restructuring.

Alec Brownlow is Assistant Professor and Acting Director of the Environmental Studies Program, Department of Geography and Urban Studies, Temple University. His research broadly explores issues of and intersections between urban ecology, public space, human-environment, and social theory.

Eliza Darling recently completed a Ph.D. in Anthropology from the City University of New York's Graduate School and University Center. She writes about political ecology, the production of nature, rural gentrification, urban social theory, and children's literature.

Matthew Gandy is Reader in the Department of Geography at University College London. His research interests are primarily concerned with cultural and environmental dimensions of urban landscape.

Stephen Graham is a Professor in the Department of Geography at the University of Durham. His research develops critical and "socio-technical" perspectives to the reconfiguration of cities, technologies, mobility systems and the relations between cities, war and terrorism. He is the co-author of *Telecommunications and the City, Splintering Urbanism* (both with Simon Marvin), co-editor of *Managing Cities*, and editor of the *Cybercities Reader* and *Cities, War and Terrorism*.

Nik Heynen is an Assistant Professor in the Department of Geography at the University of Wisconsin-Milwaukee. His research uses critical social theory to consider the production of uneven urban environments, with specific attention toward urban hunger and urban forestry.

Maria Kaika is a Lecturer in Urban Geography, University of Oxford, School of Geography and the Environment, and Fellow of St Edmund Hall. Her research interests lie with modernist urbanism and nature; representations of nature and the city in the modernist movement; governance and environmental policy; the political ecology of water supply in western cities; European water policy. She is author of *City of Flows: nature, modernity and the city* (Routledge 2005).

Roger Keil is Associate Professor in the Faculty of Environmental Studies and the Department of Political Science at York University, Toronto, Canada. His research interests lie with urban ecological politics and the politics in world cities. He is a founding member of the International Network of Urban Research and Action and has been involved in several labor and community groups. He is author of *Los Angeles: globalization, urbanization, and social struggles* (1998) and *Nature and the City: making environmental policy in Toronto and Los Angeles* (with G. Desfor 2004).

Alex Loftus is Academic Fellow of Sustainability in the Department of Geography, Royal Holloway, University of London. His research looks at the political ecology of water struggles from a historical geographical materialist perspective.

Simon Marvin is Professor and Director of the Centre for Sustainable Urban and Regional Futures, University of Salford. His research explores the changing relations between socio-technical networks and the development of cities and regions.

Will Medd is a Researcher in the Department of Sociology and Centre for Sustainable Water Management, Lancaster University. While currently developing work on sustainable water management, Will has published on complexity science, partnership working, social policy, methodology, and mobilities research.

Stuart Oliver is Senior Lecturer in Geography at St Mary's College, University of Surrey. His main research interests are in the imposition of regulation on rivers and wetlands.

David N. Pellow is Associate Professor of Ethnic Studies at the University of California, San Diego. His research and teaching focus on race and racism, corporate power, and social movements for environmental justice.

Paul Robbins is Associate Professor of Geography at the University of Arizona. His work focuses on the relationship between politics, environmental knowledge, and ecological change, including research on forests in India, elk in Montana, and lawn grasses across North America.

Greg Ruiters is a Senior Lecturer in Political and International studies at Rhodes University, Grahamstown. He did his Ph.D. in Geography at Johns Hopkins University. He has taught and written on water privatization, local government in South Africa, social movements and comparative politics. He is co-editor of "The age of commodity – Water privatization in Southern Africa", Earthscan, London.

Julie Sharp is a Ph.D. candidate in the Department of Geography at the Ohio State University. Her research interests include the gendered politics of green consumption and the differences in US and Canadian cultures of science surrounding the Great Lakes.

Laila Smith is the Director of Research and Evaluation at the Contract Management Unit for the City of Johannesburg. She is involved in providing service delivery oversight for Johannesburg Water, a public utility that provides water and sanitation to 3.5 million people. She is also involved in shaping the evolving regulatory frameworks for service delivery for the City of Johannesburg. She has researched the implications of public–private partnerships and their institutional effects in several localities across South Africa over the past five years.

Neil Smith is Distinguished Professor of Anthropology and Geography and Director of the Center for Place, Culture, and Politics at the Graduate Center, City University of New York. His research explores the broad intersection between space, nature, social theory and history.

Erik Swyngedouw is a Professor in the School of Geography and the Environment at Oxford University and Fellow of St Peter's College. His research interests cover a variety of topics although a central concern of his work is the integration of space and nature into critical social theory. His recent books include *Social Power and the Urbanization of Water: flows of power* (2004) and *Globalising Urbanization* (co-edited with F. Moulaert and A. Rodriguez, 2003).

Foreword

Neil Smith

When we eventually look back at the intellectual shibboleths of the high capitalist period – say the last three centuries – few ingrained assumptions will look so wrongheaded or so globally destructive as the common-sense separation of society and nature. Historically and geographically, most societies have avoided such a stark presumption as hubristic folly, but from physicists to sociologists, physicians to poets, the brains of the ascendant capitalist "west" not only embraced but made a virtue of society's separation from nature (and vice versa). Scientists studied a natural world, conceptually ripped from any social context except that human bodies, like "natural" ones, were just as subject to the laws of gravity or bio-chemistry; social scientists sought laws of society that didn't defy laws of nature but by exclusion assumed such laws meaningless to social questions; the best poets and artists saw humanity and nature mirrored in each other – separate enough to require a creative inferno of reconnection. In time the destructiveness of this deep-seated presumption of society separated from nature will become fully and tragically apparent; quite when this will occur depends very much on the fate of the capitalist globalism that not only fostered such a grotesque fiction but raised it almost to the level of species instinct and profited so extraordinarily from it.

Capitalism is the appropriate focus here not because capitalist societies are unique in positing such a separation. Quite the contrary. Many past and present societies recognize a disjuncture of nature and society, as in Cicero's belief more than two thousand years ago that cities really represent a "second nature", but Cicero was the perceptive exception to a general rule of early Greek thought that saw nature, society and the spiritual world as an irrevocable amalgam. His "second nature" was triumphalist in part, recognizing an increasing social power to separate itself from nature, but any celebration was circumscribed by a parallel sense that the social world was itself part of (a second) nature. In the crude materialist evolutionism of the nineteenth century, "man" continuously fought nature and separated himself from it as a means of controlling the "external" world. If that progressive vision is much too crude and teleological, there is nonetheless a sense in which the expansion of capitalist technology has accomplished a very real if limited separation of society from various natural constraints and in ways more aggressive and complete than any previous social formation. So when Marx, to some extent Hegel, and certainly the Frankfurt School of social theorists pick up Cicero's language of a second nature, the triumphalism is muted in favour of an angst about the fate of nature. "Capitalism", warned Frankfurt School theorists Adorno and Horkheimer on the eve of the atomic bomb, "is a massive racket in nature".

In observing such a discrete separation of society from nature, capitalist societies are the oddity. They have historically depended on a rampant objectification of nature, centred on the abstraction and globalization of wage labour in multifold forms. Capitalist societies externalize nature to an unprecedented extent (even as they internalize it in the commodity form). The intellectual conundrum in the Enlightenment was not to explain the entwinement of nature and society, therefore, but the opposite, namely to explain their supposed separation. If nature was to be ground into commodity form, its supposed externality was what had to be explained; it is much easier to rationalize the profit-driven rape of earth and body alike if that nature is objectified. The exploitation of nature did not begin with capitalism but it did become deepened, generalized and dramatically intensified during this period in human history. Enlightenment thinkers from Newton to Kant, Adam Smith to Montesquieu – and many others – answered the demand to understand the externality of nature. This is neither to diminish their contributions necessarily, nor to cast aspersions on their accomplishments, but it is to contextualize the ideological power that their contributions came to have. In many ways, the stunning question today – still almost unaskable – is not how to reconcile nature and society, how to understand the "interaction" of nature and society – the agenda of most in the environmental movement – but how western ideologies could have got to the point of flattering themselves so successfully that they were somehow separate from nature. Such a widespread "discourse" of separateness is thoroughly embedded in the capitalist universalization of wage labour; under wage labour not only is the raw material externalized but so too is the human potential to work. The environmental movement is not immune to this but generally complicit: the "interaction" of society and nature only makes sense if society and nature are conceived from the start as separate.

Such a powerful shibboleth inevitably conjures forth its opposite, and for all the externalization of nature binding social production to western ideologies, there is a parallel historical narrative emphasizing the oneness or universality of "man and nature". The contemporary environmental movement, especially in its ecological guise, may best embody this response. This of course poses an acute contradiction – nature is external to society yet united with it – and it is a contradiction that is highly generalized in contemporary western thought. The physicist may be a poet while the artist surely believes in the objectivity of gravity. The environmental movement actually embodies this contradiction too, variously worshipping or wanting a universal nature yet seeking to manage it as an object. American presidents now routinely claim to be environmentalists even while they open the Alaskan wilderness to "environmentally friendly" oil conglomerates. Little wonder that even environmentalist insiders now admit what socialists, radicals and anarchists have long concluded, namely that the mainstream environmental movement is dead, co-opted by the very capitalist power it once tried to fight, reincarnated as little more than green capitalism.

The left response has been varied. Some have reverted to the NGOs. Others – myself included – have stuck with academia where our voice is relatively untrussed but equally unheard. Still others have bravely struggled in environmental justice organizations, many others in unions; many of us struggle on several political fronts. As the mainstream environmental movement has been tamed, more radical voices have often resuscitated a certain apocalypticism. Peak oil! Societal collapse! Global warming! Population explosion! Natural disaster! Tsunami! Ecology of fear! War and famine! We are killing nature, is the message, and soon nature is going to exact its revenge. Without diminishing the various threats involved, I think an apocalyptic response plays into the ideological separation between nature and society that an oppositional politics needs to challenge. It also severely underestimates the ability of capitalist societies to adjust to problems of their own making, and make a profit along the way. This is not to say that capitalism is infinitely self-correcting but that it does have an extraordinary adaptability even in the face of crises of its own making. Capitalism can be quite nature-friendly in its own interests and on its own terms – witness the multi-billion dollar "natural foods" or recycling industries. We are all environmentalists now – who is not for "sustainability"? – and we will underestimate the adaptability of capitalism at our peril.

The genius of the present volume is precisely this recognition. The idea of a metabolism with nature has roots in nineteenth- and twentieth-century social thought, especially Marx. It simultaneously problematizes the relationship with nature and refuses the knee-jerk apocalypticism that marks so much left environmental response today. For Marx, it is not a question of understanding how nature and society interact; rather the point is that nature is incomprehensible except as mediated by social labour, and consequently there has to be a rethinking that posits labour as central to nature. This flies directly in the face of either an external notion of nature, which excludes labour, or a universalist notion which broadly refuses labour as the nexus of society and nature.

The notion of metabolism sets up the circulation of matter, value and representations as the vortex of social nature. But, as the original German term, "*Stoffwechsel*", better suggests, this is not simply a repetitive process of circulation through already established pathways. Habitual circulation there certainly is, but no sense of long-term or even necessarily short-term equilibrium. Rather "*Stoffwechsel*" expresses a sense of creativity in much the same way that Benjamin talks about mimesis: the metabolism of nature is always already a production of nature in which neither society nor nature can be stabilized with the fixity implied by their ideological separation. Society is forged in the crucible of nature's metabolism, for sure, but nature is equally the amalgam of simmering social change. So much is human production, "this unceasing sensuous labour and creation . . . the basis of the whole sensual world as it now exists", Marx once wrote in a polemic against Feuerbach, "that were it interrupted for only a year" not only would there be "an enormous change in the natural world", but Feuerbach would "soon find that the whole world of men and his own perceptive faculty, nay his own existence, were missing".

Sundered apart, nature and society die in reciprocal conceptual torpor. The ideology of separate and distinct social and natural spheres therefore begs the question: for what purpose? What social work does this dualism do? There are many layers to an answer, but most simply, the positing of an external nature rationalizes and justifies the unprecedented exploitation of nature (human *cum* non-human), the "massive racket" that capitalism, historically and geographically, represents.

Political ecology provides a powerful means of cracking the abstractions of this discussion about the metabolism or production of nature. Rooted in social and political theory it is also grounded in ecology and has an international scope. When complemented by an environmental justice politics, which is less internationally focused and less theoretical but more politically activist in inspiration, political ecology becomes a potent weapon for comprehending produced natures.

While all of these ideas have come together powerfully in the last two decades or more, they are only now being applied explicitly to urban nature. If they are going to have theoretical and political traction, however, these ideas need to be tested in the most produced nature of all, and that means the city. The production of urban nature is deeply political but it has also received far less scrutiny and seems far less visible, precisely because the arrangement of asphalt and concrete, water mains and garbage dumps, cars and subways seems so inimical to our intuitive sense of (external) nature. Whatever our analytical sophistication, the idea of nature as a contrivance still cuts deeply and sharply against our most engrained and peculiar prejudices. *In the Nature of Cities* helps open up this new territory and it will help to create a new structure of feeling connecting nature and the city. Radically new ideas are by definition discomfiting: only later do they seem "natural".

This book should therefore be read as a search. It is a search for ways to articulate a creative politics of nature in, of and for the city. The emphasis on urban metabolism represents a sober recognition of the power of capitalist productions of nature while winkling them open at points of opportunity for radical, even revolutionary, change. It broadly rejects the apocalyptic "death" of nature in recognition of the fact that, however perversely, societies make the natural environment they live in, to a lesser or greater degree, although not of course under conditions of their own choosing. It is therefore a search too for political possibilities. Nature is manifestly not dead but is incessantly reproduced – in ways we may detest or we may love. Nor is environmentalism dead, despite the belated recognition by some among the mainstream environmental movement that they have motored themselves to the end wall of their own political cul-de-sac.

A dialectic is at work here. As part of a broader political movement, an urban political ecology can help integrate a politics of nature into a more established "social" politics; at the same time an ecologically enhanced politics focusing on the productive metabolism of nature can further exoticize the absurd separation of nature and society while denying any anti-social universalism of nature. In the bigger picture, seeing the world so differently – outside the prism of capitalist nature – requires both analytical and poetic exploration. But insofar as the landscapes we create refract back to us a very powerful naturalization of the social assumptions that sculpted such landscapes in the first place, a revolution in our thinking may be intimately bound up with a revolution in how these landscapes are made. Seeing the world differently probably depends on making a different world from which the world itself can be seen differently.

1 Urban political ecology

Politicizing the production of urban natures

Nik Heynen, Maria Kaika,
and Erik Swyngedouw

It is in practice, hard to see where "society" begins and "nature" ends . . .
[I]n a fundamental sense, there is in the final analysis nothing unnatural about New
York City.

(Harvey 1993: 31, 28)

Urbanization as a process has constituted the city and the countryside, society and
nature, a "unity of opposites" constructed from the integrated, lived world of human
social experience. At the same time, the "urbanization of consciousness" constitutes
Nature as well as Space.

(FitzSimmons 1989: 108)

The "city" as a form of life is a specific, historically developed model of the regulation
of the societal relationship with nature . . . .. [U]rban struggles are predominantly
socio-ecological struggles, since they are always about the social and material
regulation and socio-cultural symbolization of societal relationships with nature.

(Jahn 1991: 54 – translation Keil 1995)

INTRODUCTION

In the summer of 1998, the Southeast Asian financial bubble imploded. Global
capital moved spasmodically from place to place, leaving cities like Jakarta with a
social and physical wasteland where dozens of unfinished skyscrapers were dotted
over the landscape while thousands of unemployed children, women, and men were
roaming the streets in search of survival. In the meantime, El Niño's global dynamic
was wrecking havoc in the region with its climatic disturbances. Puddles of stagnant
water in the defunct concrete buildings that had once promised continuing capital
accumulation for Indonesia became great ecological niches for a rapid explosion
of mosquitoes. Malaria and Dengue Fever suddenly joined unemployment and
social and political mayhem in shaping Jakarta's cityscape. Global capital fused
with global climate, with local power struggles, and with socio-ecological
conditions to re-shape Jakarta's urban socio-ecological conditions in profound,
radical, and deeply troubling ways.

This example is just one among many to suggest how cities are dense networks
of interwoven socio-spatial processes that are simultaneously local and global,

human and physical, cultural and organic. The myriad transformations and metabolisms that support and maintain urban life, such as, for example, water, food, computers or hamburgers always combine infinitely connected physical *and* social processes (Latour 1993; Latour and Hermant 1998; Swyngedouw 1999).

The world is rapidly approaching a situation in which most people live in cities, often mega-cities. It is surprising, therefore, that in the burgeoning literature on environmental sustainability and environmental politics, the urban environment is often neglected or forgotten as attention is focused on "global" problems like climate change, deforestation, desertification, and the like. Similarly, much of the urban studies literature is symptomatically silent about the physical-environmental foundations on which the urbanization process rests. Even in the emerging literature on political ecology (see for example Walker 2005), little attention has been paid so far to the urban as a process of socio-ecological *change*, while discussions about global environmental problems and the possibilities for a "sustainable" future customarily ignore the urban origin of many of these problems. Similarly, the growing literature on the technical aspects of urban environments, geared primarily to planners and environmental policy makers, fails to acknowledge the intimate relationship between the antinomies of capitalist urbanization processes and socio-environmental injustices (Whitehead 2003). This book seeks to address this gap and to chart the contours of a critical academic and political project that foregrounds the urban condition as fundamentally a socio-environmental process.

We were faced with two major challenges while moving this intellectual project forward. First, there is a need to revisit the overtly "sociological" nature of much of twentieth-century urban theory. If we take David Harvey's dictum that "there is nothing unnatural about New York City" seriously, this impels interrogating the failure of twentieth-century urban social theory to take account of physical or ecological processes. While late-nineteenth-century urban perspectives were acutely sensitive to the ecological imperatives of urbanization, these considerations disappeared almost completely in the decades that followed (with the exception of a thoroughly "de-natured" Chicago school of urban social ecology). Re-naturing urban theory is, therefore, vital to urban analysis as well as to urban political activism. Second, most of environmental theory has unjustifiably largely ignored the urbanization process as both one of the driving forces behind many environmental issues and as the place where socio-environmental problems are experienced most acutely. The excavation of these processes also constitutes one of the central concerns of an evolving urban political ecology.

The central message that emerges from urban political ecology is a decidedly political one. To the extent that cities are produced through socio-ecological processes, attention has to be paid to the political processes through which particular socio-environmental urban conditions are made and remade. From a progressive or emancipatory position, then, urban political ecology asks questions about who produces what kind of socio-ecological configurations for whom. In other words, urban political ecology is about formulating political projects that are radically democratic in terms of the organization of the processes through which the environments that we (humans and non-humans) inhabit become produced.

As global/local forms of capitalism have become more entrenched in social life, there are still powerful tendencies to externalize nature. Yet, the intricate and ultimately vulnerable dependence of capital accumulation on nature deepens and widens continuously. It is on the terrain of the urban that this accelerating metabolic transformation of nature becomes most visible, both in its physical form and its socio-ecological consequences.

In this introductory chapter, we chart the contours of such an ambitious urban political ecological (UPE) perspective. Obviously, our perspective is filtered through our own critical theoretical lens and political sensitivities. In the first part, we explore how urbanization is very much a process of socio-metabolic transformations and insist that the re-entry of the ecological in urban theory is vital both in terms of understanding the urban and of engaging in a meaningful environmental politics. The second part suggests how critical theory, and in particular political economy, can and should be reformulated in a way that permits taking the environment politically seriously. The third part explores the implications of urban political ecology and frames the contributions that form the core of this collection. We consider the deeply uneven power relations through which contemporary "cyborg" cities become produced. Evidently, these uneven and often outright oppressive socio-ecological processes do not go uncontested. All manner of socio-ecological activism and movements have arisen that both contest the dominant forms of urbanizing nature and chart the contours for both transforming and democratizing the production of urban natures. In the final part of this introductory chapter, the structure of the book and the main lines of the contributions are briefly outlined.

THE CITY AS SOCIO-ECOLOGICAL PROCESS

Within the last couple of decades, theorization about human/environment relations has made substantial progress. In particular, a perspective that attempts to transcend the dualist nature/culture logic and the moral codes inscribed therein has replaced this crude binary ruling of city versus the environment. Critical to this progress has been the realization that the split between humanity and environment that first became prominent during the seventeenth century (Gold 1984) has long impeded understanding of environmental issues. Along these lines Swyngedouw (1999: 445) suggests that "[c]ontemporary scholars increasingly recognize that natural or ecological conditions and processes do not operate separately from social processes, and that the actually existing socionatural conditions are always the result of intricate transformations of pre-existing configurations that are themselves inherently natural *and* social". This had of course already been recognized by Marx more than 150 years ago, and only recently regained the attention it deserves, from Marxists and non-Marxists alike (Pulido 1996; Whatmore 2002; see Swyngedouw, this volume). While the notion that all kinds of environments are socially produced is not new, the idea still holds much promise for exploration, discussion and illustration. In his landmark book, Smith (1984: xiv) suggests:

What jars us so much about this idea of the production of nature is that it defies the conventional, sacrosanct separation of nature and society, and it does so with such abandon and without shame. We are used to conceiving of nature as external to society, pristine and pre-human, or else a grand universal in which human beings are but small and simple cogs. But . . . our concepts have not caught up with our reality. It is capitalism which ardently defies the inherited separation of nature and society, and with pride rather than shame.

Despite often being neglected by urban studies, "environmental" issues have always been central to urban change and urban politics. Throughout the nineteenth century, visionaries of all sorts lamented the "unsustainable" character of early modern cities and proposed solutions and plans that would remedy the socio-environmental dystopias that characterized much of urban life. Friedrich Engels (1987 [1845]) had already noted in the mid-nineteenth century how the depressing sanitary and ecological conditions of England's great cities are related to the class character of industrial urbanization. Much later, Raymond Williams pointed out in *The Country and the City* (Williams 1985 [1973]) that the transformation of nature and the social relations inscribed therein are inextricably connected to the process of urbanization. Indeed, the urbanization process is predicated upon a particular set of socio-spatial relations that produce "an ecological transformation, which requires the reproduction of those relations in order to sustain it" (Harvey 1996: 94). The production of the city through socio-environmental changes results in the continuous production of new urban "natures", of new urban social and physical environmental conditions (Cronon 1991). All of these processes occur in the realms of power in which social actors strive to defend and create their own environments in a context of class, ethnic, racialized and/or gender conflicts and power struggles (Davis 1996).

The relationship between cities and nature has long been a point of contention for both environmentally minded social theorists and socially minded environmental theorists. Urbanization has long been discussed as a process whereby one kind of environment, namely the "natural" environment, is traded in for, or rather taken over by, a much more crude and unsavoury "built" environment. Bookchin (1979: 26) makes this point by suggesting that "[t]he modern city represents a regressive encroachment of the synthetic on the natural, of the inorganic (concrete, metals, and glass) on the organic, or crude, elemental stimuli on variegated wide-ranging ones". The city is here posited as the antithesis of nature, the organic is pitted against the artificial, and, in the process, a normative ideal is inscribed in the moral order of nature.

Although many view the notion of urban environmental landscapes as an oxymoron, Jacobs (1992 [1961]: 443) long ago already suggested that urban environments "are as natural as colonies of prairie dogs or the beds of oysters". David Harvey substantiates his claim that there is nothing intrinsically *unnatural* about New York City by suggesting that human activity cannot be viewed as external to ecosystem function (Harvey 1996: 186). "It is inconsistent", Harvey (1996: 187) continues, "to hold that everything in the world relates to everything else, as ecologists tend to, and then decide that the built environment and the urban structures that go into it are somehow outside of both theoretical and practical consideration. The effect has been to evade integrating understandings of the urbanizing process into environmental-ecological analysis." The conclusion then that there is nothing unnatural about produced environments like cities, dammed rivers, or irrigated fields comes out of the realization that produced environments are specific historical results of socio-environmental processes. This scenario can be summed up by simply stating that cities are built out of natural resources, through socially mediated natural processes.

Lefebvre's take on the notion of "second nature" provides an often-neglected platform from which to discuss the social production of urban environments. Regarding the social production of urban environments, Lefebvre (1976: 15) suggests:

> Nature, destroyed as such, has already had to be reconstructed at another level, the level of "second nature" i.e. the town and the urban. The town, anti-nature or non-nature and yet second nature, heralds the future world, the world of the generalized urban. Nature, as the sum of particularities which are external to each other and dispersed in space, dies. It gives way to produced space, to the urban. The urban, defined as assemblies and encounters, is therefore the simultaneity (or centrality) of all that exists socially.

While perhaps relinquishing some of the inherent "natural" qualities of cities, e.g. water, vegetation, air etc., Lefebvre's explanation of second nature defines urban environments as necessarily socially produced and thus paves the way to understand the complex mix of political, economic and social processes that shape and reshape urban landscapes. In addition, for Lefebvre (as well as for Harvey or Merrifield (2002)), the urban constitutes the pivotal embodiment of capitalist or "modern" social relations, and, by implication, of the wider (and often global) socio-ecological relations through which modern life is produced, materially and culturally.

While landscape architects like Olmstead and Howard are often credited with "creating" urban natural landscapes, the metabolization of urban nature has a history as long as urbanization itself (Olmstead 1895). To this end, Gandy (2002: 2) suggests that "[n]ature has a social and cultural history that has enriched countless dimensions of the urban experience. The design, use, and meaning of urban space involve the transformation of nature into a new synthesis." Still, understanding the politicized and uneven nature of this urban synthesis should be the main task.

In capitalist cities, "nature" takes primarily the social form of commodities. Whether we consider a glass of water, an orange, or the steel and concrete embedded in buildings, they are all constituted through the social mobilization of metabolic processes under capitalist and market-driven social relations. This commodity relation veils and hides the multiple socio-ecological processes of domination/ subordination and exploitation/repression that feed the capitalist urbanization process and turn the city into a metabolic socio-environmental process that stretches from the immediate environment to the remotest corners of the globe. Indeed, the apparently self-evident commodification of nature that fundamentally underpins a market-based society not only obscures the social relations of power inscribed therein, but also permits imagining a disconnection of the perpetual flows of metabolized, transformed and commodified nature from its inevitable foundation, i.e. the transformation of nature (Katz 1998). In sum, the environment of the city (both social and physical) is the result of a historical-geographical process of the urbanization of nature (Swyngedouw and Kaika 2000).

THE PRODUCTION OF URBAN NATURES

The importance of the social and material production of urban nature has recently emerged as an area of importance within historical-geographical materialist and radical geography (Benton 1996; Braun and Castree 1998; Castree 1995; Castree and Braun 2001; Gandy 2002; Grundman 1991; Harvey 1996; Hughes 2000; Keil and Graham 1998; Smith 1984; Swyngedouw 1996; 2004a,b; Desfor and Keil 2004). This presents an important departure away from the agrarian focus of much environmental history (see Worster 1993). While there is an important body of literature that focuses on urban environmental history (see Tarr 1996; Hurley 1997; Melosi 2000), urban political ecology more explicitly recognizes that the material conditions that comprise urban environments are controlled, manipulated and serve the interests of the elite at the expense of marginalized populations (Swyngedouw 2004a). These conditions, in turn, are not independent from social, political and economic processes and from cultural constructions of what constitutes the "urban" or the "natural" (Kaika and Swyngedouw 1999; Kaika 2005).

The interrelated web of socio-ecological relations that bring about highly uneven urban environments as well as shaping processes of uneven geographical development at other geographical scales have become pivotal terrains around which political action crystallizes and social mobilizations take place. The excavation of these processes requires urgent theoretical attention. Such a project requires great sensitivity to, and an understanding of, physical and bio-chemical processes. In fact, it is exactly those "natural" metabolisms and transformations that become discursively, politically and economically mobilized and socially appropriated to produce environments that embody and reflect positions of social power. Put simply, gravity or photosynthesis is not socially produced. However, their powers are socially mobilized in particular bio-chemical and physical metabolic arrangements to serve particular purposes; and the latter are invariably associated with strategies of achieving or maintaining particular positionalities and express shifting geometries and networks of social power. This social mobilization of metabolic processes, of course, produces distinct socio-environmental assemblages. This book addresses exactly this mobilization and transformation of nature and the allied process of producing new socio-environmental conditions. Roger Keil (2003: 724) has recently summarized urban political ecology (UPE) as follows:

> [T]he UPE literature is characterized by its intensely critical predisposition; critical is defined here as the linking of specific analysis of urban environmental

problems to larger socio-ecological solutions. This necessitates, as a minimum, some modicum of indebtedness to radical and critical social theory. It is no coincidence then, that the emerging field of UPE has many of its multiple roots in the intellectual traditions of fundamental social critique: eco-Marxism, eco-feminism, eco-anarchism, etc. It is also, however, indebted to a neo-pluralist and radical democratic politics that includes the liberation of the societal relationships with nature in the general project of the liberation of humanity.

Nature and humans are simultaneously social and historical, material and cultural (Smith 1996; 1998a; Castree 1995; Haraway 1997). While an understanding of the changes that have occurred within urban environments lies at the heart of political-ecology research, they must be understood within the context of the economic, political and social relations that have led to urban environmental change. It is therefore necessary to focus on the political economic processes that bring about injustice and not only on the natural artefacts that are produced through these uneven social processes (Swyngedouw and Kaika 2000). The material production of environments is necessarily impregnated with the mobilization of particular discourses and understandings (if not ideologies) of and about nature and the environment.

The social appropriation and transformation of nature produces historically specific social and physical natures that are infused by social power relationships (Swyngedouw 1996). Things like commodities, cities, or bodies, are socio-metabolic processes that are productive, that generate the thing in and through the process that brings it into being. Social beings necessarily produce natures as the outcome of socio-physical processes that are themselves constituted through myriad relations of political power and express a variety of cultural meanings (Haraway 1991; 1997). In addition, the transformation of nature is embedded in a series of social, political, cultural, and economic social relations that are tied together in a nested articulation of significant, but intrinsically unstable, geographical configurations like spatial networks and geographical scales. Indeed, urban socio-ecological conditions are intimately related to the socio-ecological processes that operate over a much larger, often global, space.

Engels (1940: 45) spoke to the complexities inherent to these socio-ecological relations when he suggested that "[w]hen we consider and reflect upon nature at large . . . at first we see the picture of an endless entanglement of relations and reactions, permutations and combinations, in which nothing remains what, where, and as it was, but everything moves, changes, comes into being and passes away". The notion of "metabolism" is the central metaphor for Marx's approach to analyzing the dynamic internal relationships between humans and nature that produces socio-natural entanglements and imbroglios referred to by Engels. In the most general sense, "labouring" is seen exactly as the specifically human form through which the metabolic process is mobilized and organized (see Swyngedouw, this volume). This socio-natural metabolism is for Marx the foundation of, and possibility for, history, a socio-environmental history through which both the nature of humans and of non-humans is transformed. To the extent that labour constitutes the universal premise for human metabolic interaction with nature, the particular social relations through which this metabolism of nature is enacted shape the form this metabolic relation takes. Clearly, any materialist approach insists that "nature" is an integral part of the "metabolism" of social life. Social relations operate in and through metabolizing the "natural" environment and transform both society and nature.

Under capitalist social relations, then, the metabolic production of use values operates in and through specific social relations of control, ownership, and appropriation, and in the context of the mobilization of both (sometimes already metabolized) nature and labour to produce commodities (as forms of metabolized socio-natures) with an eye towards the realization of the embodied exchange value. The circulation of capital as value in motion, then, is the combined metabolic transformations of socio-natures in and through the circulation of money as capital under social relations that combine the mobilization of capital and labour power. New socio-natural forms are continuously produced as moments and things in this metabolic process (see Grundman 1991; Benton 1996; Burkett 1999; Foster 2000). While nature provides the foundation, the dynamics of social relations produce nature's and society's history. Whether we consider the production of dams, the making of an urban park, the re-engineering of rivers, the transfiguration of DNA codes, the making of transgenic cyborg species like Dolly the cloned sheep, or the construction of a skyscraper, they all testify to the particular social relations through which socio-natural metabolisms are organized. Socio-ecological "metabolism" will therefore be one of the central material and metaphorical tropes that will guide the case-studies and other analyses presented in this book.

Political ecology, then, "combines the concerns of ecology and a broadly defined political economy. Together this encompasses the constantly shifting dialectic between society and land-based resources, and also within classes and groups within society itself" (Blaikie and Brookfield 1987: 17). Furthermore, Schmink and Wood (1987: 39) propose that political ecology should be used to explain "how economic and political processes determine the way natural resources have been exploited". While these broad definitions lay a sound foundation from which to begin to understand urban political ecology, these concepts are in need of further elaboration and expansion (see Forsyth 2003). The processes of urbanization, while implicit in much geographical research, often tend to simply play the role of backdrop for other spatial and social processes. While there has been work done that helps us consider the spatial distribution of limited urban environmental resources (Gandy 2002; Swyngedouw 2004a), there does not exist a framework through which to systematically approach issues of uneven urban socio-ecological change, related explicitly to the inherent spatial patterns the distribution of environmental amenities take under urban capitalism. Such a framework is an important step towards beginning to disentangle the interwoven knots of *social process, material metabolism*, and *spatial form* that go into the formation of contemporary urban socionatural landscapes (Swyngedouw and Heynen 2003). This book seeks to present urban political ecology as a theoretical platform for interrogating the complex, interrelated socio-ecological processes that occur within cities (see also Kaika 2005).

THE URBANIZATION OF NATURE, SOCIO-ENVIRONMENTAL JUSTICE AND UNEVEN GEOGRAPHICAL DEVELOPMENT

In line with seeking out a synthetic understanding of urban environments, we must point out that the social forms of urban change have been of primary interest within urban geographic research (Gober *et al.* 1991). This work, however, neglects the fact that the processes of uneven deterioration that accompany urban socio-economic restructuring also contribute to changes in the ecological forms of urban areas more broadly. While environmental (both social and physical) qualities may be enhanced in some places and for some people, they often lead to a deterioration of social and physical conditions and qualities elsewhere (Peet and Watts 1996; Keil and Graham 1998; Laituri and Kirby 1994), both within cities and between cities and other, often very distant places. A focus on the uneven geographical processes inherent to the production of urban environments serves as a catalyst for a better understanding of socio-ecological urbanization.

Issues of social justice have also explicitly entered ecological studies, most visibly through the rubric of the environmental justice movement (Wenz 1988; Bullard 1990; Szaz 1994; Dobson 1999). As a result of the political mobilization that has occurred around many environmental issues, the environmental justice literature has evolved through political praxis and focuses on the uneven distribution of both environmental benefits and damages to economically/politically marginalized people. Because it comes from praxis as opposed to theoretically driven academic research, it provides a distinctly different context through which to understand urban human/environment interactions (see Bullard and Chavis 1993; Di Chiro 1996). Because it is a *movement* rather than a research program *per se*, it must explicitly appeal to a broad coalition of either environmentally minded or social justice minded groups, thus promoting the widespread dissemination of the struggles endured. However, although much of the environmental justice literature is sensitive to the centrality of social, political and economic power relations in shaping process of uneven socio-ecological conditions (Wolch *et al.* 2002; MacDonald 2002), it often fails to grasp how these relationships are integral to the functioning of a capitalist political-economic system. More problematically, the environmental justice movement speaks fundamentally to a liberal and, hence, distributional perspective on justice in which justice is seen as Rawlsian fairness and associated with the allocation dynamics of environmental externalities. Marxist political ecology, in contrast, maintains that uneven socio-ecological conditions are produced through the particular capitalist forms of social organization of nature's metabolism.

Henri Lefebvre reminds us of what the urban really is, i.e. something akin to a vast and variegated whirlpool replete with all the ambivalence of a space full of opportunity, playfulness and liberating potential, while being entwined with spaces of oppression, exclusion and marginalization (Lefebvre 1991 [1974]). Cities seem to hold the promise of emancipation and freedom whilst skilfully mastering the whip of repression and domination (Merrifield and Swyngedouw 1997). Perpetual change and an ever-shifting mosaic of environmentally and socio-culturally distinct urban ecologies – varying from the manufactured and manicured landscaped gardens of gated communities and high-technology campuses to the ecological war-zones of depressed neighbourhoods with lead-painted walls and asbestos covered ceilings, waste dumps and pollutant-infested areas – still shape the choreography of a capitalist urbanization process. The environment of the city is deeply caught up in this dialectical process and environmental ideologies, practices and projects are part and parcel of this urbanization of nature process (Davis 2002). Needless to say, the above constructionist perspective considers the process of urbanization to be an integral part of the production of new environments and new natures. Such a view sees both nature and society as combined in historical-geographical production processes (see, among others, Smith 1984; 1996; 1998a; Castree 1995).

From this perspective, there is no such thing as an unsustainable city in general, but rather there are a series of urban and environmental processes that negatively affect some social groups while benefiting others (see Swyngedouw and Kaika 2000). A just urban socio-environmental perspective, therefore, always needs to consider the question of who gains and who pays and to ask serious questions about the multiple power relations – and the networked and scalar geometries of these relations – through which deeply unjust socio-environmental conditions are produced and maintained. This requires sensitivity to the political-ecology of urbanization rather than invoking particular ideologies and views about the assumed qualities that inhere in nature itself. Before we can embark on outlining the dimensions of such an urban political-ecological enquiry, we need to consider the matter of nature in greater detail, in particular in light of the accelerating processes by which nature becomes urbanized through the deepening metabolic interactions between social and ecological processes.

Urban political ecology research has begun to show that because of the under-lying economic, political, and cultural processes inherent in the production of urban landscapes, urban change tends to be spatially differentiated, and highly uneven. Thus, in the context of urban environmental change, it is likely that urban areas populated by marginalized residents will bear the brunt of negative environmental change, whereas other, affluent parts of cities enjoy growth in or increased quality of environmental resources. While this is in no way new, urban political ecology is starting to contribute to a better understanding of the interconnected processes that lead to uneven urban environments. Several chapters in this book attempt to address questions of justice and inequality from a historical-materialist perspective rather than from the vantage point of the environmental justice movement and its predominantly liberal conceptions of justice. Urban political ecology attempts to tease out who gains and who loses (and in what ways), who benefits and who suffers from particular processes of socio-environmental change (Desfor and Keil 2004). Additionally, urban political ecologists try to devise ideas/plans that speak to what or who needs to be sustained and how this can be done (Cutter 1995; Gibbs 2002). In other words, environmental transformations are not independent of class, gender, ethnicity, or other power struggles. These metabolisms produce socio-environmental conditions that are both enabling, for powerful individuals and groups, and disabling, for marginalized individuals and groups. These processes precisely produce positions of empowerment and disempowerment. Because these relations form under and can be traced directly back to the crisis tendencies inherent to neo-liberal forms of capitalist development, the struggle against exploitative socio-economic relations fuses necessarily together with the struggles to bring about more just urban environments (Bond 2002; Swyngedouw 2005). Processes of socio-environmental change are, therefore, never socially or ecologically neutral. This results in conditions under which particular trajectories of socio-environmental change undermine the stability of some social groups or places, while the sustainability of social groups and places elsewhere might be enhanced. In sum, the political-ecological examination of the urbanization process reveals the inherently contradictory nature of the process of socio-environmental change and teases out the inevitable conflicts (or the displacements thereof) that infuse socio-environmental change (see Swyngedouw *et al.* 2002a).

Within this context, particular attention is paid in this book to social power relations (whether material or discursive, economic, political, and/or cultural) through which socio-environmental processes take place and to the networked connections that link socio-ecological transformations between different places. It is this nexus of power and the social actors deploying or mobilizing these power relations that ultimately decide who will have access to or control over, and who will be excluded from access to or control over, resources or other components of the environment. These power geometries shape the social and political configurations under and the urban environments in which we live.

A "MANIFESTO" FOR URBAN POLITICAL ECOLOGY

Throughout this book, a series of common perspectives and approaches are presented. Although urban political ecology neither has, nor should have, a hermetic canon of enquiry, a number of central themes and perspectives are clearly discernible. We thought it would be useful to articulate these principles in sort of a ten-point "manifesto" for urban political ecology (see also Swyngedouw *et al.* 2002a,b). Although manifestos are not really fashionable these days, they nevertheless often serve both as a good starting point for debate, refinement, and transformation, and as a platform for further research.

1 Environmental and social changes co-determine each other. Processes of socio-environmental metabolic circulation transform both social and physical environments and produce social and physical milieus (such as cities) with new and distinct qualities. In other words, environments are combined socio-physical constructions that are actively and historically produced, both in terms of social content and physical-environmental qualities. Whether we consider the making of urban parks, urban natural reserves, or skyscrapers, they each contain and express fused socio-physical processes that contain and embody particular metabolic and social relations.

2 There is nothing a-priori unnatural about produced environments like cities, genetically modified organisms, dammed rivers, or irrigated fields. Produced environments are specific historical results of socio-environmental processes.

The urban world is a cyborg world, part natural/part social, part technical/part cultural, but with no clear boundaries, centres, or margins.

3 The type and character of physical and environmental change, and the resulting environmental conditions, are not independent from the specific historical social, cultural, political, or economic conditions and the institutions that accompany them. It is concrete historical-geographical analysis of the production of urban natures that provides insights in the uneven power relations through which urban "natures" become produced and that provides pointers for the transformation of these power relations.

4 All socio-spatial processes are invariably also predicated upon the circulation and metabolism of physical, chemical, or biological components. Non-human "actants" play an active role in mobilizing socio-natural circulatory and metabolic processes. It is these circulatory conduits that link often distant places and ecosystems together and permit relating local processes with wider socio-metabolic flows, networks, configurations, and dynamics.

5 Socio-environmental metabolisms produce a series of both enabling and disabling social and environmental conditions. These produced milieus often embody contradictory tendencies. While environmental (both social and physical) qualities may be enhanced in some places and for some humans and non-humans, they often lead to a deterioration of social, physical, and/or ecological conditions and qualities elsewhere.

6 Processes of metabolic change are never socially or ecologically neutral. This results in conditions under which particular trajectories of socio-environmental change undermine the stability or coherence of some social groups, places or ecologies, while their sustainability elsewhere might be enhanced. In sum, the political-ecological examination of the urbanization process reveals the inherently contradictory nature of the process of metabolic circulatory change and teases out the inevitable conflicts (or the displacements thereof) that infuse socio-environmental change.

7 Social power relations (whether material or discursive, economic, political, and/or cultural) through which metabolic circulatory processes take place are particularly important. It is these power geometries, the human and non-human actors, and the socio-natural networks carrying them that ultimately decide who will have access to or control over, and who will be excluded from access to or control over, resources or other components of the environment and who or what will be positively or negatively enrolled in such metabolic imbroglios. These power geometries, in turn, shape the particular social and political configurations and the environments in which we live. Henri Lefebvre's "Right to the City" also invariably implies a "Right to Metabolism".

8 Questions of socio-environmental sustainability are fundamentally political questions. Political ecology attempts to tease out who (or what) gains from and who pays for, who benefits from and who suffers (and in what ways) from particular processes of metabolic circulatory change. It also seeks answers to questions about what or who needs to be sustained and how this can be maintained or achieved.

9 It is important to unravel the nature of the social relationships that unfold between individuals and social groups and how these, in turn, are mediated by and structured through processes of ecological change. In other words, environmental transformation is not independent from class, gender, ethnic, or other power struggles.

10 Socio-ecological "sustainability" can only be achieved by means of a democratically controlled and organized process of socio-environmental (re)-construction. The political programme, then, of political ecology is to enhance the democratic content of socio-environmental construction by means of identifying the strategies through which a more equitable distribution of social power and a more inclusive mode of the production of nature can be achieved.

DOING URBAN POLITICAL ECOLOGY

The fifteen chapters collected in this volume explore, both theoretically and empirically, the themes, perspectives, and politics that are central to an urban political ecological analysis. Although this collection by no means aspires to be exhaustive and discusses almost exclusively the "developed" world, it brings together a rich and multi-faceted scholarship that focuses on the fusion between the social and the natural in the process of urbanization. There are a number of themes and perspectives that run through the book and that, hopefully, provide a series of coherent arguments that contribute to define both the epistemological and methodological ground on which urban political ecology rests.

Two central tropes run throughout the book, metabolism and circulation. They are mobilized as guiding vehicles that permit casting urbanization as a dynamic socio-ecological transformation process that fuses together the social and natural in the production of distinct and specific urban environments. The politicization of socio-physical circulation and metabolism processes constitutes the core of our attempt to chart an urban political ecology and its associated politics of radical democratization. Needless to say, these two metaphors are deeply contested and historically constituted in their own rights. The contributors to this collection interpret them in their own specific way. While some focus on the materiality of socio-ecological metabolic and circulatory processes, others insist on the discursive and symbolic powers associated with the foregrounding of these metaphors and how this, in turn, shapes the "nature" of the urban imaginary and urban socio-environmental politics. All agree that the production of urban "nature" is a highly contested and contestable terrain.

In the first three chapters after this introduction, the contours of an urban political-ecological project are outlined. In chapter 2, Erik Swyngedouw insists on the powerful possibilities that the mobilization of a historical-materialist framing of "metabolism" and "circulation" holds for capturing the political-ecological dynamics of urbanization. Metabolic urbanization and the production of cyborg cities are the central figures through which urban political ecology is explored in this chapter. In chapter 3, Roger Keil and Julie-Anne Boudreau mobilize urban political ecology and the metaphor of metabolics to explore how Toronto's recent urban politics and urban movements reshaped the urban agenda towards "environmentalism" in promising new directions. Matthew Gandy, in chapter 4, excavates the intricate and shifting relations between the historical dynamics of the urbanization of nature on the one hand and the transformations in ecological imaginaries on the other. All three contributions insist on the need to move away from reactionary ecological imaginaries of the past and to construct an environmental politics framed around the co-evolutionary dynamics of the social and bio-physical world. These introductory chapters provide a tapestry of the field of urban political ecology against which the other chapters of the book can be situated. In chapter 5, Eliza Darling's dazzling and whirlwind analysis of "nature's carnival" at Coney Island, New York reflects on the paradoxical carnivalesque staging of nature-as-play at the turn of the twentieth century. She explores how the tropes of nature continue to haunt urban space in an age of rapid industrialization and urbanization. For her, nature still constitutes spectacle in Gotham. From a different perspective, Stuart Oliver suggests, in his account of the disciplining of the river Thames in the UK in chapter 6, how cultural imaginaries, the desires of individuals, and the material conditions of river flows fuse together with economic imperatives in the making of a managed, engineered, and urbanized nature. The construction of distinct cultural-material urban environments is also explored in Robbins and Sharp's chapter on this quintessentially American urban nature, the lawn (chapter 7). Moving from Louis Althusser to organophosphates and back, they explore how the lawn produces a turf grass subject. Examining the array of linkages of the contemporary turf grass yard to chemical production economies and community values, they show how the lawn is a capitalized system that produces a certain kind of person, one who answers to the needs of community landscape. The fusion between the interests of the chemical industry and the constructed aesthetics of lawn-based suburbia explored in this chapter testifies to the intricate power relations, both symbolic and material, that operate at a variety of geographical scales but become materialized in the particular geographies of high-input lawns.

From the cyborg city, we move to the urban human body as the leitmotiv of chapters 8 and 9. The cyborg bodies of Nik Heynen in chapter 8 are those of the hungry, the marginalized bodies of the urban poor. The chapter charts their metabolic struggles in the context of a capitalist urbanization of food; a process that produces hunger as a socio-physical condition in the midst of the lush and abundant urbanized natures of US cities. Simon Marvin and Will Medd, in turn, excavate in chapter 9 the discursive and material politics of "fat" bodies in "fat" cities. For them, the urban metabolism and circulation of fat, both in the bodies of human as well as in the "body-work" of the city (sewers, and the like) is constructed as a threat to the circulatory and metabolic processes within bodies and cities alike. In an imaginative *tour de force* they combine the political-economy of fat with the politics of producing "lean" cities.

Chapters 10, 11, and 12 enter the political ecological metabolism of the city through the lens of water. Maria Kaika's engaging account of the politics of drought and scarcity in Athens evokes the mechanisms through which the urbanization of nature becomes an integral part of the politics and power relationships that drive the urbanization process. She suggests how the political-economy of urbanization in Athens operates, among others, in and through the interweaving of discursive and material practices with respect to the urbanization of nature, and, in particular, of water. The contested politics of urban water circulation are simultaneously the arena in which and means through which particular political-economic programmes are pursued and implemented. The geographical strategies of competitiveness and water control are also broached by Alexander Loftus who analyses in chapter 11 how the political ecology of Durban's waterscape has increasingly come to embody the contradictory tendencies of capitalism. The local waters of the city constitute a sphere in which a commercialized state entity has attempted to ensure its profitability, through fencing in something formerly considered to reside outside of capital's orbit. Simultaneously, this entity has tried to expand its operations throughout the southern hemisphere – but failed dismally. In chapter 12, Laila Smith and Greg Ruiters focus their analysis of urban water in South Africa on the choreography of public/private governance. They consider how the part-privatization of water delivery services affects the state/citizen relationship and the associated transformations in power choreographies.

The final part of the book explores socio-ecological urban politics and governance further. In chapter 13, Alec Brownlow delves into Philadelphia's contested politics to fuse a fragmented "environmentalism" with a competitive entrepreneurial strategy in the struggle to "clean-up" Philly's industrial legacy. He considers how entrepreneurial and inherited narratives of nature are both products of and responses to earlier industrial fragmentations. He shows how the new urban fragmentations and narratives of neo-liberal urbanism – be they "new" discourses of nature and eco-modernization or regimes of urban ecological governance – articulate themselves with the inherited ecologies and social geographies of the industrial city. In chapter 14, David Pellow takes the argument global. He insists that the pollution of urban areas is not fundamentally distinct from the despoliation of rural spaces because they are part of the same process and reflect the urbanization of nature on a global scale. Cities in the Global North are the point of origin for many of the world's toxic wastes. He explores the nature of activism among Global North Environmental Justice (EJ) organizations in order to construct a profile of the transnational EJ movement that combines an emphasis on challenging discursive and structural practices with sensitivity to the material and political relations between local tactics and global strategies. He also examines the changing contours and scales of urban environmental justice politics in light of the growth of transnational activism. In the final chapter, Stephen Graham chillingly explores the geo-politics of targeting urban metabolisms in new forms of warfare. In military tactics, attacking the metabolic live lines of big cities has become a "vital" and extraordinarily effective strategy of warfare. At the time of writing these lines, water distribution and electricity delivery were still not fully restored in Baghdad after they had been taken out "surgically" during the first days of the Iraq war. With the massive technical infrastructure that sustains urban metabolism becoming the target of increasingly sophisticated strategies of political violence, this chapter seeks to probe into the political ecology and political economy of forced de-modernization. That is, it explores the deliberate targeting of the "transformation of Nature into City" as a strategy of political violence. Graham analyses how the deliberate targeting of urban technics in political violence impacts on the political ecologies and urban metabolisms of targeted cities.

URBAN POLITICAL ECOLOGY: TOWARDS THE DEMOCRATIC PRODUCTION OF CYBORG CITIES

In sum, this collection seeks to suggest how urban political ecology provides an integrated and relational approach that helps untangle the interconnected economic, political, social and ecological processes that together form highly uneven urban socio-physical landscapes. Because the power-laden socio-ecological relations that go into the formation of urban environments constantly shift between groups of human and non-human actors and of spatial scales, historical-geographical insights into these ever-changing urban configurations are necessary for the sake of considering the future evolution of urban environments. An urban political ecological perspective permits new insights in the urban problematic and opens new avenues for re-centreing the urban as the pivotal terrain for eco-political action. To the extent that emancipatory urban politics reside in acquiring the power to produce urban environments in line with the aspirations, needs, and desires of those inhabiting these spaces, the capacity to produce the physical and social environment in which one dwells, the question of whose nature is or becomes urbanized must be at the forefront of any radical political action. And this is exactly what the contributions in this book attempt to illuminate. They also endeavour to open up a research agenda and a political platform that may set pointers for democratizing the politics through which cyborg cities are produced as both enabling and disempowering sites of living for humans and non-humans. "Urbanizing" the environment, therefore, is a project of social and physical environmental construction that actively produces the urban (and other) environments that we wish to inhabit today.

BIBLIOGRAPHY

Benton, T. (ed.) (1996) *The Greening of Marxism*. New York: Guilford Press

Blaickie, P. and Brookfield, P. (1987) *Land Degradation and Society*. London: Methuen

Bond, P. (2002) *Unsustainable South Africa*. London: The Merlin Press

Bookchin, M. (1979) "Ecology and revolutionary thought", *Antipode*, 10(3): 21–32

Braun, B. and Castree, N. (eds) (1998) *Remaking Reality: Nature at the Millennium*. London and New York: Routledge

Brenner, N. (2001) "The limits to scale? Methodological reflections on scalar structuration", *Progress in Human Geography*, 25: 591–614

Bullard, R. (1990) *Dumping in Dixie: Race, Class, and Environmental Quality*. Boulder, CO: Westview Press

Bullard, R. and Chavis, B. F. Jr. (1993) *Confronting Environmental Racism: Voices from the Grassroots*. Boston, MA: South End Press

Burkett, P. (1999) *Marx and Nature – A Red and Green Perspective*. New York: St Martin's Press

Castree, N. (1995) "The nature of produced nature: materiality and knowledge construction in Marxism", *Antipode*, 27: 12–48

Castree, N. and Braun, B. (eds) (2001) *Social Nature: Theory, Practice, Politics*. London and New York: Routledge

Cronon, W. (1991) *Nature's Metropolis*. New York: A.A. Norton

Cutter, S. L. (1995) "Race, class and environmental justice", *Progress in Human Geography*, 19(1): 111–122

Davis, M. (1996) "How Eden lost its garden: a political history of the Los Angeles landscape". In A.J. Scott and E.W. Soja (eds) *The City – Los Angeles and Urban Theory at the End of the Twentieth Century*. Berkeley, CA: University of California Press

Davis, M. (1998) *Ecology of Fear: Los Angeles and the Imagination of Disaster*. New York: Metropolitan Books

Davis, M. (2002) *Dead Cities*. New York: The New Press

Denton, N. and Massey, D. (1991) "Patterns of neighborhood transition in a multiethnic world: U.S. metropolitan areas, 1970–1980", *Demography*, 28(1): 41–63

Desfor, G. and Keil, R. (2004) *Nature and the City: Making Environmental Policy in Toronto and Los Angeles*. Tucson, AZ: The University of Arizona Press

Di Chiro, G. (1996) "Nature as community: the convergence of environment and social justice". In W. Cronon (ed.) *Uncommon Ground: Rethinking the Human Place in Nature*. New York: W.W. Norton and Company

Dobson, A. (1999) *Justice and the Environment: Conceptions of Environmental Sustainability and Dimensions of Social Justice*. Oxford: Oxford University Press

Engels, F. (1987 [1845]) *The Condition of the Working Class in England*. Edited by V.G. Kiernan. Harmondsworth: Penguin

Engels, F. (1940) *Dialectics of Nature*. New York: International Publishers

Engels, F. (1959) "Socialism: utopian and scientific". In L. Feuer (ed.) Karl Marx and Friedrich Engels, *Basic Writings*. London: Collins

Escobar, A. (2001) "Culture sits in places: reflections on globalism and subaltern strategies of localization", *Political Geography*, 20: 139–174

FitzSimmons, M. (1989) "The matter of nature", *Antipode*, 21: 106–120

Forsyth, T. (2003) *Critical Political Ecology – The Politics of Environmental Science*. London and New York: Routledge

Foster, J.B. (2000) *Marx's Ecology*. New York: Monthly Review Press

Foster, J.B. (2002) *Ecology against Capitalism*. New York: Monthly Review Press

Gandy, M. (2002) *Concrete and Clay: Reworking Nature in New York City*. Cambridge, MA and London: The MIT Press

Gibbs, D. (2002) *Local Economic Development and the Environment*. London: Routledge

Gober, P., McHugh, K. E. and Reid, N. (1991) "Phoenix in flux: household instability, residential mobility, and neighborhood change", *Annals of the Association of American Geographers*, 81(1): 80–88

Gold, M. (1984) "A history of nature". In D. Massey and J. Allen (eds) *Geography Matters!* London: Cambridge University Press

Grundman, R. (1991) *Marxism and Ecology*. Oxford: Clarendon Press

Haraway, D. (1991) *Simians, Cyborgs and Women – The Reinvention of Nature*. London: Free Association Books

Haraway, D. (1997) *Modest_Witness@Second_Millennium.FemaleMan©_Meets_Oncomouse™*. London: Routledge

Harvey, D. (1973) *Social Justice and the City*. Cambridge, MA: Blackwell Publishers

Harvey, D. (1993) "The nature of environment: dialectics of social and environmental change". In R. Miliband and L. Panitch (eds) *Real Problems, False Solutions*. A special issue of the *Socialist Register*. London: The Merlin Press

Harvey, D. (1996) *Justice, Nature and the Geography of Difference*. Oxford: Blackwell Publishers

Hughes, J. (2000) *Ecology and Historical Materialism*. Cambridge: Cambridge University Press

Hurley, A. (1997) *Common Fields: An Environmental History of St. Louis*. St. Louis, MO: Missouri Historical Society

Jacobs, J. (1992 [1961]) *The Death and Life of Great American Cities*. New York: Vintage Books

Jägter, J. and Raza, W.G. (2004) "Regulationist perspectives in Political Ecology: a conceptual framework for urban analysis", Mimeographed paper, Department of Economics, Vienna University of Economics and Business Administration, Augasse 2–6, A-1090 Vienna, Austria

Jonas, A. (1994) "Editorial", *Environment and Planning D: Society and Space*, 12: 257–264

Kaika, M. (2005) *City of Flows. Nature, Modernity, and the City*. New York: Routledge

Kaika, M. and Swyngedouw, E. (1999) "Fetishising the modern city: the phantasmagoria of urban technological networks", *International Journal of Urban and Regional Research*, 24(1): 120–138

Katz, C. (1998) "Whose nature, whose culture? Private productions of space and the 'preservation' of nature". In B. Braun and N. Castree (eds) *Remaking Reality: Nature at the Millenium*. London: Routledge

Keil, R. (1995) "The environmental problematic in world cities". In P. Knox and P. Taylor (eds) *World Cities in a World System*. Cambridge: Cambridge University Press

Keil, R. (2003) "Urban political ecology", *Urban Geography*, 24(8): 723–738

Keil, R. and Graham, J. (1998) "Reasserting nature: constructing urban environments after Fordism". In B. Braun and N. Castree (eds) *Remaking Reality – Nature at the Millenium*. London: Routledge

Laituri, M. and Kirby, A. (1994) "Finding fairness in America's cities? The search for environmental equity in everyday life", *Journal of Social Issues*, 50(3): 121–139

Latour, B. (1993) *We Have Never Been Modern*. London: Harvester Wheatsheaf

Latour, B. (1999) *Politiques de la Nature – Comment faire entrer les sciences en démocratie*. Paris: La Découverte

Latour, B. and Hermant, E. (1998) *Paris Ville Invisible*. Paris: La Découverte

Lefebvre, H. (1976) *The Survival of Capitalism: Reproduction of the Relations of Production*. London: Allison and Busby

Lefebvre, H. (1991) *The Production of Space*. Oxford: Blackwell Publishers

Low, N. and Gleeson, B. (1998) *Justice, Society and Nature: An Exploration of Political Ecology*. London: Routledge

MacDonald, D. H. (ed.) (2002) *Environmental Justice in South Africa*. Athens, OH: Ohio University Press

Marston, S. (2002) "A long way from home: domesticating the social production of scale". In R. McMaster and E. Sheppard (eds) *Scale and Geographic Enquiry*. Oxford: Blackwell

Marx, K. (1973) *Grundrisse*. New York: Vintage Books

Marx, K. (1975) "Early economic and philosophical manuscripts". In L. Colletti (ed.) *Karl Marx: Early Writings*. Harmondsworth: Pelican

Marx, K. (1976) *Capital Volume 1*. New York: Vintage Books

Melosi, M.V. (2000) *The Sanitary City: Urban Infrastructure in America from Colonial Times to the Present*. Baltimore, MD and London: The Johns Hopkins University Press

Merrifield, A. (2002) *Dialectical Urbanism*. New York: Monthly Review Press

Merrifield, A. and Swyngedouw, E. (eds) (1997) *The Urbanization of Injustice*. New York: New York University Press

Norgaard, R. (1994) *Development Betrayed*. London: Routledge

Olmstead, F. L. (1895) "Parks, parkways and pleasure grounds", *Engineering Magazine*, 9: 253–254

Peet, R. (1977) "The development of radical geography in the United States". In R. Peet (ed.) *Radical Geography: Alternative Viewpoints on Contemporary Social Issues*. Chicago, IL: Maaroufa Press

Peet, R. and Watts, M. (eds) (1996) *Liberation Ecologies*. London: Routledge

Pulido, L. (1996) *Environmentalism and Economic Justice: Two Chicano Struggles in the Southwest*. Tucson, AZ: University of Arizona Press

Schmink, M. and Wood, C. (1987) "The 'political ecology' of Amazonia". In P.D. Little *et al.* (eds) *Lands at Risk in the Third World: Local Level Perspectives*. Boulder, CO: Westview Press

Smith, N. (1984) *Uneven Development: Nature, Capital and the Production of Space*. Oxford: Blackwell Publishers

Smith, N. (1993) "Homeless/global: Scaling places". In J. Bird *et al.* (eds) *Mapping the Futures: Local Cultures, Global Change*. London: Routledge

Smith, N. (1996) "The production of nature". In G. Robertson *et al.* (eds) *Future Natural: Nature/Science/Culture*. London: Routledge

Smith, N. (1998a) "Antinomies of space and nature in Henri Lefebvre's 'The production of space' ". In A. Light and J.M. Smith (eds) *Philosophy and Geography II: The Production of Public Space*. London and New York: Rowman and Littlefield

Smith, N. (1998b) "El Niño capitalism", *Progress in Human Geography*, 22(3): 159–163

Spirn, A. (1996) "Constructing nature: the legacy of Frederick Law Olmsted". In W. Cronon (ed.) *Uncommon Ground: Rethinking the Human Place in Nature*. New York: W.W. Norton and Co

Swyngedouw, E. (1996) "The city as a hybrid: on nature, society and cyborg urbanization", *Capitalism Nature Socialism*, 7 (25 March): 65–80

Swyngedouw, E. (1997) "Power, nature and the city. The conquest of water and the political ecology of urbanization in Guayaquil, Ecuador: 1880–1980", *Environment and Planning A*, 29(2): 311–332

Swyngedouw, E. (1999) "Modernity and hibridity: nature, *Regeneracionismo*, and the production of the Spanish waterscape, 1890–1930", *Annals of the Association of American Geographers*, 89(3): 443–465

Swyngedouw, E. (2000) "Authoritarian governance, power and the politics of rescaling", *Environment and Planning D: Society and Space*, 18: 63–76

Swyngedouw, E. (2004a) *Social Power and the Urbanization of Water: Flows of Power*. Oxford: Oxford University Press

Swyngedouw, E. (2004b) "Scaled geographies. Nature, place, and the politics of scale". In R. McMaster and E. Sheppard (eds) *Scale and Geographic Inquiry: Nature, Society and Method*. Oxford: Blackwell Publishers.

Swyngedouw, E. (2005) "Dispossessing H_2O – The contested terrain of water privatization", *Capitalism, Nature, Socialism*, 16(1): 1–18

Swyngedouw, E. and Heynen, N.C. (2003) "Urban political ecology, justice and the politics of scale", *Antipode*, 35(5): 898–918

Swyngedouw, E. and Kaika, M. (2000) "The environment of the city or . . . The urbanization of nature". In G. Bridge and S. Watson (eds) *Reader in Urban Studies*. Oxford: Blackwell Publishers

Swyngedouw, E., Kaika, M. and Castro, E. (2002a) "Urban water: a political-ecology perspective", *Built Environment*, 28(2): 124–137

Swyngedouw, E., Page, B. and Kaika, M. (2002b) "Sustainability and policy innovation in a multi-level context: crosscutting issues in the water sector". In P. Getimis *et al.* (eds) *Participatory Governance in Multi-Level Context: Concepts and Experience*. Frankfurt: Leske & Budrich

Szaz, A. (1994) *Ecopopulism*. Minneapolis, MN: University of Minnesota Press

Tarr, J.A. (1996) *The Search for the Ultimate Sink: Urban Pollution in Historical Perspective*. Akron, OH: University of Akron Press

Titmus, R.M. (1962) *Income Distribution and Social Change*. London: Allen & Unwin

Walker, P. (2005) "Political ecology: where is the ecology?", *Progress in Human Geography*, 29(1): 73–83

Walker, R. (1981) "A theory of suburbanization: capitalism and the construction of urban space in the United States". In M. Dear and A. Scott (eds) *Urbanization and Urban Planning in Capitalist Society*. London: Methuen

Wenz, P. S. (1988) *Environmental Justice*. New York: State University of New York Press

Whatmore, S. (2002) *Hybrid Geographies*. London and New York: Routledge

Whitehead, M. (2003) "(Re)analysing the sustainable city: nature, urbanization and the regulation of socio-environmental relations in the UK", *Urban Studies*, 40(7): 1,183–1,206

Williams, R. (1985 [1973]). *The Country and the City*. London: Hogarth Press

Wolch, J., Pincetl, S. and Pulido, L. (2002) "Urban nature and the nature of urbanism". In M.J. Dear (ed.) *From Chicago to L.A.: Making sense of urban theory*. Thousand Oaks, CA: Sage.

Worster, D. (1993) *The Wealth of Nature: Environmental History and the Ecological Imagination*. New York and Oxford: Oxford University Press

2 Metabolic urbanization
The making of cyborg cities

Erik Swyngedouw

METABOLIC URBANIZATION AND CYBORG CITIES

> A cyborg is a cybernetic organism, a hybrid of machine and organism, a creature of social reality as well as a creature of fiction.
>
> (Haraway 1991: 149)

In the introductory chapter we argued that cities are constituted through dense networks of interwoven socio-ecological processes that are simultaneously human, physical, discursive, cultural, material, and organic. Circulatory conduits of water, foodstuffs, cars, fumes, money, labour, etc., move in and out of the city, transform the city, and produce the urban as a continuously changing socio-ecological landscape. Imagine, for example, standing on the corner of Piccadilly Circus in London, and consider the socio-environmental metabolic relations that come together in this global-local place: smells, tastes, and bodies from all nooks and crannies of the world are floating by, consumed, displayed, narrated, visualized and transformed. The "Rainforest" shop and restaurant play to the tune of eco-sensitive shopping and the multi-billion pound eco-industry while competing with McDonalds' burgers and Dunkin' Donuts, whose products – like burgers, coffee, orange juice, or cream cheese – are equally the result of processes that fuse together and interconnect social and biochemical relations from many places, near and far away. Consider how human bodies – of migrants, prostitutes, workers, capitalists – spices, clothes, foodstuffs, and materials from all over the world whirl by. The neon lights are fed by energy coming from nuclear power plants and from coal-, oil-, or gas-burning electricity generators. Cars, taxis, and buses move on fuels from oil-deposits (now again from Iraq) and pump CO_2 into the air, affecting peoples, forests and climates in places around the globe. All these flows complete the global geographic mappings and traces that flow through the urban and "produce" London (or any other city) as a palimpsest of densely layered bodily, local, national and global – but depressingly geographically uneven – metabolic socio-ecological processes. This intermingling of material and symbolic things produces the vortexes of modern life, combines to produce a particular socio-environmental milieu that welds nature, society, and the city together in a deeply heterogeneous, conflicting and often disturbing whole (Swyngedouw 1996).

The view that a city is a particular process of environmental production, sustained by particular sets of socio-metabolic processes that shape the urban in distinct, historically contingent ways, a socio-environmental process that is deeply caught up with socio-metabolic processes operating elsewhere, rarely grabs the headlines. Of course, the "Hygienic City" of the nineteenth century (Gandy 2004; this volume) already celebrated the making of the city as a system of circulatory conduits that would render the metabolism of the city rhyme in concert with the bio-chemical metabolisms associated with a sanitized urban life. Haussmann's opening up of Paris, King Leopold's sanitation of Brussels, the visionary construction of Vienna's Ringstrasse, and London's slum clearance also point to these combined processes of political-ecological transformation and socio-cultural reconstruction. The ecological anarchism of radical thinkers like Kropotkin or Elisee Reclus, and the various attempts at creating socially or ecologically harmonious "utopian" cities pursued with equal fervour by anarchists, socialists, liberals, and fascists, also illustrate nineteenth- and early-twentieth-century concerns with producing socially just and sustainable urban environments.

Urbanization can indeed be viewed as a process of contiguous de-territorialization and re-territorialization through metabolic circulatory flows, organized through social and physical conduits or networks of "metabolic vehicles". In this chapter, we consider how nature becomes urbanized through proliferating socio-metabolic processes. "Metabolism" and "circulation" will be the central metaphors that will guide us in this endeavour. They are not randomly selected. Both concepts have a long conceptual, cultural, social, material, and arte-factual history. They emerged as coherent concepts and materially mobilized principles in the mid-nineteenth century and both were deeply connected with projects, visions, and practices of modernization, and with the associated "modern" transformation of the city. Most importantly, in contrast to other fashionable metaphors that attempt to fuse together heterogeneous entities – like networks, assemblages, rhizomes, imbroglios, collectives – the former convey a sense of flow, process, change, transformation, and dynamism in addition to the "inner-connectedness" suggested by the other tropes. They embody what modernity has been, and will always be about: change, transformation, flux, movement, creative destruction. With its emphasis on movement, change, and process and its insistence on the socially mobilized "materiality" of life, historical materialism has been among the first social theories to embrace and mobilize "metabolism" and "circulation" as entry-points in undertaking "ontologies of the present that demand archaeologies of the future" (Jameson 2002: 215). These ontologies and archaeologies are what we shall turn to next.

HISTORICAL-GEOGRAPHICAL MATERIALISM: ENTERING METABOLISM AND CIRCULATION

Historical materialism and the remaking of environments

> Certainly we continue to have crickets and thunderstorms . . . and we continue to understand our psyches as driven by natural instincts and passions; but we have no

nature in the sense that these forces and phenomena are no longer understood as outside, that is, they are not seen as original and independent from the civil order.

(Hardt and Negri 2000: 187)

Both "metabolism" and "circulation" have long conceptual and material histories. "Circulation" gained wide currency after William Harvey's postulation of the double circulation of blood in the body. Movement, flux and conduits rapidly thereafter became formative metaphors that would shape radically new visions of and practices for acting in the world. The concept of "metabolism" arose in the early nineteenth century, particularly in relationship to the material exchanges in the body with respect to respiration. It became extended later to include material exchanges between organisms and the environment as well as the bio-physical processes within living (and non-living or decaying) entities. For example, in the writings of Jacob Moleschott (1857) and Justus von Liebig (1840; 1842), metabolism denoted not only the exchange of energy and substances between organisms and the environment, but the totality of biochemical reactions in a living thing. In fact, von Liebig's analysis turned organisms into living processes, gave them a history-as-process. Interestingly enough, von Liebig, like Edwin Chadwick, had taken the temporal/spatial separation of spaces of production and spaces of consumption through the emergence of long-distance trade and the process of urbanization (what von Liebig called the "metabolic rift") as the pivotal causes of the decline in the productivity of agricultural land on the one hand, and the problematic accumulation of excrement, sewage and garbage in the city on the other. For them, the "unsustainability" of nineteenth-century forms of urbanization was, as it is today, directly related to the spatio-temporal organization of metabolic flows and circuits. With this view of metabolism as ecological-historical process, and combined with Darwin's equally historical-metabolic views of the biological world, and Lyell's theories of the world's geological reconstruction, historical-geographical materialism could mobilize the concept of metabolism, neither as just an organic analogy to the social order (see Padovan 2000) nor as a mere metaphor to be transposed onto society, but as the very foundation of and lasting condition for the social.[1]

In social theory, the concept of metabolism was introduced in an ontological and epistemological framework in the early Marxist formulations of historical materialism. In its most general sense, materialism asserts that both origin and development of what exists is dependent on nature and "matter". Or, in other words, a certain physical Reality exists that is prior to thought, and to which thought must be related or interlinked (although it can never be identical to the Real) (Foster 2000). As Roy Bhaskar argued, "neither thought nor language form a realm of their own, they are only manifestations of actual life" (Bhaskar 1979: 100).[2] Karl Marx's historical materialism was arguably the first coherent attempt to theorise the internal metabolic relationships that shape the transformations of the earth's surface and make and remake the social and physical world. In *Grundrisse*, *Capital* and, in particular, *The German Ideology*, Marx insisted on the "natural" foundations of social development (see also Hughes 2000):

The first premise of all human history is, of course, the existence of living human individuals. Thus the first fact to be established is the physical organization of these individuals and their consequent relationship to the rest of nature . . . The writing of history must always set out from these natural bases and their modification in the course of history through the action of men . . . [M]en must be in a position to live in order to be able to "make history" . . . The first historical act is thus the production of the means to satisfy these needs, the production of material life itself.

(Marx (1974 [1846]: 42 and 48)

This environmental "production" process is conceived in the broadest possible sense. It refers to the metabolic process that is energized through the fusion of the physical properties and creative capacities of humans with those of non-humans. For Marx, this is what defines the act of "labouring", i.e. the purposeful metabolic process intended to produce and reproduce (human) life. Production is an organic process in the first instance, similar (but not reducible or identical) to the act of producing things new by other organic and non-organic "actants". What differentiates human actants from others is their organic capacity to wish differentially, to imagine different possible futures, to act differentially in ways driven and shaped by human drives, desires, and imaginations (as distinct from those of rivers, viruses, cows, or tulips). This form of acting differentiates human acting from other active "moments" or "agents" in the production and transformation of "environments". As Marx puts it:

A spider conducts operations that resemble those of a weaver, and a bee in the construction of her cells puts to shame many an architect. But what distinguishes the worst architects from the best of bees is this, that the architect arises his structure in imagination before he erects it in reality.

(Marx 1971 [1867]: Ch. 5)

Labouring is therefore nothing other than engaging the "natural" physical and mental forces and capabilities of humans in a metabolic physical-material process with other human and non-human actants and conditions. It is through the process of "transposition of labour power into human organism" (Marx 1971: 323) that this metabolic process is mobilized:

Nature builds no machines, no locomotives, railways, electric telegraphs, self-acting mules, etc. These are products of human industry; natural material transformed into organs of the human will over nature, or of human participation in nature. They are organs of the human brain, created by the human hand.

(Marx 1973 [1858]: 706)

These products of transformed nature and embodied "dead" labour take on a thing-like character, which, like any other actant, is enrolled again in subsequent assemblages. In fact, "[A]ny product can take on a 'life' of its own, and may come to dominate the living labour that makes it. The 'nature of things' is indeed to become non-human actors" (Kirsch and Mitchell 2004: 23). If the act of labouring, broadly conceived, constitutes a socio-ecological process, then the particular relational frame through which this labour is socially organized has to become an integral part of understanding the continuous (re-)making of what we can now discern as socio-natural entities (Castree 2000; 2002). The circulation of goods, or of entities, is evidently directly associated with the notion of metabolism, which involves precisely such a process of transformation-in-movement. In other words, metabolic circulation fuses together physical dynamics with the social regulatory and framing conditions set by the historically specific arrangement of the social relations of appropriation, production, and exchange – in other words, the mode of production. The things, the products used by labour in production always enter the metabolic processes as already configured assemblages, collectives, networks that, in turn, through socio-metabolic circulatory processes, mobilize new human and non-human "actants" and produce new assemblages or collectives. As Timothy Luke (1999: 39) notes:

> Marx can be seen as an extended critique of Latour's sense of collectivization, inasmuch as he uses the notion of the commodity to describe the association of humans and nonhumans. Since Marx's examination of the commodity form under capitalism looks at ways in which human labor is mixed with nonhuman things to create value, much of his analysis is a careful study of who dominates whom in the process of such collectivization, with commodification leading to the endless "co-modification" of human and nonhuman beings in both nature and culture. These ties now define coevolution.

These "collectives" are those proliferating objects that Donna Haraway calls "cyborgs" (Haraway 1991) or that Bruno Latour refers to as "quasi-objects" (Latour 1993); these hybrid, part social, part natural – yet deeply historical and thus produced – objects/subjects are intermediaries that embody and express nature *and* society and weave networks of infinite liminal spaces. These assemblages, like commodities, are simultaneously real, like nature; narrated, like discourse; and collective, like society (Latour 1993: 122). They take on cultural, social, and physical forms and enter social and ecological processes in new and transformed manners. The city, in its parts and as a whole, is a kaleidoscopic socio-physical accumulation of human/non-human imbroglios. In the production of these assemblages and entanglements, the figures of "metabolism" and of "circulation" take centre stage in a historical materialist and dialectical account. In the next section, we shall delve deeper into the origin and mobilization of "metabolism" and "circulation" within historical materialism.

Metabolism as metaphor and practice

Marx and Engels were among the first to engage the term "metabolism" to grapple with the dynamics of socio-environmental change and evolution (Fisher-Kowalski

1998; 2003). In fact, "metabolism" is the central metaphor for Marx's definition of labour and for analyzing the relationship between human and nature:

> Labour is, first of all, a process between man and nature, a process by which man, through his own actions, mediates, regulates, and controls the *metabolism* between himself and nature. He confronts the materials of nature as a force of nature. He sets in motion the natural forces which belong to his own body, his arms, legs, head, and hands, in order to appropriate the materials of nature in a form adapted to his own needs. Through this movement he acts upon external nature and changes it, and in this way he simultaneously changes his own nature. . . . [labouring] is the purposeful activity aimed at the production of use-values. It is an appropriation of what exists in nature for the requirements of man. It is the universal condition for the metabolic interaction between man and nature, the ever-lasting nature-imposed condition of human existence, and it is therefore independent of every form of that existence, or rather it is common to all forms of society in which human beings live.
>
> (Marx 1971 [1867]: 283 and 290)

For Marx, this socio-natural metabolism is the foundation of history, a socio-environmental history through which the natures of humans and non-humans alike are transformed (see also Godelier 1986). To the extent that labour constitutes the universal premise for human metabolic interaction with nature, the particular social relations through whom this metabolism of nature is enacted shape its very form. Clearly, any materialist approach insists that "nature" is an integral part of the "metabolism" of social life. Social relations operate in and through metabolizing the "natural" environment, and transform both society and nature. For historical materialism, then, ecology is not so much a question of values, morals, or ethics, but rather a mode of "understanding the evolving material interrelations (what Marx called 'metabolic relations') between human beings and nature . . . From a consistent materialist standpoint, the question is . . . one of coevolution" (Foster 2000: 10–11) (see also Norgaard 1994; Levins and Lewontin 1985). Foster (2000: 15–16) continues to argue that:

> [A] thoroughgoing ecological analysis requires a standpoint that is both materialist and dialectical . . . [A] materialist sees evolution as an open-ended process of natural history, governed by contingency, but open to rational explanation. A materialist viewpoint that is also dialectical in nature (that is, a non mechanistic materialism) sees this as a process of transmutation of forms in a context of interrelatedness that excludes all absolute distinctions . . . A dialectical approach forces us to recognize that organisms in general do not simply adapt to their environment; they also affect that environment in various ways by affecting change in it.

In other words, non-human entities act in their metabolic exchange – in their "enrolment" as Latour (1993) would call it – with other human and non-human actants. This materialist view is decidedly "constructionist" in the sense that it considers socio-natural processes as historically specific, produced, and contingent. However, it does not foreground a notion of "social construction", as the non-human plays a pivotal and foundational role in the process; it merely evokes the view of nature as "produced".

Marx undoubtedly borrowed the notion of "metabolic interaction" from Justus von Liebig,[3] the founding theoretician of modern agricultural chemistry. In contrast to other sociologists avant-la-lettre, like Comte and Spencer, who used the concept of metabolism as an analogy to grapple with social metabolism and for whom "nature offered the gnoseological structures to survey the workings of society" (Padovan 2000: 7), Marx, Engels, or Adam Schäffle, mobilized "metabolism" in an ontological manner in which human beings, like society, were an integral, yet particular and distinct, part of nature.

The original German word for metabolism is *Stoffwechsel*, which translates literally as "change of matter". This simultaneously implies circulation, exchange *and* transformation of material elements. As matter moves, it becomes "enrolled" in associational networks that produce qualitative changes and qualitatively new assemblages. While the newly produced "things" embody and reflect the processes of their making (though a process of internalization of dialectical relations – see Harvey (1996)), they simultaneously differ radically from their constituent relational parts. For von Liebig, chemical metabolism was a process of "creative destruction" in which the new irrevocably transformed the old. Metabolism as a biochemical process is a contradictory one, predicated upon fusion, tension, conflict, and ultimately transconfiguration, which, in turn, produces a series of new "entities", often radically different from the constituting components, yet equally re-active. Metabolism (with a few rare exceptions), consequently, is a historical process, it has a time arrow. Labour (itself an organic metabolic procedure), then, becomes the organic activity through which this metabolic process is mobilized in a purposeful, human manner by enrolling heterogeneous things into specific metabolic interactions:

> Actual labour is the appropriation of nature for the satisfaction of human needs, the activity through which the metabolism between man and nature is mediated.
> (Marx 1861–1863)

While every metabolized thing embodies the complex processes and heterogeneous relations of its making at some point in the past, it enters (or becomes enrolled), in its turn and its own specific manner, into new assemblages of metabolic transformation. These dynamic heterogeneous assemblages form a circulatory (although not necessarily closed) process. Under conditions of generalized commodity production, the process takes on the form of circulation of commodities and the circulatory reverse flow of capital (as embodied dead labour, that is past metabolic transformations). This processual metabolism is, according to Foster (2000), central to Marx's political economy and is directly implicated in the circulation of commodities and, consequently, of money: "[t]he economic circular flow then was closely bound up, in Marx's analysis, with the material exchange (ecological circular flow) associated with the metabolic interaction between human beings and nature" (Foster 2000: 157–158). Indeed, under capitalist social relations, the metabolic production of use values operates in and through specific control and ownership relations, and in the context of the mobilization of both nature and labour to produce commodities (as forms of metabolized socio-natures) with an eye towards the realization of the embodied exchange value. The circulation of capital as value in motion is, therefore, the combined metabolic transformations of socio-natures in and through the reverse circulation of money as capital under social relations that combine the mobilization of capital, nature or dead labour, and labour power. New socio-natural forms, including the transformation of labour power as living labour, are continuously produced as moments and things in this metabolic process (see Grundman 1991; Benton 1989; 1996; Burkett 1999; Foster 2000). Whether we consider the production of dams, the re-engineering of rivers, the management of biodiversity hotspots, the transfiguration of DNA codes, the cultivation of tomatoes (genetically modified or not) or the construction of houses, they all testify to the particular associational relations through which socio-natural metabolisms are organized (in terms of property and ownership regimes, production or assembly activities, distributional arrangements, and consumption patterns).

Of course, the ambition of classical Marxism was broader than reconstructing the dialectics of historical socio-natural transformations and their contradictions. Historical materialism also questioned and critiqued the process of discursive (or ideological in Marxist terms) purification, of separation and binarization of the world into things "social" and things "natural" that, in Latour's vocabulary, produced the modern "constitution" and derailed the project of becoming "modern" (while, in the process, filling this symbolic void with all manner of socio-natural imbroglios). Historical-geographical materialism as a dialectical (that is, non-teleological) evolutionary (that is, actively produced history) organicism (that is, the unity of the heterogeneous social and the heterogeneous natural) not only addresses the cultural, discursive, "ideological", moral/ethical constructions of nature that were as prevalent in the nineteenth century as they are today, but offered a view of the world that unified the natural and the social while critiquing radically the "modern" separation of "society" from "nature".[4] In fact Marx had already prefigured Bruno Latour's clarion call to "re-modernise", to re-connect the two poles that have been severed by modernity, in *Grundrisse*:

> It is not the unity of living and active humanity, the natural, inorganic conditions of their metabolic exchange with nature, and hence their appropriation of nature, which requires explanation, or is the result of a historic process, but rather the separation between these inorganic conditions of human existence and his active existence

> (Marx 1973 [1858]: 489)

However, by concentrating on the labour process as mere social process (as was and is the case for most of modern sociology, Marxist sociology included), some

Marxist analysis – particularly during the twentieth century – tended to replicate the very problem it meant to criticize. The "void" referred to above was silenced rather than problematized, ignored rather than taken as the "space" for politics, for struggle, for pre-figuring radical socio-ecological transformation, and realizing alternative socio-natural relations. In other words, while mainstream economics forgot the natural foundations of economic life[5] (only to rediscover them recently, under the guise of environmental economics), much of Marxist theory equally became an exclusively "social" theory, rather than a socio-ecological one. Put simply, the over-emphasis on the social relations under capitalism that characterized much of Marxist (and other) social analysis tended to abstract away from or ignore the material and socio-physical metabolic relationships, their phantasmagorical representations and symbolic ordering. This resulted in a partial blindness in the social sciences of the twentieth century to questions of political ecology and socio-ecological metabolisms.

Some recent approaches to the society–nature problematic, such as Actor Network Theory or (political-) ecological theories of a variety of kinds, have provided a new grammatical apparatus that has "profoundly revitalized empirical studies of human–nature–technology relations . . . But . . . it remains important that we incessantly raise the question . . . why are 'things as such' produced in the way they are – and to whose potential benefit" (Kirsch and Mitchell 2004). While a historical-materialist mobilization of metabolism might begin to shed light on the production of socio-natural entities, this has to be fused together with another equally central metaphor and material condition, one that is closely related to metabolism, namely, circulation.

The invention of circulation

> Enlightened planners wanted the city in its very design to function like a healthy body, freely flowing as well as possessed of clear skin. Since the beginnings of the Baroque era, urban planners had thought about making cities in terms of efficient circulation of the people on the city's main streets. The medical imagery of life-giving circulation gave a new meaning to the Baroque emphasis of motion. Instead of planning streets for the sake of ceremonies of movement toward an object, as did the Baroque planner, the Enlightenment planner made motion an end in itself.
>
> (Sennett 1994: 263–264)

Alongside the emergence of the notion of "metabolism" in the natural and social sciences (an emergence not wholly disassociated with the rising "metabolic rift" caused by industrialization and urbanization), the notion of "circulation" began to gain greater and wider currency. For example, the idea of "water circulation", that water piped into the city must leave the city by its sewers is not older than the nineteenth century (in the west). Circulating water, following a given path and finally returning to its source, remained foreign to western urban imaginations, spatial representations and engineering systems until then. Modern urbanization, highly dependent on the mastery of circulating flows, was linked with the representation of cities as consisting of and functioning through complex networks of circulatory systems (Kaika and Swyngedouw 1999).

Before the "discovery" of circulatory systems, the movement of water was seen merely as evaporation: the separation of the "spirit" from the "water" (Goubert 1989). This view that things happen, appear, or disappear through "extraction" was widely held before circulatory views began to replace them. In chemistry, for example, phlogiston theory of the seventeenth century, formulated by Johann Becher and still defended by Priestley, rested on the basis of extractionist views. Such theories prevailed until Antoine Lavoisier's eighteenth-century discovery, which postulated chemical reactions as (metabolic) transfigurations or re-arrangements of components that in the process produced qualitatively new assemblages, but in which nothing was lost or disappeared. Together with phlogiston theory, the representation of the respiratory system, plant growth, the Physiocrats' view of the production of material wealth from the given natural conditions of the soil, even the Malthusian unidirectional flow of food, all indicate the incapacity of early post-renaissance people to conceive of "circulation" as an infinite cyclical process.

When William Harvey (1628) promulgated his ideas of the double circulation of blood in the vascular system of the human body in 1628, a revolutionary insight came into being which would begin to permeate and dominate everyday life, engineering, and intellectual thought for centuries to come, both metaphorically and materially.[6] By the end of the century, medical practice had accepted the idea of the circulatory (metabolic) system, leading to a profound re-definition of the body. In the nineteenth century, the metabolic circulation of chemical substances and organic matter (see von Liebig's contribution above) became increasingly accepted, and would form the basis of modern ecology. The "circulation" and the "metabolism" of matter became fused together as the two central metaphors through which to capture processes of socio-natural change, and of modernity itself.

Indeed, the use of the word "circulation" to refer to the movement of money within a national economy established itself within a generation of William Harvey's claim (Harvey 1999 [1628]). Thomas Hobbes, in *Leviathan* (1651), for example, had already compared the problems of a government that was unable to raise sufficient tax revenue to "an ague; wherein, the fleshy parts being congealed, or by venomous matter obstructed, the veins which by their natural course empty themselves into the heart, are not, as they ought to be, supplied from the arterie, whereby there succeedeth at first a cold contraction, and trembling of the limbs; and afterwards a hot, and strong endeavour of the heart, to force the passage of the blood" (cited in Harvey 1999 [1628]). Francis Bacon, in his essay *Of Empire*, wrote that merchants "are vena porta; and if they flourish not, a kingdom may have good limbs, but will have empty veins, and nourish little" (cited in Harvey 1999 [1628]).

At the beginning of the eighteenth century, the term "circulation" had become established in many sciences, referring to the flow of sap in plants and the circulation of matter in chemical reactions (Teich 1982). "Circulation" becomes a dominant metaphor after the French Revolution: ideas, newspapers, gossip and – after 1880

– traffic, air, and power "circulate". From about 1750, wealth and money begin to "circulate" and are spoken of as though they were liquids, flowing incessantly to become a process of accumulation and growth. Society begins to be imagined as a system of conduits (Sennett 1994). Montesquieu in *Lettres Persanes* (p. 117) speaks of "[T]he more 'circulation' the more wealth" and in *l'Esprit des Lois* of "[M]ultiplying wealth by increasing circulation". Rousseau (1766) refers to "[T]his useful and fecund circulation that enlivens all society's labour" and to "a circulation of labour as one speaks of the circulation of the money" (cited in Illich 1986). Of course, by the mid-nineteenth century, the *flâneur* – dandy, artist, detective, and stroller, the favourite literary characters of Baudelaire and, later, with Walter Benjamin, of the *passages* – has been well represented and theorized as an object of circulation within this urban space. Of course, in the process, "circulation" became less and less identified with closed circular movement, and more with change, growth, and accumulation. Similar to the way von Liebig discovered the mechanisms of metabolism through considering the "metabolic rift", "circulation" gained greater socio-ecological currency exactly when it became seen as an integral part of a process of change and transformation.

Adam Smith and Karl Marx conceived of a capitalist economy as a metabolic system of circulating money and commodities, carried by and structured through social interactions and relations. Accumulation is dependent on the swiftness by which money circulates through society. Each hiccup, stagnation or interruption of circulation may unleash the infernal forces of devaluation, crisis and chaos. Society's wealth and the relationships of power on which wealth is constructed are seen as intrinsically bound up with and expressed by the "circulation speed" of money in all its forms (capital, labour, commodities). Later, David Harvey (1985) would analyze the circulation of capital and its urbanization as a perpetual mobile channelled through a myriad of ever-changing production, communication and consumption networks. The development and consolidation of circulating money as the basis of material life, and the relations of domination and exclusion through which the circulation of money is organized and maintained shapes this "urbanization of capital".

By the mid-nineteenth century some British architects also begin to speak of the inner city mobilizing the metaphor of circulation. Sir Edwin Chadwick formulated the ideology of circulating waters effectively for the first time in his 1842 *Report into the Sanitary Conditions of the Labouring Population of Great Britain*. In his report, Chadwick imagined the new city as "a social body through which water must incessantly circulate, leaving it again as dirty sewage". Water ought to "circulate" through the city without interruption to wash it of sweats and excrements and wastes. The brisker this flow, the fewer stagnant pockets that breed congenital pestilence there are and the healthier the city will be. Unless water constantly circulates through the city, pumped in and channelled out, the interior space imagined by Chadwick can only stagnate and rot. This representation of urban space as constructed in and through perpetually circulating flows of water is conspicuously similar to imagining the city as a vast reservoir of perpetually circulating money.

Viollet-le-Duc introduced circulation as a bodily metaphor for the organization of the urban villa. In fact, Chadwick's papers were published under the title *The Health of Nations* during the centenary commemoration for Adam Smith (Chadwick 1887). Like the individual body and bourgeois society, the city was now also described as a network of pipes and conduits. The brisker the flow, the greater the wealth, the health and hygiene of the city would be. Just as William Harvey redefined the body by postulating the circulation of the blood, so Chadwick redefined the city by "discovering" its needs to be constantly washed (Illich 1986: 45). New principles of city planning and policing were emerging based upon the medical metaphors of "circulation" and "flow". The health of the body became the comparison against which the greatness of cities and states was to be measured. The "veins" and "arteries" of the new urban design were to be freed from all possible sources of blockage (Sennett 1994: 262–265; Corbin 1994).

With circulation as a metabolic process firmly established as practice and as solid representation of the process of socio-ecological change, attention quickly moved from metabolism and circulation to "speed" or, in other words, to the "movement of movement". Metabolic circulation of the kind analyzed by Marx, and now firmly rooted in generalized commodity production, exchange, and consumption, is increasingly subject to the socially constituted dynamics of a capitalist market economy in which the alpha and omega of the metabolic circulation of socio-ecological assemblages is the desire to circulate money as capital. As Douglas (2004) notes:

> Not only now would political rationality understand the motion of matter, and of bodies, it would seek above all to perfect the mechanisms of producing it. The "movement-of-movement", or "speed", as a technical achievement, emerges at this time (the early nineteenth century) as a societal principle, reordering the whole of the modern world. In the most radical way possible Virilio begins to answer the question of how efficiency was established in the modern urban landscape . . . The power of movement was subject to a spatial codification (in the city, in the workhouse, in the hospital, in the manufactory). By the beginnings of the nineteenth century this "codification" had been achieved, and a second "reordering" could now be effected. This reordering, rather than charting the middle ground between rapidity and stasis, aimed to "release" the full productive, dynamic efficiency of the (national) population in and through time. Motion had emerged as the destiny and law of a new politics of order. The full equivalence of Virilio's "metabolic vehicles" to Foucault's "bearers of order" becomes clear. Dromological power – or in the words of Foucault, "capillary power" – had emerged as the practical basis and first principle of the "free society" and "coded individual" established simultaneously with the apparatus of modern "governmentality". Mobility, in other words, had become simultaneously the means to liberation and the means to domination; the "accumulation of men" running simultaneously with "the accumulation of movement", and – one might add – the "accumulation of capital".

For Paul Virilio (1986), the freedom for people to come and go was replaced by an obligation to move. The creation of urban space as space of movement of people, commodities, and information radically altered the choreography of the city. Places and spaces became less and less shared, motion devalues or threatens to devalue place; connections are lost, identities reconfigured, and attachments broken down. While the urbanization of nature led to a spiralling accumulation of unstable socio-natural assemblages, the components of these assemblages became radically disassociated from their geographical origin as speed, movement and mobility ironically rendered the fields of vision and connections more opaque, transient, and partial. Although the city turned into a metabolic vehicle, the rift between the social and the natural became engrained deeper than ever in the modern urban imagination.

(HYBRID) NATURES AND (CYBORG) CITIES

> The metabolic requirements of a city can be defined as the materials and commodities needed to sustain the city's inhabitants at home, at work and at play ... The metabolic cycle is not completed until wastes and residues of daily life have been removed and disposed of with a minimum of nuisance and hazard.
>
> (Wolman 1965: 179)

> A barrel of crude oil sold for about $13 in 1998. The same quantity of whole blood, in its "crude" state, would sell for more than $20,000 [in Manhattan, NY].
>
> (Starr 1998)

When mobilizing the twin vehicles of "metabolism" and "circulation" from a historical-materialist epistemological perspective, the modernist tropes of "nature" and "society" transform radically. Modernity's bifurcation, separation, and binarization is recognized by historical materialism as exactly what it is: an image, a metaphor, a trope; one that can be and is mobilized for all manner of cultural, social, or political projects (Kaika 2005). A dialectical approach recognises both the radical non-identity of actants (human and non-human) enrolled in socio-metabolic processes within an assemblage, while recognising the social, cultural, and political power relations embodied relationally in these socio-natural imbroglios. The production of (entangled) things through metabolic circulation is necessarily a process of fusion, of the making of "heterogeneous assemblages", of constructing longer or shorter networks. In fact, both "hybridity" and "cyborg" are misleading as tropes, and may even be implicated in radically reproducing the underlying binary representation of the world. Hence, the bracketing of "hybrid" and "cyborg" in the title of this section refers exactly to the "excess of meaning" inscribed in coding the city as either "hybrid" or "cyborg".

Metabolic circulation, then, is the socially mediated process of environmental, including technological, transformation and trans-configuration, through which all manner of "agents" are mobilized, attached, collectivized, and networked. The heterogeneous assemblages that emerge, as moments in the accelerating and intensifying circuitry of metabolic vehicles, are central to a historical-geographical materialist ontology:

As plants, animals, minerals, air, light, etc., in theory form a part of human consciousness, partly as objects of natural science, partly as objects of art . . . so they also form in practice a part of human life and human activity. Man lives physically only by those products of nature; they may appear in the form of food, heat, clothing, housing, etc. The universality of man appears in practice as the universality which makes the whole of nature his inorganic body: (1) as a direct means of life, and (2) as the matter, object, and instrument of his life activity. Nature is the inorganic body of man, that is nature insofar it is not the human body. Man lives by nature. This means that nature is his body with which he must remain in perpetual process in order not to die.

(Marx 1982: 63)

As Luke (1999: 43) argues, "the conditions of associating humans and nonhumans in ancient, Asiatic, feudal, or capitalist relations of collectivization can thus be used to understand how power, knowledge, and conflict co-modified people and their things in any given society". These assemblages of humans and non-humans, of dead labour and inert materials, are reminiscent of the "hybrids" and the "cyborgs" of Latour and Haraway, respectively (see Luke 1999). However, while Haraway asks penetrating questions as to why "cyborgs" are produced the way they are and the relations of power inscribed in these imbroglios, this question remains silent in Latour's work. For him, the key issue centres on transforming the "constitutional" arrangements through which human and non-human actants become mobilized or enrolled (Latour 2004). In sum, while Latour defends a democratic republic of heterogeneous associations, Haraway maintains a perspective that emerges from a radically different ontological position. A deep ontological divide opens here. As Benedikte Zitouni (2004) convincingly argues:

Haraway views any entity as an *embodiment* of relations, an *implosion*, the threads of which should be teased apart in order to understand it. Whereas Latour views any entity as *a piece of matter* that is continuously affected and that contracts links with a larger networks *that allows it to live, to be*. On the one hand, the entity *crystallizes* the network; on the other hand the entity is *supported* by the network. Haraway studies the network in order to define the entity; Latour studies that same network in order to define the entity's consistency and persistence . . . Dialectics, congealment, crystals, prisms, representations are not possible tools any longer for urban studies but instead we view pieces of matter, of any kind, that act, react and interact with one another, that gain their consistency, persistence and existence or lose them through the affects and links to other agents. Power differences and inequality can no longer be stated as such, as a departure point into the city but have to be explained through the many actions and relations between objects, humans and non humans. There is nothing *behind* any space or agent, only attachments *aside* of it that make it stronger or weaker, allow it to exist or lead it to perish.

(Zitouni 2004)

It is in this latter sense that we wish to see the city as a metabolic circulatory process that materializes as an implosion of socio-natural relations, a process which is organized through socially articulated networks and conduits whose origin, movement, and position is articulated through complex political, social, economic, and cultural relations. These relations are invariably infused with myriad configurations of power that saturate material, symbolic, and imaginary (or imagined) practices.

Studies on urban metabolism have often uncritically pursued the standard industrial ecology perspective based on some input–output model of the flow of "things" (see Table 2.1 on London's metabolism). Such analysis merely poses the issue, and fails to theorize the making of the urban as a socio-environmental metabolism (see, for example, Weisz *et al.* 2001). While insightful in terms of quantifying the urbanization of nature, it fails to theorize the process of urbanization as a social process of transforming and reconfiguring nature. It would not be too much of an exaggeration to state that most processes of transformation of nature are intimately linked to the process of urbanization and to the urbanization of nature. From this perspective, it is surely strange to note that relatively little empirical or theoretical work has been undertaken that explicitly attempts to theorise environmental change and urban change as fundamentally interconnected processes.

Modern urbanization or the city can be articulated as a process of geographically arranged socio-environmental metabolisms. These are mobilized through relations

Table 2.1 The metabolism of Greater London (7,000,000 inhabitants)

Inputs	Tonnes per year
Fuel (oil equivalents	20,000,000
Oxygen	40,000,000
Water	1,002,000,000
Food	2,400,000
Timber	1,200,000
Paper	2,200,000
Glass	360,000
Plastics	2,100,000
Cement	1,940,000
Bricks, blocks, sand, tarmac	6,000,000
Metals	1,200,000

Wastes	Tonnes per year
Industry and demolition	11,400,000
Household, civic and commercial	3,900,000
Wet digested sewage sludge	7,500,000
Carbon dioxide gas	60,000,000
Sulfur dioxide gas	400,000
Nitrogen oxide gas	280,000

Source: www.global-vision.org/city/metabolism.html (H. Girardet).

that combine the accumulation of socio-natural use and exchange-values, which shape, produce, maintain, and transform the metabolic vehicles that permit the expanded reproduction of the urban as a historically determined but contingent form of life. Such socially driven material processes produce extended and continuously reconfigured intended and non-intended spatial (networked and scalar) arrangements and are saturated with heterogeneous symbolic (representational) and imaginary (wish images) orders, albeit "overdetermined" (Althusser 1969) by the generalized commodity form that underpins the capitalist "nature" of urbanization. The phantasmagorical (spectacular) commodity-form that most socio-natural assemblages take not only permits and facilitates a certain discourse and practice of metabolism, but also, perhaps more importantly, "naturalize" the production of particular socio-environmental conditions and relations. For example, it seems much easier to imagine an apocalyptic environmental future of humankind (of the kind perpetuated by global climate change pundits, bio-diversity preservation activists, or GM-warriors) than to imagine a political change in the actually existing social ordering of the metabolic process, one that would imply a re-construction of the produced environments.

The urbanization of nature is largely predicated upon a commodification of parts of nature while, in the process, producing new metabolic interactions and shaping both symbolic and material socio-natural interactions. The urbanization of nature necessitates both ecological transformation *and* social transformation. Urbanized nature propels the diverse physical, chemical, and biological "natural" flows and characteristics of nature into the realm of commodity and money circulation with its abstract qualities and concrete social power relations. Produced nature becomes legally defined and standardized, according to "scientific" politically and socio-culturally defined norms that are enshrined in binding legislation. Homogenization, standardization, and legal codification are essential to the commodification process. The urbanization process makes nature enter squarely into the sphere of money and cultural capital and its associated power relations, and redraws socio-natural power relations in important new ways. Indeed, the political-ecological history of many cities can be written from the perspective of the need to urbanize and domesticate nature and the parallel necessity to push the ecological frontier outward as the city expanded (Swyngedouw 2004). As such the political-ecological process produces both a new urban and rural socio-nature. The city's growth, and the process of nature's urbanization are closely associated with successive waves of ecological conquest and the extension of urban socio-ecological frontiers. Local, regional, and national socio-natures are combined with engineering narratives, economic discourses and practices, land speculation, geo-political tensions, and global money flows. This metabolic circulation process is deeply entrenched in the political-ecology of the local and national state, the international divisions of labour and power, and in local, regional, and global socio-natural networks and processes.

CONCLUSIONS

"Metabolism" and "circulation" permit excavating the socio-environmental basis of the city's existence and its change over time. The socio-naturally "networked" city can be understood as a giant socio-environmental process, perpetually transforming the socio-physical metabolism of nature. Nature and society are in this way combined to form an urban political ecology, a hybrid, an urban cyborg that combines the powers of nature with those of class, gender, and ethnic relations. In the process, a socio-spatial fabric is produced that privileges some and excludes many, that produces significant socio-environmental injustices. Nature, therefore, is an integral element of the political ecology of the city and needs to be addressed in those terms. Urbanizing nature, though generally portrayed as a technological-engineering problem is, in fact, as much part of the politics of life as any other social process. The recognition of this political meaning of nature is essential if sustainability is to be combined with a just and empowering urban development; an urban development that returns the city and the city's environment to its citizens. Being modern, as the poet Arthur Rimbaud (1873) captured it in the nineteenth century, is exactly about the active creation of situations and events, and participating in the production of our natures in so doing. Urban modernity as a particular set of processes of socio-metabolic transformations promises exactly the possibility of the active, democratic, and empowering creation of those socio-physical environments we wish to inhabit. In this sense, modernity is not over; it has not yet begun.

ACKNOWLEDGEMENTS

I would like to thank the British Academy Research Grant *Los Pantanos o la Muerte! Contested modernization, the production of nature and the hydraulic imperative in Fascist Spain, 1938–1974*. This chapter is adapted and extended from a paper published in *Science as Culture* (2006).

NOTES

1   Ernst Haeckel, who coined the term ecology (1866), mobilized organic metaphors to describe social conditions, and started a long lineage of human ecological analysis, one that would ultimately drive a wedge between the natural sciences and the social sciences as the legitimacy of such unmediated trans-formulations was increasingly questioned. Human ecology would subsequently bifurcate into a dematerialized social ecology, primarily through the Chicago School, on the one hand, and industrial ecology on the other. The latter, moving increasingly in the direction of a variety of types of commodity chain or goods-flow analysis, would increasingly distance itself from relational social theory (Fisher-Kowalski 1998; 2003; Fisher-Kowalski and Hüttler 1999; Newcombe 1977).

2   This statement, of course, does not mean that thought or languages are simply the epiphenomenon of "material" relations. On the contrary, very complex dialectical arrangements infuse the articulation of the real, the symbolic, and the imaginary (for different ways of exploring these articulations, see, for example, Žižek (Žižek and Daly 2003) or Lefebvre (1991) in the construction of the real.

3   Although Schmidt (1971) and Fisher-Kowalski (1998) maintain that Moleschott (1857) provided the influential insights, this is convincingly rebuked by Foster (2000), who maintained that von Liebig (1840) was of central importance. In any case, the use of "metabolism" was widespread in the emerging social sciences at the time and both Marx and Engels were familiar with the ongoing scientific debates in biology.
4   This has become engrained in social theory since its founding fathers Durkheim, Weber, and a "socialized" Marx.
5   While the Physiocrats were radically and correctly critiqued, the rational kernel of their mythical theorization was equally dismissed radically.
6   The first person apparently to suggest the circulation of blood in the arterial system was Ibn-al-Nnafiz (physician, born in Baghdad and died in Cairo in 1288) (Illich 1986: 40). The idea of circulation remained alien to the imagination of sixteenth-century Europeans. Two sixteenth-century scientists suspected what Harvey would later discover: Servetus (a Spanish genius and heretic burnt by Calvin — he also edited Ptolemy's geography in Lyon — and student of Vesalius in Paris) and Realdus Colombus of Padua (also student of Vesalius). Harvey was a student of Vesalius in 1603.

BIBLIOGRAPHY

Althusser, L. (1969) *For Marx*. London: Verso
Benton, T. (1989) "Marxism and natural limits: an ecological critique and reconstruction", *New Left Review*, 178: 51–86
Benton, T. (ed.) (1996) *The Greening of Marxism*. New York: Guilford Press
Bhaskar R. (1979) *The Possibility of Naturalism*. Atlantic Highlands, NJ: Humanities Press
Burkett, P. (1999) *Marx and Nature – A Red and Green Perspective*. New York: St Martin's Press
Castree, N. (2000) "Marxism and the production of nature", *Capital and Class*, 72: 5–37
Castree, N. (2002) "False anthitheses? Marxism, nature and actor-networks", *Antipode*, 34: 111–146
Chadwick, E. (1842) *Report on the Sanitary Conditions of the Labouring Population of Great Britain*. London: B.P.P., Vol. XXVI
Chadwick, E. (1887) *The Health of Nations*, 2 vols. Ed. R.W. Richardson. London: Longmans Green & Co.
Corbin, A. (1994) *The Foul and the Fragrant*. London: Picador
Douglas, I.R. (2004) "The calm before the storm: Virilio's debt to Foucault, and some notes on contemporary global capital". Online. Available HTTP: <http://proxy.arts.uci.edu/~nideffer/_SPEED_/1.4/articles/douglas.html> (accessed 15 May 2004)
Fischer-Kowalski, M. (1998) "Society's metabolism. The intellectual history of material flow analysis, Part I, 1860–1970", *Journal of Industrial Ecology*, 2(1): 61–78
Fischer-Kowalski, M. (2003) "On the history of industrial metabolism", in D. Bourg and S. Erkman (eds) *Perspectives on Industrial Ecology*. Sheffield: Greenleaf Publishing
Fischer-Kowalski, M. and Hüttler, W. (1999) "Society's metabolism: The state of the art. The intellectual history of material flow analysis, Part II, 1970–1998", *Journal of Industrial Ecology*, 2(4): 107–137
Foster, J.B. (2000) *Marx's Ecology – Materialism and Nature*. New York: Monthly Review Press
Gandy, M. (2004) "Rethinking urban metabolism: water, space and the modern city", *City*, 8(3): 371–387
Godelier, M. (1986) *The Mental and the Material*. London: Verso

Goubert, J.P. (1989) *The Conquest of Water: The advent of health in the industrial age.* Cambridge: Polity Press

Grundman, R. (1991) *Marxism and Ecology.* Oxford: Clarendon Press

Haeckel, E. (1866) *Generelle morphologie des organismen.* Berlin: G. Reimer

Haraway, D. (1991) *Simians, Cyborgs and Women – The reinvention of Nature.* London: Free Association Books

Hardt, M. and Negri, A. (2000) *Empire.* Cambridge, MA: Harvard University Press

Harvey, A.D. (1999) "The body politic: anatomy of a metaphor", *Contemporary Review,* August. Online. Available HTTP: <http: //articles.findarticles.com/p/articles/mi_m2242/is_1603_275/ai_55683940> (accessed 12 June 2004).

Harvey, D. (1985) *The Urbanization of Capital.* Oxford: Blackwell

Harvey, D. (1996) *Justice, Nature and the Geography of Difference.* Oxford: Blackwell

Harvey, W. (1628) *Exercitatio anatomica de motu cordis et sanguinis in animalibus.* Francofurti: Sumptibus Gulielmi Fitzeri

Hughes, J. (2000) *Ecology and Historical Materialism.* Cambridge: Cambridge University Press

Illich, I. (1986) *H₂O and the waters of Forgetfulness.* London: Marion Boyars

Jameson F. (2002) *A Singular Modernity.* London: Verso

Kaika, M. (2005) *City of Flows.* London and New York: Routledge

Kaika, M. and Swyngedouw, E. (1999) "Fetishising the modern city: the phantasmagoria of urban technological networks", *International Journal of Urban and Regional Research,* 24(1): 120–138

Kirsch, S. and Mitchell, D. (2004) "The nature of things: dead labor, nonhuman actors, and the persistence of Marxism", *Antipode,* 36(4): 687–706

Latour, B. (1993) *We Have Never Been Modern.* London: Harvester Wheatsheaf

Latour B. (2004) *Politics of Nature: How to bring the sciences into democracy.* Cambridge, MA: Harvard University Press

Lefebvre, H. (1991) *The Production of Space.* Oxford: Blackwell

Levins, R. and Lewontin, R. (1985) *The Dialectical Biologist.* Cambridge, MA: Harvard University Press

Liebig von, J. (1840) *Principles of Agricultural Chemistry, with Special Reference to the Late Researches Made in England.* English Edition. London: Walton & Maberly, 1855

Liebig von, J. (1842) *Animal Chemistry: or, Organic Chemistry in its Application to Physiology and Pathology.* Edited from the author's manuscript by William Gregory. With additions, notes, and corrections by Dr. Gregory and John W. Webster. A facsimile of the Cambridge edition of 1842. New York: Johnson Reprint

Luke, T.W. (1999) *Capitalism, Democracy, and Ecology.* Urbana-Champaign, IL: University of Illinois Press

Marx, K. (1861–1863) *Economic Manuscripts, 1861–1863,* Online. Available HTTP: <http: //www.marxists.org/archive/marx/works/1861/economic/ch13.htm> (accessed 30 March 2005)

Marx, K. (1971) *Capital,* Volume I. New York: Penguin (first published 1867)

Marx, K. (1973) *Grundrisse.* New York: Vintage Books (first published 1858)

Marx, K. (1974) *The German Ideology* (ed. C.J. Arthur). London: Lawrence and Wishart (first published 1846)

Marx, K. (1982) *Economic and Philosophic Manuscripts, Selected Writings.* London: Lawrence & Wishart

Mayer, J.R. (1845) *Die organische Bewegung in ihrem Zusammenhange mit dem Stoffwechsel.* Heilbronn: C. Drechsler

40   Erik Swyngedouw

Moleschott, J. (1857) *Der kreislauf des lebens*. Mainz: Von Zabern

Montesquie, C. de Secondat, baron de, 1689–1755 (1973) *Lettres Persanes*, Edition établie et présentée par Jean Starobinski. Collection Folio 475. Paris: Gallimard

Montesquie, C. de Secondat, baron de, 1689–1755 (1995). *l'Esprit des lois*. Paris: Nathan

Newcombe, K. (1977) "Nutrient flow in major urban settlements: Hong Kong", *Human Ecology*, 5(3): 179–208

Norgaard, R. (1994) *Development Betrayed: the end of progress and a coevolutionary revisioning of the future*. New York and London: Routledge

Padovan, D. (2000) "The concept of social metabolism in classical sociology", *THEOMAI*, 2

Rimbaud, A. (1873) *Une saison en enfer*. Bruxelles: M.J. Poot & Co

Schmidt, A. (1971) *The Concept of Nature in Marx*. London: New Left Books

Sennett, R. (1994) *Flesh and Stone*. London: Faber and Faber

Starr, D. (1998) *Blood: an epic history of medicine and commerce*. New York: Alfred A. Knopf

Swyngedouw, E. (1996) "The city as a hybrid – on nature, society and cyborg urbanization", *Capitalism, Nature, Socialism*, 7(1): 65–80

Swyngedouw, E. (2004) *Social Power and the Urbanization of Water: flows of power*. Oxford: Oxford University Press

Teich, M. (1982) "Circulation, transformation, conservation of matter and the balancing of the biological world in the eighteenth century", *Ambix*, 29: 17–28

Virilio, P. (1986) *Speed and Politics: an essay on dromology*, Semiotext(e). Cambridge, MA: MIT Press

Weisz, H., Fisher-Kowalski, M., Grünbühel, M., Haberl, H., Krausman, F. and Winiwarter, V. (2001) "Global environmental change and historical transitions", *Innovation*, 14(2): 117–142

Wolman, A. (1965) "The metabolism of cities", *Scientific American*, 213(3): 178–193

Zitouni, B. (2004) "Donna Haraway and Bruno Latour: an ontological divide", Paper presented at the "Technonatures II" conference, School of Geography and the Environment, Oxford University, 24 June 2004

Žižek, S. and Daly, G. (2003) *Conversations with Žižek*. Oxford: Polity Press

3 Metropolitics and metabolics

Rolling out environmentalism in Toronto

Roger Keil and Julie-Anne Boudreau

INTRODUCTION: URBAN POLITICAL ECOLOGY

This chapter is based on the theoretical and conceptual approach summarized as Urban Political Ecology (UPE) and specifically its aspect of urban metabolism. Specifically, we argue that in Toronto, during the past decade, something we call "metabolic metropolitics" has taken hold. Largely in spite of, or in the back of dramatic neoliberalizing processes, which the urban region underwent during that time, there has been a process of "roll-out-environmentalism". This development has had particular visibility in those areas of urban society–nature interactivity, where a massive physical redistribution of *material flows* in the urban fabric has taken place, often with a concomitant change in the mode of social regulation that accompanies these flows.

We have taken the inspiration for the notion of "roll-out-environmentalism" from Jamie Peck and Adam Tickell's recent paper on the switch from roll-back to roll-out neoliberalism in the 1980s and 1990s (Peck and Tickell 2002). We argue that the establishment of a neoliberal regime in Toronto during the 1990s had the unexpected and somewhat paradoxical side-effect of producing a strengthened urban ecological agenda – so it seems. Conservative elites in the urban region tried to rearrange governance systems and processes to their economic, cultural and political advantage after the election of a provincial Tory government in Ontario in 1995, the amalgamation of the city of Toronto into a "megacity" of 2.5 million people, and the tenure of an explicitly conservative mayor (Mayor Mel Lastman, 1998–2003) (Keil 2000; 2002). During this time of aggressive neoliberal restructuring in the urban social welfare state and in the political economy of Toronto, these elites seemed to have left their environmental flank unprotected from a surging environmental activism, which effectively used its free space to make major changes to the way material streams and metabolic relationships in the city were structured. This activism used both the existing progressive environmental basis of the inner city and the expanded playing field of the megacity to their advantage in the political struggle for an expanded municipal and regional environmentalism. The latter includes a deliberate broadening of the urban political ecological field to include "suburban" issues and suburban constituencies and a clever use of the more centralized and powerful decision-making structures in the expanded municipality. This development created a new "sustainability fix" (While *et al.* 2004) and included a significant jump in scale for metropolitan environmental politics from the two solitudes of inner city and exurban issues and actors. In effect, it also entailed a democratization of the societal relationships with nature across the urban region as the governance of the metropolitan environment left the "subpolitical" realm of much of conventional urban ecology in favor of a "public ecology" (Luke 2003) constructed around issues such as waste, water, air pollution, and pesticide use.

UPE – roughly understood as a discourse and practice floating on Marxist urban theory, constructivist political ecology, urban ecology, social ecology, environmental justice theory and practice, ecological modernization theory and others – has become an important home for intellectual debates on these questions (Keil 2003; Swyngedouw and Heynen 2003). Swyngedouw, Kaïka and Castro have recently summarized this field in ten concise points that circumscribe an incipient yet already fairly contoured project (2002: 124–125).

The major point made by proponents of UPE is that the domination of nature and the domination of humankind are connected processes and that these processes come together in the urban. Most participants in the debates around UPE would agree that the idea of liberation, as Lefebvre would have us believe, must come about through the urban societies in which we inevitably live. But it also must go through the natural – physical and symbolic – metabolisms that we equally, unavoidably belong to. This makes the "materiality" of nature a central concern of UPE (as opposed to urban political *economy*) (Bakker 2003). The biophysical reality of most urban metabolic processes makes them subject not just to symbolic, discursive deliberations of all sorts but also of quite physical engineering practice (and discourse). Whether it is water, waste, wetlands, waste or energy, there is always a tangible reality to the processes in question (Görg, 2003). Swyngedouw has added to this with his notions of socionature, hybridity, and the particularly powerful concept of quasi-objects – influenced equally by Latour and Marx (Swyngedouw 2004). This metabolic and material relationship of the natural and the social that now comes together in urban life is the topic of urban political ecology.

Central to this definition of UPE is the notion of "metabolism" and the "interwoven knots of *social process, material metabolism* and *spatial form* that go into the formation of contemporary urban socionatural landscapes (. . .) [I]t is on the terrain of the urban that [the] accelerating metabolic transformation of nature becomes most visible, both in its physical form and its socioecological consequences" (Swyngedouw and Heynen 2003: 906–907). The notion of urban metabolism has been present in discussions of urban environments at least since its first widespread use in the 1960s following the seminal article by Abel Wolman (1965), "The metabolism of cities". Often cited as a principle to understand the position of urban regions in a larger world (Girardet 1992), this notion shares with a similar concept, the "ecological footprint" (Wackernagel and Rees 1996), the curious lack of much empirical follow-up. While it is relatively easy to grasp that cities depend on inflows and outflows of materials, energy, etc., it is more complicated (and rarely undertaken) to do a full empirical study of such outflows. A recent study on the metabolism of Toronto, for example, claims that it is "the first urban metabolism of a Canadian urban region, and possibly the first for a North American city. It also makes a first attempt at comparing the urban metabolism models of a few cities worldwide" (Sahely *et al.* 2003: 469). This work builds on previous, and rare, similar studies, the most well-known of which have been the studies of Hong Kong's metabolism (Newcombe *et al.* 1978; Warren-Rhodes and Koenig 2001). We will take up the results of the Toronto study below. At this point, we would merely like to point to a certain restrictiveness of much existing urban metabolism analysis. Sahely, Dudding and Kennedy define urban metabolism as "a means of quantifying the overall fluxes of energy, water, material, and wastes in and out of an urban region. Somewhat analogous to human metabolism, cities can be analyzed in terms of their metabolic flow rates that arise from the uptake, transformation, and storage of materials and energy and the discharge of waste products" (Sahely *et al.* 2003: 469; Warren-Rhodes and Koenig 2001). While such quantification leads to impressive results and suggests comparability with other urban regions, it has a few weaknesses that need to be addressed: 1. Beyond reference to policy changes (introduction of recycling, for example), there is little attention paid in these studies to the political changes in the study area; 2. While economic changes are being registered, a fundamental critique of the capitalist economy that underlies such changes is missing; 3. Social factors (modes of regulation; habits of consumption, etc.) are rarely factored into the equation (apart from noting differences such as the auto dependency of North American cities versus the pedestrian nature of Hong Kong's mobility system); 4. Nature is seen as relatively static: material streams are described as mostly unchanging in character and itself not with a sense of agency but – in good engineering tradition – as an object of human ingenuity. We believe that the notion of urban metabolism can be usefully applied if one keeps these four caveats in mind and it is in this more comprehensive sense that, below, we employ it.

Sahely *et al.* (2003: 478) conclude in their Toronto metabolism study:

> The most noticeable feature of the GTA metabolism is that inputs have generally increased at higher rates than outputs over the study years [1987–1999]. The inputs of water and electricity have increased marginally less than the rate of population growth (25.6%), and estimated inputs for food and gasoline have increased by marginally greater percentages than the population with the exception of diesel fuel. With the exception of CO_2 emissions, the measured output parameters are growing slower than the population. The outflows of residential waste and wastewater loadings have even reduced in absolute terms.

The authors also admit that, "Several of the improvements to the efficiency of metabolism can be attributed to enlightened policy and wise investment" (p. 478). The reduction of waste due to recycling is a case in point. But the study does not detail the measures taken by recent municipal governments, nor does it problematize the social and cultural changes that led to the "efficiencies". As we will argue below, the "efficiencies" may be more rising "effectiveness" of straight environmental policies applied at various scales of the urban region. Let us look at some of the recent changes in environmental policies affecting the urban metabolism of Toronto.[1]

The technological model (or: bringing the incinerator back in?)

Given the extreme challenges of fiscal and economic realities in a neoliberal regulatory environment, progressive urban ecological initiatives and policies can easily be contextualized as unintended overall reorganization of socio-ecological relationships that they can be turned on their head. The tremendous successes of more progressive environmental policies are now running up against the material and discursive limits of reform. This is most clearly expressed in the area of waste. While the City introduced so-called Green Bins citywide in order to capture the so-called wet waste of households, it is struggling to find solutions to the continuingly growing mountain of common household waste, which escapes recycling and is being shipped by caravans of trucks along a more than 400 km long road to Michigan. As ecological modernization has gripped the material streams and waste habits of all Torontonians, it remains unclear whether the spirit of reduction can outlast the nagging aggressiveness of the proponents of more conventional methods of waste management and of technologically advanced forms of incineration.[2] Waste streams have reached an impasse as old dumps are closing, new ones are not available, shipment of waste runs into public opposition along the trucks' route and in the recipient location; on one hand, this has led to sophisticated new waste diversion strategies, which also rely on compliance within households throughout the city. The city has a plan for 100 percent waste diversion by 2010. The Waste Diversion Task Force 2010, comprised of the entire Toronto City Council, was charged with "finding a "Made in Toronto" solution for waste diversion from landfill" (City of Toronto 2001a). After public consultation a report to that effect was released and the new policy was passed in 2001 (City of Toronto 2001a). On the other hand, given the enormous pressure on the municipality due to huge waste output, less benign and more conventional technology-based proposals are back. Among them is the renewed call for incineration. At the last City Council meeting before the latest municipal election in 2003, an environmental assessment for a pilot program was improved for new and emerging technologies like gasification, etc., which is just another term for incineration (New Tech Waste 9/03).[3]

Water networks and delivery systems

These are being scrutinized for their ability to deliver clean and healthy water to the region in an efficient and socially just way (Debbané and Keil 2004; Young and Keil 2005); the sewerage system, which drains the cities households and industries, must be updated all the time as the system ages, capacities reach limits and demand grows. Public workers, who are the backbone of the water system's flawless and reliable performance, have linked the discourse on ecological modernization to a social and environmental justice claim that involves both the security of their

Figure 3.1 "Watering the road again?" Municipal advertising campaign against water
waste

Source: Produced by Axmith McIntyre Wicht Ltd for the City of Toronto

Figure 3.2 "Relax, it's just a weed!" Municipal advertising campaign for the new
by-law that bans pesticides from private ornamental gardens

Source: Produced by Axmith McIntyre Wicht Ltd for the City of Toronto

jobs and the maintenance of high standards of skill and scrutiny in the system. Marketization, privatization and commodification of water are seen as potential threats to a time-honored system of public service delivery in water services. The threat of privatization has not been the only important issue related to water. The city has pledged to invest hundreds of millions of dollars into its failing water and sewer infrastructure; a Water Efficiency Plan was passed by City Council in 2003 with the goal of reducing water and wastewater flows in an attempt to meet the demands of a growing regional population (City of Toronto 2003). A Works Best Practices Program has been introduced. Storm water quality management and sewage flow monitoring have been improved and re-regulated.

Among the big ways in which socio-ecological material streams have been re-regulated has been the Deep Lake Water Cooling project, a PPP with Enwave District Energy Limited in the lead. This scheme, which had been discussed in environmental activist circles in Toronto since the early 1990s, will eventually use pumped-in water from the bottom of Lake Ontario to cool the office towers in the central city in a much more efficient and ecologically sustainable way than its current oil-dependent method (City of Toronto 2005a). It is clear that this scheme, which was ridiculed less than a decade ago as environmentalist fantasy, has now gained the attention of the development elites of the inner city, who are keen on lowering their energy costs in an age of $50 barrel oil.

Smog policies

These are another area of rapid and drastic change. The Toronto Atmospheric Fund (TAF) of the City of Toronto was established in 1991 to finance local initiatives to combat global warming and improve air quality in Toronto. Highly successful and renowned world-wide, the TAF survived amalgamation and devolution and has been an important institution of local environmental policy making (City of Toronto, 2005c). Air pollution control policies are necessarily devices for scale-jumping as municipal governments ostensibly reach beyond their limits in creating clean air programs.

Similar incursions into the conventional way of dealing with its material streams have been made regarding the issue of road salt, a formidable issue in the yearly rhythm of socio-ecological management in a winter city. While the City has not stopped using salt, it has a plan minimizing it. While in official parlance, safety and cost concerns are still prioritized, in February 2002 the City Council "endorsed a recommendation to encourage the Toronto school boards and all City agencies to follow the City's lead in reducing the use of salt on their properties for de-icing purposes" (City of Toronto 2002).[4]

Pesticide policies

An important step forward for environmentalists was taken with the passing of a municipal by-law in 2003 that forbids future use of pesticides in urban gardens and front yards. This law symbolizes a major victory of the environmental movement to ban a significant amount of chemicals from passing through the material streams of the city. Since 1998, when city council expressed its commitment to "phase out pesticide use on public green spaces" and subsequently reduced such use by 97 percent to the current by-law that governs private green spaces, the municipality has made big changes to the political ecology of toxins in Toronto. A related by-law was passed in May 2003. It "will restrict outdoor use of pesticides in the City of Toronto". The law, which allows some pesticide use for public health reasons, came along with the establishment of a Pesticide By-Law Advisory Committee to regulate and oversee the policy's implementation (City of Toronto 2005b). Following the analysis of the local state as vulnerable to civil society intervention (see Desfor and Keil 2004: ch. 2), it is possible to argue that the success of this initiative can be partly explained as a consequence of the territorial and scalar contradictions of how nature is organized in capitalism. Taking advantage of local conditions and urban political dynamics allowed TEA and its allies to win a surprising victory over a giant chemical industry which was organizing a well-funded campaign to stop the initiative. Roberts explains:

> There are many industries that are absolutely absent here – so when it comes to pesticides for instance there are lots of pesticide applicators but there is no pesticide industry here. And they can handle citizen lobbies in Ottawa and at Queen's Park but they can't do it across the country – city by city – they'll go broke. So the fossil fuel industry is not represented here – only users of fossil fuels. And this is why I believe radicals should orient to municipal politics. They're never going to get an opportunity to move the agenda without opposition – other than ideological – there's no one that's going to lose their factory or plant or workers that will lose their job. You know like the CAW supported the Island Bridge and everything else because their members' jobs are on the line. So I always strategically pick issues that there's not a vested interest that will lose out so that was one of the definitions of a quickstart.
>
> The reason why I go into it is that it's not as limiting a factor as you might think. Because we have no pulp and paper here – there's no oil – it's a long list of things we don't have. Other than the fact that a person is an instinctive conservative they're not funded by – there's not pesticide companies – to my knowledge – that are funding councilors to run again to take on the demon environmentalists or anything like that. It would be worth while if we could get somebody to just check their corporate donations but I would be totally surprised if anybody's receiving money from national corporations because of their role in defending their local interest. And to the best of knowledge that whole pesticide thing as an example was only the pesticide applicators – like the national industry didn't even send people. So there's high consensus – low resistance; multiple beneficiaries.
>
> (Personal interview, 3 December 2003)

A NEW ENVIRONMENTAL REGIME?

The environmental regime after roll-out environmentalism shapes up to be a consensus regime built on the strong pillars of a locally defined ecological modernization project, propped up constantly by the driving energy of a significantly more radical urban ecological movement, which has also moved more directly into the realm of the local state. The new consensus position is somewhat of a contested hegemony as Wayne Roberts explains elegantly:

> First of all when you have a consensus position it doesn't mean that we all agree – we all have found our own way to this position. So there is a consensus in Toronto that our future lies as a clean green city.
>
> (Personal interview, 3 December 2003)

Even before the election of progressive mayor David Miller in November 2003, the conditions had been created to fundamentally redraw the material streams from – to borrow from Herbert Girardet (1992) – a more *linear* to a more *circular* urban metabolism. While this change has clearly been on the agenda of many large municipalities, the Toronto case is particularly interesting because this still ongoing shift has occurred under political and social conditions which, at face value, would have not suggested such changes to be possible: the condition of roll-out neoliberalism set in motion by consecutive Tory provincial governments and a compliant Toronto mayor, who was a representative of the city's development regime.

To begin with, we can posit that some of these changes were possible precisely because the elites left a large chunk of the metropolitanization of Toronto's political mechanisms to "ordinary people" and their progressive institutional and political representatives. At least, this is what the President of the Toronto Labour Council, John Cartwright, too, seems to suggest happened in past decades in Toronto:

> while we can talk about the vision of key public figures like David Crombie or Jane Jacobs, there was something else happening back then. There was a tremendous involvement of ordinary people challenging things they didn't like. They organized, worked hard, and came up with real alternatives that are nowadays celebrated as elements of our city's greatness . . . All of those people – ordinary people – made Toronto the city that we cherish and celebrate. Their struggles were often opposed by powerful elites. But their struggles were successful, and we need to look at how they built enough power to achieve success.
>
> (Personal interview, 19 May 2004)

It is in this vein that we will argue below that environmentalism was "rolled out" following the initiative of ordinary activists during neoliberal times. Yet, we are not arguing that Toronto has become a strange ecological oasis in a neoliberal environment. Rather, the "mainstreaming" of local environmentalism (Churley, personal interview, 28 May 2004; Tabuns, personal interview, 19 May 2004) had somewhat predictable effects on the political astuteness of the environmentalists themselves: they became more professionalized as their importance in municipal politics grew and much of their political program and strategy – while highly successful in changing the urban metabolism of Toronto – became more accommodationist in the process. The protest/activist mode of urban political ecological groups was slowly moulded into a more policy/consultant mode. The growing professionalism of Toronto environmentalists was at once a necessary outcome of the changed political landscape after amalgamation, a function of the career of environmentalist organizations, and the willingness of urban bureaucracies – which at that point contained many environmentalists themselves – to lend an open ear to the complaints of the movement (Perks, personal interview, 3 December 2003; Roberts, personal interview, 3 December 2003). In what follows, we contextualize the metabolic changes in Toronto in the economic and political frame in which they have taken place: neoliberalism, ecological modernization and the struggle for a more sustainable city.

MAKING NATURE COMPETITIVE IN TORONTO

The 1990s in Toronto were a period in which two distinct regimes of urban ecological modernization followed each other (for a recent overview of an application of this concept see Desfor and Keil 2004). First, coming immediately out of the exuberant and growth-intoxicated 1980s but based on the historical compromise of the liberal 1970s was an "ecosystem approach". This approach was promoted in particular by the Royal Commission on the Toronto Waterfront since 1988 and became the operative ideology for all urban development discourse in the early 1990s. It remained at least partly committed to the core values of Toronto's liberal growth regime; not surprisingly, the head of the Royal Commission was ex-mayor David Crombie, who was one of the figureheads of the reform movement. Since the recession tested Toronto's growth regime in general and urban development in particular after the crash of 1989, a second ecomodernization discourse started to form, which was more aggressive and neoliberal than its predecessor and soon freed itself from the constraints that may have been placed on free market reign by traditions of redistributional politics, conservationist discourse and democratization. This second, more neoliberalized ecomodernization project made itself relevant to a region in economic crisis and fell in line with the more conservative politics that beset the Province after 1995. The symbolic benchmark of this period was the elaboration of a soil remediation policy, which allowed for economic-ecological win-win situations all around and provided the basis for a thorough revitalization of real estate market once the economy picked up in the late 1990s (Desfor and Keil, 2004; Stewart, 1999).

The countervailing discourse to the predominant practice of ecological modernization was a growing popular and sometimes radical urban environmental politics that developed around a few prominent issues, most notably around ecological spaces such as the Leslie-Street-Spit, the Don Valley and the Rouge River. But there were also significant struggles around issues of toxic soil hotspots, local air pollution through incineration, and green work (Hartmann 1999; Desfor and Keil 2004; Keil 1994). In this alternative discourse, the Toronto Environmental Alliance[5] became the dominant voice for a progressive and democratic – if somewhat centre-city and middle-class biased – urban ecological movement in the city. The Task Force to Bring Back the Don, a more moderate organization of mostly conservationist activists in the neighborhoods abutting the Don River in the East of Toronto, was the most visible showpiece of Toronto's civic environmentalism and the clearest example of the emergence of a new system of urban ecological governance that integrated civil society concerns directly into the architecture of the local state (Desfor and Keil 2004).

The political centrepiece to this alternative and critical discourse has been provided by a set of progressive city councilors and provincial representatives, particularly those associated with the New Democratic Party. Of those, Jack Layton appears as the most representative of both a persistent catalyst of radical ecological politics in the City and conduit into the administrative and political process. One long-time observer of the political scene in Toronto notes:

> Jack Layton who was an incredible Councilor – the year after he, when he lost the Mayoralty election he was footloose and fancy free and the two of us set up a coalition for a green economy with Gary Gowan from the Liberals and he went deep green in that period. It began as a sort of Keynesian economics essentially like people who believed in government spending to get out of the depression and they may as well spend it on an environmental project as anything else. Some went deep green in the process and he was one of them and he really has a profound understanding of it and wanted to do something about it and I think that's why he eventually left municipal politics – he wanted to play with those ideas in a bigger arena.
>
> (Roberts, personal interview, 3 December 2003)

Both discourses – *ecomodernist* and *radical ecological* – were active at the time when Toronto was amalgamated in 1997. They actually both played heavily into the amalgamation debate where one side spoke of economic efficiency and growing out of the ecological crisis through more consolidated economies of scale; and where the other side insisted on the value and virtue of the local communities and ecological topologies of the diverse and globalized city. The question of the size of government and the size of municipal area was closely connected to environmental sensibilities in at least four ways: 1. The "civic environmentalism" of Toronto was closely linked to the downtown reform politics of the post-1970s; 2. The split in political culture between the central city and the suburbs was perhaps nowhere felt as much as in the field of environmental politics; 3. Most environmentalists were convinced that amalgamating Metro Toronto was too little or too much centralization: the bioregional scale may have been more appropriate for ecological regulation; 4. The neoliberal impetus of the consolidation nurtured fears among environmentalists that programs would be cut, regulation of polluters would be weakened and more suburban settlement forms would outweigh sustainability concerns.

The general expectation among environmental activists in 1997 was that amalgamation would have negative effects on the tremendous progress achieved since the beginning of the 1990s. This pessimism was largely due to a political analysis that contrasted the progressive inner city with the more conservative suburbs. Fowler and Hartmann write: "The aftermath of the first mega-city election, which saw the reduction of elected officials from 106 to 57, not only shifted the political balance to the right, it also shifted political power from the downtown core to the generally more conservative suburbs. The new mayor, former North York mayor Mel Lastman, came to power on a Tory-style agenda featuring a tax freeze. And he could count on a majority of councilors – mostly from the suburbs – to support his neo-liberal agenda" (2002: 157). This "suburbanization of local politics" (Keil 1998) was a distinct reality and a shock to the liberal-progressive political regime that had existed in the downtown core for the past two decades. It was generally and correctly assumed that this regime had provided a comfortable shell in which very successful urban ecological projects such as the work of the Task Force to Bring Back the Don River could be moved forward (Desfor and Keil 2004: ch. 4). A dissociation of development politics from social and environmental concerns seemed typical of this period as spectacular megaprojects, mostly in the entertainment and sports sectors, were built and planned for, and condominium towers went up along the waterfront, while homeless youth were pushed from squats, squeegee kids were criminalized, and environmental concerns were driven into the background of the public agenda. Fowler and Hartmann conclude that in the context of the new governance structure of the megacity, "the Toronto environmental movement and other allies struggled to keep environmental issues from falling off the public policy radar screen. After a decade of growing success, environmental advocates found themselves in a new political terrain that left little room for any environmental initiatives that could not be presented as 'good for business'" (2002: 157). The contextualization of all things environmental in a development regime that was at best ecomodernist, and at worst indifferent to ecological affairs, indeed, created much anxiety in the environmental community in Toronto. Interestingly, though, the changes in the regime did not come as expected. One reason, following Fowler and Hartmann, was the attention the Toronto Transition Team (TTT), responsible for setting up the newly amalgamated municipal government, paid to the "many and difficult environmental issues facing the new city" (2002: 158). One of the TTT's recommendations was the establishment of an Environmental Task Force (ETF), which was indeed created by City Council in March 1998.

The TTT's recommendation was not the only reason for the birth of the ETF. The new municipal regime was also born in a set of specific crises that needed more than the usual attention. As a consequence, the environment, homeless activism and diversity policies again became important areas of municipal activity in Toronto in the late 1990s. The outcome of the dissocation of the mid-1990s was the sectioning off of certain responsibilities from the main development discourses and their subordination under it when and where needed (Olympic bid, waterfront). The new mayor Mel Lastman, as oblivious as he had been to issues such as homelessness, poverty or the environment, realized quickly that in order to govern both the City with its complex socio-ecological structures and dynamics and the wildly unruly City Council, he would have to compromise and delegate. Some progressive downtown councilors played a key role in this politics of compromise. They were all associated with the social democratic New Democratic Party, in particular the left-wing power couple Jack Layton and Olivia Chow as well as gay councilor Kyle Rae. While the former two became the environmental, housing/homelessness and children advocates of the new city, Rae became an unlikely ally for Lastman's regime in the promotion of glitz, culture, entertainment, central-city gentrification and revanchist measures in the downtown core.[6]

ROLL OUT ENVIRONMENTALISM?

Work of the Environmental Task Force led to a remarkable new Environmental Governance regime built most importantly around an innovative document called the Environmental Plan for the City of Toronto published in February 2000. The plan marked a shift from struggles of the environmental community against the roll-back neoliberalism of the mid-1990s and the emergent competitive regime (Kipfer and Keil 2002) to a "roll-out environmentalism" that accepted an institutional compromise with the ecological modernization regime of the day but moved ahead in strident ways in substantive areas, mostly due to the leadership of councilor Jack Layton, who also became the head of the Federation of Canadian Municipalities and began to use the urban green agenda as a plank in his platform to become the national leader of the NDP in 2003. The scale jumping of the urban environmental agenda in this remarkable way in an unlikely moment of neoliberal retrenchment was unpredictable and unexpected. The politics of scale engaged by Layton and his allies was immediately successful in breaking out from both the downtown ghetto of left-liberal environmentalism and the specific restrictions that had taken hold of Toronto politics in the face of fiscal austerity and ideological suburbanism. Taking the "downtown" agenda of social and environmental sustainability national created a frame of legitimacy for an activist and policy agenda that was under severe attack locally.

The establishment of the ETF was reported to have occurred under a specific set of tensions between "more traditional environmentalists [who] wanted the task force to act as a vehicle for entrenching the best environmental protection initiatives from the former municipalities throughout the new mega-city" and "other members [who] saw environmental protection as intimately tied to economic prosperity and social health" (Fowler and Hartmann 2002: 158). The ETF launched so-called "Quick Start" initiatives in order to spread established good environmental practices throughout the city and established multi-stakeholder working groups on various environmental topics.[7] The notions of "environment" and "sustainability" remained contested in the process and the result a governance structure for environmental matters built around a "Sustainability Roundtable" and a governance document in the form of the Environmental Plan. Sustainability criteria were written firmly into all city policy processes at that point: "In some ways, the city's first-ever

Environmental Plan is as much a sustainability plan as an environmental plan. The first part of the plan contains more conventional recommendations aimed at improving the health of the city's air, land, and water. The latter part of the plan contains a number of recommendations that promote sustainability in the transportation, energy use, and economic development sectors" (Fowler and Hartmann 2002: 159). The plan is a remarkably upbeat and powerful document, which frankly presents some of the main environmental challenges faced by Canada's largest city and makes sensible and practical recommendations for improvements mostly through a mix of governance innovations, spread of best practices and moderate ecological modernization measures, often couched in a language mostly as compatible with the sensitivities of Toronto's civic environmentalism as with the lexicon of the neoliberal zeitgeist of the Lastman regime. Fowler and Hartmann ultimately propose that "the Toronto Environmental Plan suggests an alternative: a particular type of economic growth that is both financially frugal and also sensitive to environmental and social concerns", leading the city down a "path that, if followed, will lead to a much healthier, happier, environmentally benign, and vibrant economic future for Torontonians" (2002: 161).

But the Lastman years were really just a time of institutional progress "behind the scenes": "roll-out environmentalism" could only really take hold in the bureaucracy after the election of David Miller, when problematic practices of the administration in the context of a computer leasing contract were brought to light in a highly publicized mismanagement inquiry, and the path was cleared for a different administrative style, more in line with the social democratic mayor's idea of a "clean" bureaucracy.[8] It was also a period of modest success for the Toronto environmental movement. This can be exemplified in the work of the Environmental Alliance (TEA). During the past few years, TEA sunk its teeth into a variety of issues both related to the emerging expansion of the urban economy and to more marginal or niche issues related to the Toronto environment. It seems that the dramatic change in urban governance priorities during the Lastman years towards a more competitive and revanchist city created precisely the niches or venues in which urban environmental policy could thrive.

The environment was brought to the municipal government agenda in the early 1990s through a pressure politics led by mostly left-leaning local politicians and activists. As the decade went on, and as environmental politics became more institutionally and procedurally integrated into the political process at City Hall, support for environmental issues started to move across the political spectrum. While conservatives tended to be against many environmental initiatives which often were perceived to endanger, for example, existing contracts in the hard-lobbying construction and engineering industries, it is not unusual to now find support for ecological concerns among fiscal or social conservatives. Wayne Roberts talks about the enthusiasm of one right-wing Toronto councilor for rooftop gardens and that of another – who had grown up on a farm – for progressive and bioregional food policies (personal interview, 3 December 2003). Another long-term environmental activist (albeit a conservative one) and now city councilor, Glenn De Baeremaeker, speaks to the same point when he observes that there is now somewhat of an environmental hegemony in Toronto, where even some of his more conservative colleagues would not openly question the necessity of certain environmental measures that are before Council, but would merely ask about cost and efficiency: the conflict may be about the technique used in a specific case but not about whether something should be done at all (De Baeremaeker, personal interview, 3 March 2004).[9]

A version of a regime of ecological modernization, which is less tied to the promise of Toronto's civic regime and more neoliberal in character, has taken hold in the exurban 905 telephone area north of Toronto, where development is now principally sold as green, smart and in harmony with "nature". In what De Baeremaeker (personal interview, 3 March 2004) has called "an Orwellian approach to language" a programmatic "live with nature" approach has now emerged, in which even those people who have lived after the principle "the more sprawl the better" have come around to use the smart growth language that is now pervasive everywhere. This strategy often reverts to the notion that development actually "produces" a better environment as it plants trees in subdivisions where there were empty farmers fields filled with soybeans. Support for public transit, such as the Light Rail Transit lines in the making for suburban growth centres Markham or Vaughan, also plays with the general theme of the hegemonic ecological modernization discourse. But while this support may look environmental at first glance, it is really generated as a thinly disguised attempt to create a better infrastructure for more sprawl.

AMALGAMATING URBAN NATURE

Surprisingly, then, amalgamation has had positive net effects in terms of the urban environmental discourse. Whether it has led to better "actually existing" sustainability remains to be seen (Krueger and Agyeman forthcoming). Amalgamation was criticized for its scalar failure, i.e. its inability to address urban and environmental issues at the most appropriate scale: instead of bringing in a more locally democratic governance structure with clear bottom-up democratic procedures, it saddled the municipality with devolution and downloading to the detriment of the kinds of local spaces that make experiment and change possible; instead of consolidating the greater urban area at the bioregional scale in order to make at least an attempt at creating an integrated socio-ecological policy area, it created an artificially demarcated polity at the metropolitan level and left the regional contradictions between inner- and outer-city intact. None of these structural weaknesses of governance restructuring under the provincial Tories have been significantly altered, but the polity of this newly created jurisdiction started to work around these structural deficiencies. There is reason to believe that the newly created municipal governance structure, for example, became more transparent, and surprisingly more open to the desires of environmentalists. Some of it had to do with the expanded and much more unpredictable council, which necessitated new and constantly renegotiated alliances and created windfall efficiencies in the bureaucracy. Wayne Roberts notes: "Which is the incredible thing with amalgamation is if you can get it through a committee you get it through in minutes, seconds because they say, well, the right, left and centre are on the committee and if they've okayed it then its OK" (personal interview, 3 December 2003).

Amalgamation had meant the merging of a number of important middle management positions across the City. In addition, there was a predominance of politicians and power brokers from the old suburbs and from the Metropolitan Toronto Government in the transition process to the "megacity". This had been largely interpreted as resulting in a marginalization of the historically very progressive municipal bureaucrats from the old core city. In addition, the ideological conservatism and political nepotism of former mayor Mel Lastman, who brought many of his former North York staff with him into leading positions at City Hall created a chilly climate for many reform oriented bureaucrats. Some would assume that the municipal governance culture at City Hall changed, during these years, irreversibly towards a managerial culture led by ideas of efficiency and market orientation. Still, we argue that there is a sustained progressive impetus among the newly amalgamated City's 40,100 employees based on their history of social engagement – from the original trade union organizing in the nineteenth century to the more current social, (multi)cultural and environmental civic activism that motivates people to become public workers. What seems to have happened, ironically, is that the provincial Tories, despite their attempt to purge the City of radicals and progressives, and instead of weakening their worst political enemies in the core of Toronto, gave them a larger playing field for their activities. The local state showed both its resilience in the face of external attack, an incredible flexibility in handling its affairs and an ability to rebound. The Toronto bureaucracy actually strengthened its presence in environmental affairs precisely at the time when the macro-environment suggested an ecological Armageddon at the provincial level.[10]

Moreover, the success of roll-out environmentalism in hard times was grounded to a large degree on the support from large parts of City staff. Roberts remembers:

> What happened with the Environmental Task Force was that Layton pulled together I think a pretty good team of politicians – pretty good in terms of citizen representatives who would again run the gamut from left to right and be people who would command some general respect and there were staff people who were looking for a chance to do something in this area. If you went to the early meetings you would see that however many people were on the task force I can't remember but the chairs around it were always packed and mostly – at least half of them were staff.
>
> So that tells you two things: one is that there was staff interest and that people were looking for an opportunity to get one of these programs and start to work on them and two that there are people who are interested in a career track and they smelled this as an area that's going to become important. It had a lot of staff interest.
>
> (Personal interview, 3 December 2003)

After Mel Lastman was gone, a newly reinvigorated debate about democratization and transparency took hold in the city. While still a city councilor, the current mayor, David Miller, headed a discussion series on civic engagement for the CAO's office of Toronto. Miller carried the mandate of this group over into his election campaign and into office. Once in power, Miller espoused various policies of citizen involvement such as a Porto Alegre-style budgeting process and a series of roundtables as advisory bodies to the elected council (City of Toronto Chief Administrator's Office 2001).

WHERE ARE WE GOING FROM HERE? OR: HAS AMALGAMATION BEEN GOOD FOR NATURE?

The neoliberal ecological modernization model, which made roll-out environmentalism possible, has now reached an entirely new stage. With the election of a new mayor and council in November 2003, a potentially new regime was born: a regime of *social-democratic modernization*. Wayne Roberts comments on the consequences of Miller's election for food policy in Toronto:

> When David Miller got elected, the next morning I got up and I felt like I had a hundred pound weight lifted off my shoulders, For the last eight years we have been fighting to keep programs alive that could be cut at any moment at anybody's whim.
>
> Now all of a sudden we're talking about a whole new Council. The Food Policy Council has two Councilors that are assigned to it. They have never once come to a meeting of the Food Policy Council in the three years that I've been on it. We had six members on the Food and Hunger Action Committee and we had to literally write every speech they did – you name it.
>
> (Personal interview, 3 December 2003)

A senior bureaucrat in City Hall expressed similar experiences of the shift. In explaining that the city was leaving the (neoliberal) age of efficiencies above all behind, a new era of "effectiveness" had seemed to begin with Miller's election:

> Just drive around the city. Have a look at the parks. People come back to the – this is the starkest analogy – people that have lived in the city 10 years ago and they've moved away and they came back and they went "what happened to my city?" You know? And its been ten years, and I *can't* blame it on amalgamation, because amalgamation was built on the premise of cost saving and *efficiencies*. Where I was going with this thought is, this is a new era we feel that there's a new . . . energy around the city, that we're going into an *effectiveness*. What are we effective at? How do we make the services better to meet the demands and expectations of the public. So this is a new era for us, and I think it's just started since the latest mayoral election.
>
> (Librecz, personal interview, 25 May 2004)

Roll-out environmentalism and roll-out neoliberalism are embraced in a set of new policies at the urban level, which reorganize major areas of urban economic, political and social life. Whereas changing the capitalist foundations of Toronto's existence are not up for debate by any means, the city's major material streams and their social regulation are under constant redefinition.

CONCLUSION

In all these cases (here just considered for the amalgamated core City of Toronto), behavioural changes, material re-regulation and changes to the political economy (e.g. public or private; outsourced or unionized) are being fought over by political actors of all sectors. To each major proposed change to the socio-ecological relationships – both material and discursive – in the urban region, there are two sometimes converging but often oppositional subtexts of expertization, techno-logization, and marketization/commodification on one hand, and of what Andrew Light (2003) calls "civic environmentalism", on the other hand. It needs to be seen to what degree, for example, the growing expertization, which has brought eco-plogical activists directly into the local state, has increased or decreased democratic involvement in these processes as it has meant de-legitimization of more grassroots forms of environmental knowledge production (Fischer 2000). It also remains to be seen whether expertization and the march of the environmentalists through the institutions of the local state leads to the effective disappearance of opposition and the cooptation of all potential critique. It needs to be asked, as well, whether the new governance of water, air and waste has led to new air, waste and water actor networks and ultimately a new urban political ecology of air, waste and water in Toronto.

The obstacles to creating a cleaner urban region remain formidable. On the one hand, it remains undeniable that Toronto remains an automobile region. As Sahely, Dudding and Kennedy note: "It is apparent that the parameters of greatest concern are the increasing inputs of gasoline and diesel and the associated output of CO_2," (2003: 478–479). Indeed, the Ontario auto industry, of which Toronto is the geographical centre, is reported to have produced about 2.7 million cars in 2004, more than automobile powerhouse Michigan (Hakim 2004). The region is also home to the world's largest car parts manufacturer, Magna Corporation. In the face of such overwhelming industrial power, any municipal metabolic politics must be kept in perspective. Toronto remains, as Lefebvre would have had it, a "society of controlled consumption" (2003) fired by a Fordist core that seems immune to any greening. The gap between the "cleaning" of the city and the continued automobile dependency is obvious and cannot easily be closed.

The mode of regulation of metabolic metropolitics remains contested. A rather surprising editorial in the influential *Toronto Star* of 21 November 2004 argued vehemently against the spread of recycling at the expense of "regular" garbage pick-up (and was met with equal vehemence by letter writers in the general public, many of whom criticized the *Star* for its lack of sensitivity to Toronto's ecological necessities). This intervention reminded Torontonians of the divisive politics around garbage, which the urban region experienced in the early 1990s, when the then

NDP provincial government created a much criticized Interim Waste Authority to deal with the allegedly looming landfill crisis in the region; similarly, there still existed memories of the more recent struggle to prevent garbage from being shipped to the Adams Mine site in Northern Ontario (Perks, personal interview, 3 December 2003; Grier, personal interview, 2 December 2004).

Lastly, the environmentalists themselves are keeping up the pressure. As the ever-active TEA has pointed out repeatedly, despite the improvements in many areas, much needs to be done as air pollution, ground and water contamination and other systemic problems persist (see, for example, TEA 2005). And a recent study by a group called Pollution Watch found that pollution in Canada on the whole actually got dramatically worse in the past decade, with the Greater Toronto Area standing out as being relatively more polluted than other parts of the country (Pollution Watch 2004).

So, while not all is well in Miller's world of roll-out environmentalism, the principle of the City turning its attention to environmental issues has made inroads in the everyday life of the urban bureaucracy and is beginning to make a difference in people's homes and public spaces. Metabolic metropolitanism is here to stay. It will be difficult – at least in the area of collective consumption and individual households – to return to the days of linear metabolism and supply side thinking. Whether there is real attention paid to the civic activism that has brought ecological issues into the core of the governance system will have to be seen. In Toronto after neoliberalism, a tender peace has been struck between ecological modernizers, radical environmentalists and other interested groups (including the various industries). The new "sustainability fix" which has been proposed for the region by way of this compromise entails an incipient rethinking of the metabolic streams on which the city is built. It has now become the challenge for all involved to reconnect the focus on material streams with concerns about democracy, participation and social justice.

ACKNOWLEDGEMENTS

We are extremely grateful for the research assistance of Douglas Young and Punam Khosla for this article. Funding for this research has been provided by SSHRC Standard Grant # 410–2003–1207, "Gouvernance métropolitaine et compétitivité internationale: les examples de Montreal et Toronto" (with PI Pierre Hamel, University of Montreal, and Bernard Jouve, UQUAM).

NOTES

1 Note that virtually all of these were implemented after the end of the metabolism study by Sahely *et al.* (2003).
2 It was rumored during the Fall 2004 US presidential election, after Democratic candidate John Kerry had threatened to close the border to Canadian garbage if elected, that the City was quietly looking for alternative dump sites in Southern Ontario.
3 For a good discussion of the complexities of the garbage crisis in today's cities see Gandy 1994.

4   Next to these examples, we could also name the new provincial policy to create a green belt in Southern Ontario. Taking huge areas of land under development restrictions results in tangible reductions of material streams and changes the spatial mode of regulation significantly in the region.

5   On their website, the Toronto Environmental Alliance presents itself in this way: "The Toronto Environmental Alliance was formed in 1988 in order to provide an activist voice to local Toronto issues. TEA's mandate is to promote a greener Toronto. Our vision of a healthy community is based on equity, access, safety and a clean environment. We work with concerned individuals, public health agencies, local governments and grassroots organizations in order to encourage the participation of Toronto citizens on local issues and to provide a forum for citizens to be heard on environmental issues.

  "TEA focuses on six major campaign areas, where we work on a variety of projects and programs related to Smog and Climate Change, Urban Pesticides, Waste Reduction, Sustainable Transportation, Water, and Involving Youth. (. . .)

  "TEA speaks out on issues that shape our local environment so that, with the help of community volunteers, TEA has made progress on critical environmental issues while expanding the skills and knowledge of our broader community. Over the past decade TEA has grown to play a well-respected role both in the community and in municipal governments" (http://www.torontoenvironment.org/abouttea/index.html; accessed 23 November 2004).

6   One observer remembered how the politics of compromise in the environmental area took shape after amalgamation: "So a couple of months after the election Layton said to Lastman – he bumped into him in the hall – [. . .] you know we don't have an environmental policy – of course we didn't have any policy cause it was an amalgamated city – so Lastman says why don't you get one – I'll put you in charge – you'll be chair of a committee and we'll set up environmental policies for the new city – so that's how he became head of The Environmental Task Force" (Roberts, personal interview, 3 December 2003).

7   Quick Start became a widely used policy tool in other areas than the environment later. Wayne Roberts talks about its application in the food policy field: "Out of the Task Force we developed this concept that we've been working on ever since called the 'quickstart'. And we developed three criteria for a quickstart and I've since developed ten and I'm probably going to try to write it up sometime before the new year which is – partially so people know whether they have an issue that is a quickstart or not – like is child poverty an issue that you can move on quickly? You might think it is because everybody loves children – they're not guilty of being lazy and they're not responsible for their poverty or anything like that – so this is a checklist for you to know – and if you have four of them but you have the potential – you know these are the six things you have to work on. But we developed three criteria at the time. That is that it has a high degree of consensus and low degree of resistance – if there's a high degree of resistance you're not going to have a quickstart by definition – you're going to have a long haul. So the fantastic thing about municipal politics is that there are many circumstances where that is true because the bourgeoisie is not equally represented geographically across the country" (Roberts, personal interview, 3 December 2003).

8   Mayor Miller had campaigned using a broom as a symbol for sweeping the city clean of corruption. Next to the so-called MFP inquiry, which he spearheaded already as a councilor, Miller also moved quickly in his new job as mayor to make city hall more transparent to citizens.

9   De Baeremaeker relates an anecdote from recent memory: it used to be that in the more suburban political circles mention of the existence of white tail deer in city parks would not be greeted with amazement and wonder or lead to an impulse to protect habitat but to the impulse to get out the rifles to shoot the animals instead. Today, even opponents to growth control in the exurban development regime would try to cushion their interests in a rhetoric of ecology.

10 Both Wayne Roberts and Glenn De Baeremaeker suggest that this development has not yet taken place in the suburbs/905. Why? "Because people in the NGOs don't live in the 905" (Roberts 2003).

BIBLIOGRAPHY

Amin, A. (2004) "Regions unbound: towards a new politics of place," *Geografiska Annaler*, 86B (1): 33–44

Bakker, K. (2003) "A political ecology of water privatization," *Studies in Political Economy*, 70 (Spring): 35–48

City of Toronto (2001) *Waste Diversion Task Force*, Online. Available HTTP: < http: //www.city.toronto.on.ca/taskforce2010/2010_report.htm> (accessed 17 April 2005)

City of Toronto (2002) Highlights of City Council Meeting of February 13, 14, and 15. Online. Available HTTP: http://www.city.toronto.on.ca/council_highlights/2002/ 021302.htm (accessed 18 April 2005)

City of Toronto (2003) *Toronto's Water Efficiency Plan*, Online. Available HTTP: <http: //www.city.toronto.on.ca/watereff/plan.htm> (accessed 17 April 2005)

City of Toronto (2005a) *Deep Lake Water Cooling Project cools Toronto . . . naturally*. Online. Available HTTP: <http://www.city.toronto.on.ca/water/deep_lake/ > (accessed 17 April 2005)

City of Toronto (2005b) *Pesticides: Brochures and Fact Sheets*, Online. Available HTTP: <http://www.city.toronto.on.ca/health/hphe/pesticides_reducing.htm> (accessed 17 April 2005)

City of Toronto (2005c) *Toronto Atmospheric Fund*, Online. Available HTTP: <http: //www.city.toronto.on.ca/taf/index.htm> (accessed 17 April 2005)

City of Toronto – Chief Administrative Officer (2001) *Building the New City of Toronto: Reflections on Civic Engagement*. Toronto: City Hall.

Debbané, A.-M. and Keil, R. (2004) "Multiple disconnections: environmental justice, urban sustainability and water", *Space and Polity*, 8(2): 209–225

Desfor, G. and Keil, R. (2004) *Nature and the City: Making Environmental Policy in Toronto and Los Angeles*. Tuscon, AZ: University of Arizona Press

Fischer, F. (2000) *Citizens, Experts and the Environment: The Politics of Local Knowledge*. Durham, NC: Duke University Press

Fowler, Edmund P. and Franz Hartmann (2002) "City environmental policy: connecting the dots," in Edmund P. Fowler and David Siegel (eds) *Urban Policy Issues: Canadian Perspectives*. Second Edition. Don Mills: Oxford University Press

Gibbs, D., Jonas, A. and While, A. (2002) "Changing governance structures and the environment: theorising the links between economy and environment at the local and regional scale", *Journal of Environmental Policy and Planning*, 5(3): 235–254

Girardet, H. (1992) *The Gaia Atlas of Cities*. New York: Anchor Books

Görg, C. (2003) *Regulation der Naturverhältnisse: Zu einer kritischen Theorie der ökologischen Krise*. Munster: Westfalisches Dampfboot

Hakim, D. (2004) "Ontario to overtake Michigan as auto kingpin", *The Globe and Mail*, November 29: A1-A8

Hartmann, F. (1999) "Nature in the city: urban ecological politics in Toronto". Ph.D. Dissertation, Political Science, York University

Keil, R. (1994) "Green work alliances: the political economy of social ecology," *Studies in Political Economy*, 44: 7–38

Keil, R. (1998) "Toronto in the 1990s: dissociated governance?," *Studies in Political Economy*, 56: 151–167

Keil, R. (2000) "Third way urbanism: opportunity or dead end?," *Alternatives* (Rob Walker, ed.) (commissioned), 25(2) (April–June)

Keil, R. (2002) "'Common sense' neoliberalism: progressive conservative urbanism in Toronto, Canada", *Antipode*, 34(3): 578–601

Keil, R. (2003) Progress report: urban political ecology, *Urban Geography*

Kipfer, S. and Keil, R. (2002) "Toronto Inc? Planning the competitive city in the new Toronto," *Antipode*, 34(2): 227–264

Krueger, R. and Agyeman, J. (forthcoming) "Sustainability schizophrenia or 'Actually existingsustainabilities': the politics and promise of a sustainability agenda in the US", *Geoforum*

Lefebvre, H. (2003) *The Urban Revolution*. Minneapolis, MN: University of Minnesota Press

Light, A. (2003) "Urban ecological citizenship," *Journal of Social Philosophy*, 34(1): 44–63

Luke, T.W. (2003) "Global cities vs.global cities rethinking contemporary urbanism as public ecology", *Studies in Political Economy*, 70: 11–34

Newcome, K., Kalma, J. and Aston, A. (1978) "The metabolism of a city: the case of Hong Kong", *Ambio*, 7(1): 3–15

Peck, J. and Tickell, A. (2002) "Neoliberalizing space". In N. Brenner and N. Theodore (eds) *Spaces of Neoliberalism*. Oxford: Blackwell

Peet, R. and Watts, M. (eds) (1996) *Liberation Ecologies*. London: Routledge

Pollution Watch (2004) *Shattering the Myth of Pollution Progress in Canada: A National Report. Toronto, Pollution Watch Online*. Online. Available HTTP: <www.pollution watch.org> (accessed, 18 April 2005)

Sahely, H., Dudding, S. and Kennedy, C. (2003) "Estimating the urban metabolism of Canadian cities: Greater Toronto Area case study", *Canadian Journal of Civil Engineering*, 30: 468–483

Stewart, K. (1999) "Greening social democracy? Ecological modernization and the Ontario NDP", Ph.D. Dissertation. Political Science. York University, Toronto

Swyngedouw, E. (2004) *Social Power and the Urbanization of Water. Flows of Power*. Oxford: Oxford University Press

Swyngedouw, E., Kaika, M. and Castro, E. (2002) "Urban water: a political-ecological perspective", *Built Environment*, 28(2): 124–137

Swyngedouw, E. and Heynen, N. (2003) "Urban political ecology, justice and the politics of scale", *Antipode*, 35(5): 898–918

TEA (Toronto Environmental Alliance) (2005) *Smog, Climate Change and Energy*, Online. Available HTTP: <http://www.torontoenvironment.org/climate?PHPSESSID=2d970 a8652db2ec461615077477c6ba0> (accessed 18 April 2005).

The Toronto Star (2004) "Editorial: Toronto's messy garbage collection", *The Toronto Star*, 21 November 2004

Wackernagel M. and Rees, W. E. (1996) *Our Ecological Footprint: Reducing Human Impact on the Earth*. Gabriola Island, BC; Philadelphia, PA: New Society Publishers.

Warren-Rhodes, K. and Koenig, A. (2001) "Escalating trends in the urban metabolism of Hong Kong: 1971–1997", *Ambio*, 30(7): 429–438

While A., Jonas, A. and Gibbs, D. (2004) "The environment and the entrepreneurial city: searching for the urban 'sustainability fix' in Manchester and Leeds," *International Journal of Urban and Regional Research*, 28(3): 549–569

Wolman, A. (1965) "The metabolism of cities", *Scientific American*, 213(3): 178–193

Young, D. and Keil, R. (2005) "Urinetown or morainetown? Debates on the re-regulation of the urban water regime in Toronto", *Capitalism, Nature, Socialism*, 16: 2 (June)

4 Urban nature and the ecological imaginary

Matthew Gandy

INTRODUCTION

The artist Lucian Freud is perhaps best known for his striking figurative representations of the human body. In the recent retrospective of his work held at the Tate Gallery in London, however, we find an intriguing exception to these studio portraits represented by a painting entitled *Wasteground with Houses, Paddington* (1970–2). This intricate tableau, which reveals a remarkable glimpse of London from the window of his studio, is framed by the rear elevation of a typical Victorian terrace. The drab greyish-brown brickwork and stained cornices are enlivened by ranks of chimney stacks with their jumble of fulvous earthenware chimney pots. Cutting through the middle of the scene is a mews of former stables now appropriately converted into a row of smart garages and in the foreground is an expanse of rubble-strewn waste ground. Despite the twisted remains of abandoned furniture and rusted metal this former bomb site is now brimming with botanical interest: the faded spikes of the ubiquitous *Buddleia davidii* are interspersed with other characteristic colonizers of London's post-war landscape such as ground-elder *Aegopodium podagraria* and rosebay willowherb *Chamaenerion angustifolium*. This, then, is an urban landscape, a seemingly unremarkable fragment of urban nature yet a critical reminder of the intricate combination of nature and human artifice which has produced urban space. An "urban ecology" is by definition a human ecology and is no more or less "natural" than any other kind of modern landscape whether it be a managed fragment of wild nature in a national park or those accidental pockets of nature of the type that Freud observed from the window of his studio in West London.

The interaction between nature and the modern city raises a series of conceptual complexities. If we understand the city to be a special kind of nodal point within an extending hyphal mesh of urbanization this still leaves the idea of urban nature as a somewhat ill-defined entity. The urbanization of nature, a transformation that has gained accelerated momentum over the last few decades, is clearly much more than a gradual process of appropriation until the last vestiges of "first nature" have disappeared. The production of urban nature is a simultaneous process of social and bio-physical change in which new kinds of spaces are created and destroyed, ranging from the technological networks that give sustenance to the modern city to new appropriations of nature within the urban landscape. The word "nature" is used here to encompass two somewhat different clusters of ideas: on the one hand, the term nature is used to denote a menagerie of concrete forms ranging from the human body to parks, gardens or complete ecosystems; and on the other hand, nature is evoked as an ideological and metaphorical schema for the interpretation of reality. In practice, however, these abstract and concrete elements are often interwoven to produce a densely packed urban discourse within which the origins and implications of different conceptions of nature are often afforded only cursory reflection.

The rise of the modern industrial city necessitated a refashioning of relations between nature and culture. Yet to refer to this transformation simply as the production of "urban nature" does not fully capture the complexity of this transition. The term "metropolitan nature" is probably more apposite since it can be deployed to signal recognition of the specific ways in which cultures of nature evolved in response to the socio-economic development and technological complexity of the modern city (see, for example, Green 1990). The urbanization of nature – and the concomitant rise of a metropolitan sensibility towards nature – encompasses not just new approaches to the technical management of urban space such as improved housing and sanitation but also extends to different kinds of cultural interactions with nature as a source of leisure. The transformation of nature in the modern city thus extends from new modes of urban governance or "governmentality" to use Foucault's term to changing modes of cultural perception so that both the strategies and techniques of negotiating urban space become inseparable.

Conceptions of the modern city have often been framed in terms of degrees of deviation from a supposed "natural" mode of living or in terms of analogies made with the body of a living organism. Ideas drawn from nature have played a significant role in developing an "ecological imaginary" in which ideas or metaphors drawn from the bio-physical and medical sciences have been used to understand the form and function of the modern city. The dynamics of urban change have been conceived, for example, in terms of processes such as ecological succession, the metabolic transmutation of nature or even the post-industrial impetus towards putrescence and decay. Underlying many formulations of the ecological imaginary, however, there is an implicit naturalization of urban processes so that urbanization is no longer conceived as the outcome of historical change but rather as a cyclical dynamic alterable through technological modifications rather than by political contestation. By developing a conception of urban nature as a medley of different elements we can begin to critically dissect some of the nature-based metaphors which have played such an influential role in the development of critical urban discourse. This chapter seeks, therefore, to explore a hiatus between the conceptual stasis emanating from organicist conceptions of urban form and an alternative set of readings of urban space which place greater emphasis on the malleable, indeterminate and historically specific dimensions to the urban experience.

ABERRATIONS AND UTOPIAS

From the middle decades of the nineteenth century onwards, the urban experience became increasingly synonymous with the experience of modernity itself. In Britain, for example, the urban population increased from just under a quarter in 1800 to over 75 percent in 1900 so that urban life had changed from a minority experience to the majority experience. London, for instance, had a population of around 1 million in 1800 rising to 4.5 million by 1881. Similarly, Berlin saw its population of 200,000 in 1800 rise to 1.5 million by 1890 and Paris experienced a five-fold increase in population over the course of the nineteenth century to reach over 2.5 million by 1900. Other cities growing dramatically during the nineteenth century include Chicago, Glasgow, Manchester, New York, Naples, Rome, St Petersburg, Vienna and Moscow. These industrial cities necessitated a new synthesis between nature and culture extending from the construction of urban technological networks to the establishment of new modes of municipal governance. The modern metropolis that emerged out of the chaos of the nineteenth-century city was driven by a combination of factors: advances in the science of epidemiology and later microbiology which gradually dispelled miasmic conceptions of disease; the emergence of new forms of technical and managerial expertise in urban governance; the innovative use of financial instruments such as municipal bonds to enable the completion of ambitious engineering projects; the establishment of new policy instruments such as the power of eminent domain and other planning mechanisms which enabled the imposition of a strategic urban vision in the face of multifarious private interests; and the political marginalization of agrarian and landed elites so that an industrial bourgeoisie, public health advocates and other voices could exert greater influence on urban affairs (Gandy 2004).

From the nineteenth century onwards, we find that the urban experience begins to take an increasingly dominant place in modern culture. It is paradoxical that although modern cities were frequently evoked in organicist terms as bodies or organisms in their own right, cities were at the same time widely perceived to reside outside nature or the "natural order" as parasites or monsters. Thomas Hardy, for example, described London as "a monster whose body had four million heads and eight million eyes" and a spate of nineteenth-century novels such as James Greenwood's *The Wilds of London* (1874) dwelled on the poverty, darkness and danger associated with the industrial metropolis. Yet, developing in parallel with these eschatological responses to urbanization, we can also detect a different set of discourses focused on the implications of urbanization for modern consciousness. By the early twentieth century we find a proliferation of interpretations of the jarring and disorientating quality of urban life: Georg Simmel, for example, explores the "blasé outlook" of the city dweller in Berlin as a means to handle the "rapidly shifting stimulations of the nerves"; Virginia Woolf describes the atomism of urban life in the seemingly vast and alienating expanse of metropolitan London; and James Joyce conceives of early-twentieth-century Dublin as a series of fragmentary encounters between different characters struggling to make sense of their lives. In these literary and sociological evocations of the modern city the urban experience enables the development of new forms of social, political and sexual awareness. As the city played a role in the enlargement of identity, we find that modern consciousness finds expression through intensified pleasures of nature within the industrial metropolis. With the gradual distancing of the body from the fatigue, illness and malnourishment of the past, new cultures of metropolitan nature developed including excursions into semi-wild fragments of nature at the urban fringe and the development of new aesthetic sensibilities towards landscape. As the social and political possibilities engendered by the modern metropolis unfolded, however, the progressive potential of the city began to increasingly conflict with traditional conceptions of urban order. The idea of "nature" was to play a defining role in this emerging tension between modernity and tradition as an ambiguous motif capable of underpinning both radical approaches to urban design and at the same time questioning the very foundations for urbanism itself.

The increasing association of the industrial city with the destruction of rural life sharpened the perceived antinomy between "city" and "country". The modern city was widely characterized as an aberrant spatial form that threatened to undermine existing ties of social and communal solidarity (see Williams 1973). The perceived superiority of rural life forms part of a powerful anti-urban sentiment connecting between the Jeffersonian ideals of small-town America and a succession of later writers on cities from Ferdinand Tönnies in the 1870s to Louis Wirth in the 1930s and Jane Jacobs in the 1960s. In a contemporary context, anti-urban views have resurfaced as part of an ecological critique of modernity which has a wide ranging influence on architecture and urban design. The so-called "New Urbanism", for example, owes much to the perceived superiority of small-town life within which ideological motifs of stability and sustainability draw heavily on nature-based conceptions of urban design. A determinist conception of spatial form as a dominating influence over human behaviour is combined with a form of ecological nostalgia for an imagined past.

In contrast with reactionary visions of the modern city as an aberration we can find an alternative lineage of urban thought originating within the Renaissance ideals of the city-state which drew inspiration from the designs of Hippodamus, Vitruvius and other early advocates of symmetrical urban form. Renaissance scholars such as Leon Battista Alberti asserted that beauty in architecture was derived from the mimesis of nature but he went beyond neo-Platonic conceptions of creativity to emphasize the critical role of human skill in the full realization of aesthetic perfection (see Bacon 1967; Forty 2000). Urban design becomes an extension of beauty in nature whether reflected in the geometric arrangement of space or the embellishment of urban life through gardens, fountains and other meticulous appropriations of nature within the fabric of the city. The cultural utilization of nature is thus both an aesthetic quest to change the urban landscape but also an attempt to foster greater degrees of social and spatial order. The emergence of the nineteenth-century urban beautification movements, for example, sought to reintroduce nature into cities in order to prevent urban space from becoming an uninterrupted vista of development. A myriad of new organic spaces began to appear as a means to re-establish contact between nature and urban society.

With the introduction of features such as public parks, botanical gardens and tree-lined boulevards we find the explicit inclusion of a designed nature within the heart of the modern city.

The city beautiful movement evolved into the garden city movement of the early twentieth century and the search for a more ambitious synthesis between nature and urban form. The garden city brought together what was at best an inchoate mix of different ideas ranging from the utopian planning ideals of Ebenezer Howard to the naturalistic landscape designs of Frederick Law Olmsted who drew a stark contrast between the cultural vibrancy of industrial America and the "rustic vice" of plantation agriculture. The various approaches to the garden city as it diffused through Europe and North America combined an eclectic mix of influences including romantic and Beaux Arts traditions but transcended the earlier ad hoc interventions through the articulation of a comprehensive approach to urban planning and design. Yet this apparent reconciliation between "city" and "nature" masked the actual transformation of nature under the impetus of capitalist urbanization. The principal legacy of the city beautiful and garden city movements was not the creation of utopian fragments in the urban landscape, important though these were, but the linking of landscape design and city planning ideals with burgeoning middle-class aspirations. In subsequent decades this earlier attempt to find a synthesis between nature and culture would in fact lead towards ever greater degrees of spatial polarization through the growth of suburbs, peripheral housing estates and other twentieth-century efforts to dismantle the inner core of modern cities.

FROM DESIGN TO FUNCTION

Developments such as the urban beautification and garden city movements were largely tangential to the underlying dynamics of capitalist urbanization, yet they remain one of the most influential dimensions to urban design. If we shift our attention, however, towards the function rather than the design of urban space we find that the transformation of nature is far more pervasive and complex than it might first appear. The production of urban nature is inseparable, for example, from the development of urban technological networks which served to bind the modern city into a more integrated spatial form. The central cores of older cities were modernized to make way for roads, railways and speculative land development forcing the working classes into ghettos and industrial districts of intense poverty. The "Haussmann approach" of comprehensive reconstruction pioneered in Second Empire Paris was also extended to Amsterdam, Barcelona, Cologne and many other cities (see Centre de Cultura Contemporánia de Barcelona 1994). Yet this impetus towards spatial rationalization was not without its critics. In Vienna, for example, Otto Wagner attempted to create a modern city based around light, space and ease of movement, but his vision was challenged by Camillo Sitte with his rejection of utilitarian rationalism (see Harvey 1989; Schorske 1981). Wagner's disavowal of nature as an organizational impetus for urban design was not shared, however, by all of the leading figures within the modernist movement: Le Corbusier and Frank

Lloyd Wright, for example, continued to place nature at the centre of their work even if their conception of the modern city conflicted with the more vernacular urbanisms of the past.

From the late nineteenth century onwards urban planning emerged as a clearly defined discipline accompanied by the growing role of technical elites in the institutions of modern governance. From the 1880s onwards, for example, innovations such as land use zoning and regional planning gathered momentum. Technical and administrative expertise played an increasingly significant role in the modern repertoire of "governmentality" and the management of complex urban societies. By the early twentieth century we find increasing emphasis on the "scientific management" of cities in the segregated and hierarchical ordering of urban space. Earlier attempts to create a utopian synthesis of nature and culture were gradually supplanted by a more radical technologically inspired vision. Progressively greater emphasis was placed on the radical separation of land uses in the "hygienic city" so that light, air and movement took precedence over the congested mingling of land uses in the nineteenth-century city. The idea of "speed" became the focal point for a new urban imaginary rooted in the creative destruction of the past. These technological fantasies reached their apogee in the designs of Italian futurists such as Antonio Sant'Elia with his emphasis on multi-level roadways as a means to perfect the circulatory dynamics of urban space. The sketches and plans for these technological utopias depict towering new buildings and virtually empty roads in an era before post-war congestion and the grassroots political challenge to the excesses of technological modernism.

The ideology of "hygienism" in twentieth-century urban planning belied the persistence of environmental and miasmic conceptions of the healthy city in combination with the post-bacteriological revolution in the scientific management of space: free circulation of both air and people was both a utopian gesture towards the horrors of the nineteenth-century city but also an attempt to create a new kind of organic unity within the modern metropolis. The early decades of the twentieth century saw attempts to forge closer links between urban nature and the public realm so that the earlier innovations of municipal parks, improved sanitation and pedagogic displays of nature in zoos and museums could be extended to encompass a more ambitious conception of the role of nature in the modern city. Under the American New Deal, for example, we can discern a shift towards an expanded conception of urban nature to encompass a wider programme of social reform including improvements in health, housing and urban infrastructure. The reshaping of nature on behalf of the modern city also encompassed vast engineering projects to provide water and power so that the new landscapes of dams and aqueducts in the American West, for example, cannot be conceived independently from the vast urban agglomerations with which they are connected. Yet the earlier associations of water engineering projects with a progressive political agenda, whether in Roosevelt's America or Nehru's India, have now waned to the extent that many of these engineering projects have become a leitmotif for the rapacious impact of modern cities on impoverished and politically marginalized rural communities (see, for example, Cutler 1985; Roy 2002).

In the twentieth century the changing relationship between nature, technology and urban space was driven to a significant degree by the spread of car ownership. This technological dynamic transcended national differences to the extent that we can discern striking similarities between the landscaped highways of Germany, Italy and the United States. In Martin Wagner's plans for 1920s Berlin, for example, the need for regional mobility was combined with the development of new peripheral housing estates. Wagner attempted to re-organize urban space in order to promote the greatest possible human happiness so that the rationalization of social and economic life and the rationalization of space became inseparable facets of the same process (see Scarpa 1986). Similarly, in Fritz Schumacher's plans for Hamburg (1909) and Cologne (1920) the centres of these cities were to be opened out with parks and public spaces to foster a new kind of leisure-oriented metropolitan culture (see, for example, Centre de Cultura Contemporània de Barcelona 1994). And in the United States, Robert Moses brought a distinctively car dominated vision to the modernization of the New York metropolitan region within which middle-class aspirations would play a decisive role. During the late 1950s and early 1960s, however, the construction of urban highways began to open up a conflict between the centralized engineering dominated ethos behind infrastructure development and growing demands for greater public participation in urban planning. The ideal metropolis conceived by technical experts and urban managers was increasingly in conflict with the lived reality of the modern city. Urban planning faced the disintegration of the kind of putative "public interest" which had sustained the ideal of comprehensive urban renewal. Planners themselves increasingly recognized that the ideal of "master planning" was illusory and began to explore ways of bolstering their legitimacy through wider public consultation. Patterns of infrastructure investment that had previously been conceived as integral to urban revitalization had now become directly implicated in post-war urban decline and the destruction of city life (see Gandy 2002). In the US, for example, the collapse of the consensus over highway construction in the 1960s mirrors the broader dissolution of the New Deal bipartisan consensus in public policy. The close interrelation between discourses of urban planning and the progressive impulses of modernist thought gradually began to unravel in the face of combined fiscal and political challenges. As urban planning became increasingly dominated by massive state subventions for corporate sectors such as cars and real estate we find an increasing polarization in space between grim housing projects for the working classes and the burgeoning suburbia of middle-class consumer aspirations. With the rise of the fragmentary metropolis the designed landscape was increasingly an adjunct to corporate atria, speculative housing developments and other market-led responses to the urban crisis of 1960s and 1970s. In a sense, therefore, the ideological resonance of nature had come full circle to emulate the ad hoc interventions of the past: the role that metropolitan nature had played in the building of a functional public realm had been gradually supplanted by a more piecemeal emphasis on the decorative contributions of nature to the design of urban space.

ECOLOGY, MODERNITY AND THE POST-INDUSTRIAL METROPOLIS

Urbanization has now become synonymous with the globalization of economic and cultural life. In 1900 there were no more than a dozen cities in the world with more than a million people and agriculture remained the dominant economic activity except for a relatively small number of industrialized nations. By the end of the twentieth century, however, over 500 cities had populations exceeding 1 million people and over half of the world's population was urban. The contemporary "urbanization revolution" dwarfs the experience of nineteenth-century Europe and North America yet is distinct from this earlier transition in a number of critical respects. The so-called "brown agenda", which dominated the rancorous UN environmental summit held in Johannesburg in 2002, reflects the scale of the public health challenge facing contemporary cities but the current housing and sanitation crisis has originated in a fundamentally different context to that of the nineteenth-century city. These rapidly growing cities in the global South reflect an urban dynamic which is unrelated to the classic paradigms of city governance and planning whether in the sprawling slums of São Paulo or the construction frenzy underway in China's Pearl Delta. It is increasingly difficult to talk in terms of any general or identifiable model for urban development as each element takes shape within its specific context and parameters. The place of technical expertise has been superseded by a new entrepreneurial vista ranging from the most precarious slum settlements to the latest generation of immense skyscrapers that dwarf those of twentieth-century Europe or North America. The scale and complexity of this global urban transformation militates against any teleological extension of past experience and necessitates new insights into the urban process. The need to connect policy deliberation with the establishment of effective and legitimate forms of urban governance remains as important now as it was in the past but such arguments can no longer rely on either the scientific logic of public health advocacy or rationalist conceptions of urban space promoted by a coterie of technical experts.

The gathering critique of modernist planning and design from the 1970s onwards has fostered new intersections between urban design and the bio-physical sciences. In the place of a cogent critique of the inequities engendered by capitalist urbanization we find a growing engagement with socio-biological ideas such as "defensible space" which were eagerly incorporated into critiques of public architecture and urban design. Increasing emphasis on individual property rights and demands for fiscal independence from the urban poor gradually coalesced around a new kind of urban agenda exemplified by the latest surveillance strategies and the rise of gated communities. These developments have intensified the ambiguity of urban nature as both an inherent element within a functional public realm but also as a means to enhance property values as the management of hitherto public spaces has been increasingly taken over by quasi-public agencies or private foundations dependent on the whim of individual or corporate benefactors.

The post-war crisis in the rationale and impact of urban planning has been a central element in the ecological critique of modernity yet a closer inspection of urban environmental discourse reveals the innate ambiguity of "ecological politics" as the basis for any progressive response to urban problems. The emergence of anti-nuclear movements in Europe and the environmental justice movements in the USA reflect a very different appropriation of ecological and environmental discourses to the reactionary anti-urbanism of the past. The impact of urban environmental disasters such as Seveso (1976) and Bhopal (1984) as well as the chronic ill health experienced in the poisoned cities along the US–Mexican border and other toxic locales has spurred a new synergy between the politics of social and environmental justice (see, for example, Hofrichter 1993; Hurley 1995). Though these radical political challenges to militarism, industrial negligence and the productivist logic of consumer capitalism share important elements with the ecological critique of modernity they nonetheless embrace a more dialectical, inclusive and culturally determined conception of nature.

The return to nature in the post-industrial metropolis also denotes a conscious rejection of the kind of aridity engendered by the concrete landscapes associated with technological modernism. The understanding and utilization of urban eco-systems has become more sophisticated to embrace a more holistic conception of the interaction between bio-physical processes and urban society. The development of new approaches to "ecological restoration", for example, marks a self-conscious attempt to recreate the bio-diversity of ecosystems that preceded the growth of the industrial metropolis in order to foster a different kind of synthesis between nature and culture. In the case of river channels, for instance, we can find examples of ecological restoration efforts which not only add aesthetic interest to the landscape but also contribute towards improvements in flood control and waste water treatment to produce a post-industrial or late modern synthesis between advances in ecological science and new approaches to landscape design (see Gauzin-Müller 2002; Gumprecht 1999).

These developments have in part been fostered by the return of nature to post-industrial cities so that the inner areas of some formerly industrial cities such as Baltimore, Detroit or Pittsburgh have taken on an increasingly Arcadian feel. In the photographic essays of Camilo José Vergara, for example, we can observe how inner urban areas have been reclaimed by nature through a mix of abandonment, neglect and structural change to produce "green ghettos". "In many sections of these ghettos", notes Vergara, "pheasants and rabbits have regained the space once occupied by humans, yet these are not wilderness retreats in the heart of the city" (1995: 16). The growing presence of nature within former industrial landscapes can be conceived as a kind of urban entropy whereby the distinction between human artifice and ecological succession becomes progressively blurred. In the literature of J.G. Ballard, for instance, the fragility of the modern city is repeatedly portrayed through a tendency towards dilapidation and decay. In the post-industrial landscapes of Ballard, Iain Sinclair and other authors we find that elaborate highway interchanges, hi-rise apartments and other characteristic features of the twentieth-century city take on the form of urban ruins set amidst a complex palimpsest of new social and technological structures (see Davis 2002; Picon 2000). A similar topographical trope of urban decay is also reflected in cinematic representations of the post-industrial metropolis. In Terry Gilliam's *12 Monkeys* (1995), for example, we encounter a post-apocalyptic Baltimore that has been taken over by elephants, lions, spiders and other organisms. This eerie spectacle is hardly an example of ecological restoration but rather a futuristic zoöpolis where urban space is controlled by animals rather than by human beings. The post-industrial metropolis, and its cultural representations, is suggestive of a very different kind of city to that of the nineteenth-century metropolis but it is a city for which we are still searching for an appropriate conceptual vocabulary. The characterization of urban segregation in terms of "ecological zones" by the Chicago School of urban sociology, for example, has more recently been reworked, albeit somewhat ironically, for example, in Mike Davis's exploration of the "ecology of fear" in contemporary Los Angeles. The nineteenth-century metabolic insights of Karl Marx and Justus von Liebig have been reprised in order to provide a counterfoil to the functionalist emphases of "industrial metabolism", "ecological footprints" and other static conceptions of the modern city (see Swyngedouw 2004a; 2004b). And the dystopian genres of monstrous urbanism originating in romanticist reactions towards the nineteenth-century industrial city have been widely appropriated within more savvy examples of science fiction cinema and literature as a means to provide allegorical critiques of contemporary social and political developments.

CONCLUSIONS

The politics of urban nature is characterized by a range of "political ecologies" that can be differentiated from one another on the basis of their contrasting approaches to the conceptualization of nature. Any critical engagement with urban environmental change must contend with problems of terminology and historicity so that although many aspects of contemporary urban discourse derive from the nineteenth-century city we can nonetheless identify a critical break since the 1960s in which the "ecological imaginary" has played an enhanced yet deeply problematic role. The ecological imaginary, which comprises a cluster of dichotomous, ethological and neo-romantic readings of nature, remains rooted in organicist conceptions of urban space. The dynamics of urban change are widely conceived in terms of an adjustment towards a notional "equilibrium state" or as a set of processes that must be forcibly realigned towards a putative set of "natural" parameters. Yet this appeal to nature as something that resides outside of social relations is a corollary of fragmentary conceptions of cities as discrete entities that remain unconnected with wider processes of social and political change.

Ranged against the organicist lineage of the "ecological imaginary" we can identify alternative approaches to the understanding of urban nature that recognize the cultural and historical specificities of capitalist urbanization. The urban ecology of the contemporary city remains in a state of flux and awaits a new kind of environmental politics that can respond to the co-evolutionary dynamics of social and bio-physical systems without resort to the reactionary discourses of the past. By moving away from the idea of the city as the antithesis of an imagined bucolic ideal we can begin to explore the production of urban space as a synthesis between nature and culture in which long-standing ideological antinomies lose their analytical utility and political resonance. Thus far, however, the development of more fluid and mutually constitutive conceptions of urban nature have had relatively little impact on popular discourses of "ecological urbanism" where the emphasis has tended towards the functional dynamics of metabolic pathways or the promotion of new forms of bio-diversity as a corollary of social and cultural complexity. It is perhaps only through an ecologically enriched public realm that new kinds of urban environmental discourse may emerge that can begin to leave the conceptual lexicon of the nineteenth-century city behind.

BIBLIOGRAPHY

Bacon, E.N. (1967) *Design of Cities*. London: Thames and Hudson

Centre de Cultura Contemporània de Barcelona (1994) *La Visions Urbaines*. Barcelona: Centre de Cultura Contemporània de Barcelona

Cutler, P. (1985) *The Public Landscape of the New Deal*. London and New Haven, CT: Yale University Press

Davis, M. (1998) *Ecology of Fear*. New York: Metropolitan Books

Davis, M. (2002) *Dead Cities and Other Tales*. New York: The New Press

Forty, A. (2000) "Nature", in *Words and Buildings: A Vocabulary of Modern Architecture*. London: Thames and Hudson

Gandy, M. (2002) *Concrete and Clay: Reworking Nature in New York City*. Cambridge, MA: The MIT Press

Gandy, M. (2004) "Rethinking urban metabolism: water, space and the modern city", *City*, 8: 371–387

Gauzin-Müller, D. (2002) *Sustainable Architecture and Urbanism: Concepts, Technologies, Examples*. Basel: Birkhäuser

Green, N. (1990) *The Spectacle of Nature: Landscape and Bourgeois Culture in Nineteenth-Century France*. Manchester: Manchester University Press

Gumprecht, B. (1999) *The Los Angeles River: Its Life, Death and Possible Rebirth*. Baltimore, MD: The Johns Hopkins University Press

Harvey, D. (1989) *The Condition of Postmodernity: An Inquiry into the Origins of Cultural Change*. Oxford and Cambridge, MA: Blackwell

Hofrichter, R. (ed.) (1993) *Toxic Struggles: The Theory and Practice of Environmental Justice*. Philadelphia, PA: New Society

Hurley, A. (1995) *Environmental Inequalities: race, class, and pollution in Gary, Indiana, 1945–1980*, Chapel Hill, NC: University of North Carolina Press

Picon, A. (2000) "Anxious landscapes: from the ruin to rust", *Grey Room*, 1: 64–83

Roy, A. (2002) *The Algebra of Infinite Justice*. London: Flamingo

Scarpa, L. (1986) "Martin Wagner oder die Rationalisierung des Glücks", in *Martin Wagner 1885–1957. Wohnungsbau und Weltstadtplanung: Die Rationalisierung des Glücks*. Berlin: Akademie der Künste

Schorske, C.E. (1981) *Fin-de-siècle Vienna: Politics and Culture*. Cambridge: Cambridge University Press

Swyngedouw, E. (2004a) *Social Power and the Urbanization of Water: Flows of Power*. Oxford: Oxford University Press

Swyngedouw, E. (2004b) "Circulations and metabolisms: hybrid natures and cyborg cities", paper presented to the research colloquium *Re-naturing Urbanization*, held at the University of Oxford, 30 June 2004 (see Chapter 2)

Vergara, C.J. (1995) *The New American Ghetto*. New Brunswick, NJ: Rutgers University Press

Williams, R. (1973) *The Country and the City*. Oxford and New York: Oxford University Press

5 Nature's carnival
The ecology of pleasure at Coney Island

Eliza Darling

INTRODUCTION

Coney Island boasts a peculiar history of smouldering elephants. In early January of 1903, a performing pachyderm named Topsy was electrocuted in an off-season publicity stunt staged by Frederick Thompson and Skip Dundy, the masterminds behind Coney Island's magnificent Luna Park. Topsy became the unfortunate victim of poor management and a power struggle between Thomas Edison and his archrival, George Westinghouse, after developing a penchant for killing irksome humans (the last of whom, legend has it, met his violent end after feeding Topsy a lit cigarette) in her waning years. Seeking to rid themselves of the seditious beast while simultaneously garnering a little press, Thompson and Dundy, crafty show-men through and through, sought Edison's assistance after the ASPCA objected to a public elephant-hanging as excessively cruel. Edison, engaged at the time in a fierce battle to discredit the alternating current electrical system pioneered by Westinghouse, promptly sent a team of electricians to Coney Island to demonstrate its deadly effects on Topsy. He even made a film of the execution, preserving for posterity the visually compelling (yet ultimately futile) evidence of the menace of AC power as it coursed through the electrodes attached to the great beast's feet.

Topsy was not the first elephant to go up in smoke on Coney Island. Less than a decade before her gruesome demise, Coney's Elephant Hotel – magnificent local landmark and spectacular financial failure – burned to the ground on a September evening in 1896 after a brief and rocky history of loans, liens and ultimate abandonment. Built in 1884, the enormous and remarkably life-like Elephant Hotel had been built by James V. Lafferty as novelty lodging, but never saw a profit and was all but deserted by the time it was claimed by conflagration (*Brooklyn Daily Eagle* 1896: 14). Despite its failure as a pecuniary venture, the elephant colossus quickly burned its way into the public imagination as a rightful and indispensable part of the Coney landscape, immortalized in the phrase "visiting the elephant", a popular local euphemism for having an illicit affair (Carlin 1989: 2).

Penned some seven years apart, the following elephant obituaries by the *New York Times* and the *Brooklyn Daily Eagle* are eerily similar, evoking the sense of melodramatic spectacle that has long characterized "nature" on Coney Island, even in death:

The pride of Coney Island, its big elephant, is gone. At 10:30 o'clock last night it took fire, and in twenty minutes the huge beast of wood and tin collapsed, first falling to its knees with what sounded almost like a groan of agony, and then rolling over into a shapeless mass, where it smoldered and burned until it was finally drowned into submission by the fire department.

(*Brooklyn Daily Eagle* 1896: 14)

At 2:45 the signal was given, and Sharkey [employee of the Edison Company] turned on the current. There was a bit of smoke for an instant. Topsy raised her trunk as if to protest, then shook, bent to her knees, fell, and rolled over on her right side motionless . . . In two minutes from the time of turning on the current [veterinarian] Dr. Brotheridge pronounced Topsy dead.

(*New York Times* 1903: 1)

Although these events ring with an air of carnivalesque absurdity through the haze of time, in fact neither the electrocution of a live elephant nor the combustion of an elephant-shaped hotel were particularly remarkable events in the heyday of Coney Island, whose claim to fame was its extraordinary capacity to turn a profit from spectacular nature – in particular, the bizarre, the extreme, the grotesque, the sensual, and the dangerous. Nature is rather like the elephant in the living room when it comes to the Coney Island literature: everyone knows it's there, but nobody really talks about it – or rather, no one gives it a label, a handle for grasping the loose but ubiquitous nature tropes that are often overshadowed by the characterization of Coney Island as the archetypical "artificial" space and which, through their eclipsed and silent presence, help to reify the sense of "unreality" that is Coney's hallmark.

This chapter looks at the old material with new eyes – examining the familiar icons of the Nickel Empire through the window of nature, or, more precisely, the production of nature in both a material and a discursive sense. In this chapter, I examine Coney Island in a highly circumscribed manner in terms of time, space, and meaning, concentrating on the amusement industry in the first half of the twentieth century – particularly its four original amusements parks, Sea Lion, Steeplechase, Luna and Dreamland – and treating the island primarily as a place to play, setting aside for the moment its characteristics as a place to live and a place to work. Far from an attempt at a comprehensive political ecology, this chapter looks specifically at the paradox of consciousness which made the tropes of what Neil Smith calls "external" nature (Smith 1996; 1984) such an integral part of urban amusement at the most classic of American playgrounds, drawing on both Michel Foucault's categorization of carnivals as "temporal heterotopias" (Foucault 1967) and Richard White's analysis of American nature as a space that is increasingly associated with human play and decoupled from human labour (White 1996).

Why Coney Island? Why not Disneyland, Sea World, Mardi Gras, the Chicago Columbian Exhibition, Barnum and Bailey Circus, the county fair, or any one of dozens of iterations of the common carnival? There are several reasons. In the first place, while Coney Island did not, by a long shot, give birth to the amusement industry, its significance in the subsequent development of such spaces cannot be overestimated, particularly in the context of American history. Coney Island, "where America learned how to play", as Coney historian Edo McCullough so aptly put it (McCullough 1957: 4), was a crucible for the pioneering of many of the tactics, technologies, and spatial forms that were quickly emulated by the purveyors of carnivalesque entertainment across the country – among them, the enclosed amusement park and the one-price cover charge (Adams 1991: 43–44), as well as the roller coaster (Denson 2002: 286). In the second, Coney Island was (and is) a quintessentially urban space whose growth and development has been inextricably bound up with the growth and development of the burgeoning metropolis which looms over its shoulder. It was this proximity which not only made Coney Island more heavily patronized (in proportion to population) in 1909 than Disneyworld was in 1989 (Nasaw 1993: 3), but made it qualitatively different from its more rural cousins, such as the rodeo, the livestock show, the county fair – forms which not only emerged from a separate historical trajectory, but related differently to nature precisely because of their uniquely agrarian pedigree. Finally, the ability of Coney Island's savvy showmen to make a profit by turning the artifact of wild, weird and dangerous nature into the penultimate urban spectacle has been emulated but never matched. Although Coney Island served as the prototype for amusement parks across the United States, the singular sense of boisterous, bawdy, sordid absurdity that was The People's Playground has never been replicated – least of all by Disney, whose hyper-sanitized spaces pale pitifully in comparison. As Carlin notes:

> It is enlightening to compare how Disneyland, the great American amusement park of the late twentieth century, replaced Coney Island's earthy symbolism with sterile, technological perfection and presexual infantile obsessions. Disneyland is a monument to sublimation and control. Coney Island was a monument to bodies and the potential for abandonment into pure libidinous pleasure.
>
> (Carlin 1989: 4–5)

Coney Island regularly garnered the ire of those tiresome defenders of American "decency" who continue to plague the nation with their moral indignation – including, of course, Robert Moses, whose dislike for the sensual vulgarity of the Coney Island amusement industry has been described by at least one historian as "pathological" (Denson 2002: 66) and whose mid-century campaign to "clean up" the beachfront, boardwalk and concessions are strikingly redolent of Rudolph Giuliani's ham-handed attempt to sanitize urban space some fifty years later. Coney was similarly reviled by many purveyors of amusement across the nation who sought to emulate its formula for success but not its reputation for iniquity (Adams 1991: 65; Snow 1984: 12). They have paid accordingly for their pandering self-righteousness, in artistic integrity if not profit: Bambi is no trade-off for the magnificent and marauding Topsy.

SODOM BY THE SEA: FLESH AND FANTASY AT CONEY ISLAND

As Edo McCullough wrote in his immensely entertaining 1957 history *Good Old Coney Island*, "Nature has been uncommonly kind to Coney Island . . . It was as if an intelligence had decreed that just here, convenient to what would become the world's largest city, there should be a splendid place for bathing" (McCullough 1957: 7). He speaks here of Coney's magnificent five-mile beach, where Walt Whitman famously swam naked in the waves and ambled on the sand (Adams 1991: 42; McCullough 1957: 23) in the quiet interlude between the colonization of the island by a group of excommunicated English Anabaptists in the mid-seventeenth century and its transformation by developers, speculators, politicians and showmen into the boisterous prototype for the American amusement park at the end of the nineteenth. It was the Coney Island beaches which first drew commercial investment to the desolate shore in the early 1800s, beginning with the Coney Island House in 1829 and burgeoning quickly into the array of seaside hotels, bathhouses and concessions that crowded the dunes like weeds by the turn of the century, and it was the Coney Island beaches which would continue to draw tourists a hundred years later, when the last of the original great amusements parks had crumbled into ruin (Zukin *et al.* 1998: 637; Weinstein 1992: 272).

Historically, the irresistible appeal of Coney Island has derived from the convenient propinquity of two apparently antipodal attractions, one the epitome of "natural" recreation (the beach), the other the quintessence of "artificial" diversion (the amusement park). The Whitney Museum of American Art has captured this inimitable sense of duality in what it called the "two Coney Islands, the one of flesh and the other of fantasy" (Carlin 1989: 1). But it would be a mistake to see nothing but organic on the sea-side of the boardwalk and nothing but synthetic on the other: the beach, of course, has been materially "improved" by humans several times over in its history, but more compellingly, there has long been a good deal of "nature" to be found on the city-side of the divide, and it was placed there every bit as strategically as the truckloads of sand which have widened the shorefront. Indeed, nature (both organic and iconographic) so thoroughly and fluently pervaded the Coney Island amusement industry that it often seemed to impudently mock the vast expanse of grey Atlantic that hemmed it in – as if the nature Coney Island created sought to surpass the nature Coney Island inherited. It is no accident that Luna Park, arguably the most magnificent of the four, has been compared to an "enchanted garden", hailed as an "electric Eden" (Snow 1984: 14). Coney was an artificial arcadia, but with an air of the demonic, the corrupt, the fallen: "Perhaps Coney Island is the most human thing that God ever made, or permitted the devil to make" quipped Richard Le Gallienne in 1905 (quoted in Snow 1984: 9).

At first glance, Coney Island's amusement parks would appear to be the last place one would search for anything akin to what we colloquially call "nature". Indeed historians have remarked that the success of Coney Island at the turn of the century fairly hinged upon the development of that penultimate apotheosis of artificiality, the machine:

The essence of Coney Island was its juxtaposition of mechanical amusement devices with an atmosphere of illusion and chaos. The precision and predictability of gears, wheels, and electricity created a fantasyland of disorder, the unexpected, emotional excess, and sensory overload . . . Coney . . . allowed members of the growing urban working class . . . to assimilate and participate in a culture ever more dominated by the machine.

(Adams 1991: 41)

While Coney Island began as a seaside resort which drew throngs of tourists to its beaches, the delights afforded by nature in its so-called "natural" state came quickly to be overshadowed by a growing preoccupation with the entertaining propensities of technology, that increasingly complex congealment of labour which served to alienate human beings from both the product of their labours and their environment under the auspices of capitalistic industrialization.[1]

Machines made the fun here. That, in itself, may have had a potent appeal. People who were struggling to cope with growing technological complexities in their jobs could spend an afternoon with the tables turned: at Coney, the machinery worked to divert them . . . In a decade, these elaborate mechanical diversions completely changed the nature of a day at Coney Island. Now people came to one of the finest beaches on the Atlantic coast and never even thought of going swimming; they might catch only a quick blue flash of ocean over the park wall before the roller coaster plunged.

(Snow 1984: 18)

Indeed both Judith Adams and Richard Snow note that this increasing mechanization of Coney Island, along with its eventual proletarianization with the demographic shift in patronage that resulted in the emergence of the "Nickel Empire", produced a paradoxical space where machinery became a source of revelry, abandon and enjoyment for the very class it otherwise enslaved on the factory floor.

And yet, despite this domination of the Coney landscape by the automated, the simulated, and the imitated, "external" nature – what many people in the (post) industrial world have come to understand as that part of the universe which stands apart from and external to humanity and its works (Smith 1996; 1984) – was everywhere to be found on Coney Island, not only in the continuing presence of the beach, but in the naturalistic emblems, objects, and imagery that pervaded Coney's four largest beachfront amusement parks: Sea Lion (1895–1902), Steeplechase (1897–1964), Luna (1903–1946) and Dreamland (1904–1911).

Sensual nature: Coney Island's animals

I will begin with the most obvious manifestation of external nature: Coney Island's fauna. The great amusement parks were full of living animals, from Sea Lion's wolves to Dreamland's polar bears. Luna Park was especially known for live animal acts. Fred Thompson and Skip Dundy were particularly fond of elephants (the latter was rumoured to have considered them a lucky charm), their herd of which in 1909 gave rides to some ten thousand people a week (Berman 2003: 57) but also kept camels, diving horses, ostriches, alligators, monkeys, pigs and goats, among others. All of the parks hosted circus acts or menageries at one time or another, usually in addition to several individual and more highly specialized animal shows. Live animals were used not merely for entertainment, but for hard labour: when Thompson and Dundy moved their cornerstone "Trip to the Moon" ride from Steeplechase to their newly created Luna Park on the former site of Sea Lion, it was Luna's elephants, including Topsy, which did most of the heavy lifting. And of course, Coney Island was the site of three horse racing tracks until 1910.

In addition to living beasts, Coney Island was fairly riddled with animal imagery. From the Elephant Hotel to the mechanical cow that dispensed five-cent glasses of milk at Culver Plaza, animal effigies abounded at Coney Island. At Luna, Heppe's Kandy Meat Market sold sweet confections in the shape of farm animals and pets. Steeplechase of course boasted its namesake ride, which sported wooden racehorses. Dreamland had Andrew Mack's Fish Pond, with mechanical tin fish for the catching. Sea Lion sported a plethora of aquatic symbols in addition to its 40 genuine sea lions. And of course, park rides were frequently name for various fauna, from the Sea Serpent at Steeplechase to the Butterfly at Luna to the Leap Frog Railroad at Dreamland.

fire, flood, and fury: dangerous nature

As Richard Snow has noted, "Nothing stirred a showman more than a truly devastating catastrophe" (Snow 1984: 46). The savvy architects of Coney Island were quick to recognize the lucrative potential of nature's capacity to shock and awe with stunning displays of spectacular power. Natural disaster exhibits were a favourite with Coney Island crowds from the beginning of the twentieth century, usually recreated with life-like miniature models of famously doomed localities which were subsequently destroyed by fire, flood, earthquake, twister, or volcanic eruption to the appreciative gasps of spectators. Floods were a favourite subject. One of the earliest exhibits re-created the Galveston Flood, in which thousands of people perished in a hurricane-spawned tidal surge in the resort town of Galveston, Texas in 1900. The famously gruesome Johnstown Flood of 1889 was also captured in a Coney Island diorama. The 1902 eruption of Martinique's Mount Pelee, the 1906 San Francisco earthquake, the destruction of Pompei and the cyclones of Kansas also provided the basis for similar natural-disaster shows which demonstrated the awesome power of uncontrolled and uncontrollable nature.

In addition to those exhibits which *re-enacted* natural disaster as spectacle, Coney Island itself often inadvertently *became* disaster-as-spectacle, particularly when catastrophic fires broke out, drawing slack-jawed crowds from the local neighbour-hood to gape at the futility of human intervention in that most mercurial of nature's forces. Devastating conflagrations swept through Steeplechase in 1907, Dreamland in 1911, and Luna in 1945, shortly before its permanent demise. Following the fire that destroyed the first incarnation of Steeplechase Park in 1907, owner George Tilyou, in a now-legendary attempt to turn a profit from misfortune, erected a sign in front of the still-smouldering ruins reading:

> To inquiring Friends: There was a lot of trouble yesterday that I have not had to-day, and there is lots of troubles to-day that I did not have yesterday. – Geo. C. Tilyou
>
> On this site will be erected a bigger, better Steeplechase Park.
> Admission to the Burning Ruins – 10 cents.

Due to the large number of beasts often housed at the great amusements parks, the fiery demise of Coney's animals also often comprised part and parcel of the spectacle of disaster. In his online history of Coney Island, Jeffrey Stanton gives the following account of the agonizing public death of a lion caught up in the 1911 Dreamland fire:

> The lion climbed up the incline of the railroad as two trainers followed the cat's bloody footprints in near darkness. They were followed by [owner] Ferrari, [lion-tamer] Bonavita and two armed policemen. They could hear the great cat's roars ahead, and they the trainers kept firing their blanks to keep the cat on the move. When they finally reached the top and the open air, Black Prince was standing, outlined against the sky. He made a splendid target and Ferrari gave the cops the orders to fire. They emptied their guns into them as thousands watched from below. He fell and still twitched. Someone threw up an axe and policeman Coots split open his skull. They found 24 bullets in the lion's head alone. The crowd below roared and Black Prince's body was dragged down the incline and into the street for display at 3:25 A.M.
>
> (Stanton 1997b)

As with Topsy, even in death Coney Island could turn nature into gruesome spectacle.

Sea, space, and science: nature as frontier

One of the most common ways in which Coney Island's showmen packaged nature for public consumption would later become something of an archetype in the amusement business: nature as a frontier of geographical exploration. The precursors to such modern amusement industry standards as Space Mountain, Sea World, Busch Gardens, and Pirates of the Caribbean could be found in many of Coney's early rides, from Luna's "Trip to the Moon" and "Twenty Thousand Leagues Under the Sea" to Dreamland's "Trip in an Airship" and "Over the Great Divide" – cycloramas which produced the illusion of spectacular, exotic, and sometimes perilous travel to the far reaches of known or imagined geographical space through a combination of light, sound, and mechanical motion. Thompson and Dundy's "Trip to the Moon", which debuted at the 1901 Pan-American Exposition in Buffalo before moving to Steeplechase the following year and eventually providing the cornerstone for the opening of Luna Park in 1903, took patrons on a spectacular journey to the moon on a winged ship called "Luna". In a similar vein, Luna Park's "Twenty Thousand Leagues Under the Sea" comprised a submarine ride to the North Pole, replete with live seals and polar bears reclining on an iceberg, as well as an "Eskimo Village".

One of the most extraordinary side shows in Coney Island's history hinged upon nature not as a fantastical or geographical frontier, but as a scientific and technological one. From 1904 to 1943, first Dreamland and later Luna Park hosted an exhibit of Premature Baby Incubators, a phenomenon that was founded at the 1896 Berlin Exposition as the *Kinderbrutanstalt* ("child hatchery") by Dr Martin Arthur Couney, who eventually moved it to its permanent home in Coney Island. Couney turned to expositions as a venue for displaying (and funding) his invention, the first mechanical incubator for human infants, after encountering little enthusiasm for the project among the European or American medical communities (Adams 1991: 50–51). Despite an initial outcry from the Brooklyn Society for the Prevention of Cruelty to Children, Couney's exhibit was approved by the American Medical Association and took up residence in Luna Park in a fully-equipped preemie ward furnished with nurses to oversee daily operations, wet nurses to feed the infants, and lecturers to explain the technical details to the visiting public, in addition to the machines which regulated incubator temperature and filtered in a constant supply of clean air. Such units were not common in American hospitals at the time, even when the technology was available, and as a consequence some 8,000 premature babies were brought to Coney Island to be nursed through their precarious first months. Over the course of its 39-year run, Couney's exhibit saved the lives of over 6,500 of them (many of whom held reunions at Coney Island for years afterward), including his own daughter, and was a smash hit with the public – particularly childless women, many of whom would return again and again to follow the fortunes of particular infants that had caught their eye (Adams 1991: 51–52).

Freaks, geeks, and foreigners: "grotesque" nature

One of the most enduring tropes in the exhibition of nature at Coney Island is the phenomenon of "freaks": human beings whose morphological or behavioural attributes place them outside the spectrum of prescriptive human physiology, including the extremely large or the extremely small, those of ambiguous gender, those with extra appendages, unique skin textures or extraordinary physical abilities – or those who could be made to *appear* like any of the above (fake freaks or "gaffes"). Because normative definitions of what constitutes the "conventional" human form tend to be historically contingent, both within societies and between them, early freaks at Coney Island included those considered strange merely by virtue of their relative cultural exoticness as well as those who were culturally familiar but physically exotic.

It was Samuel Gumperz who first introduced the freak show to Coney Island in 1904 with the installation of Lilliputia, a village populated by some 300 Little People, or those born with one of the several varieties of dwarfism, at Dreamland. Lilliputia was a fully-functioning half-scale replica of fifteenth-century Nuremburg, Germany, replete with its own fire department (which contributed valiantly if futilely to the battle to save Dreamland from conflagration in 1911) and cadre of lifeguards. The popularity of Lilliputia spurred Gomperz to create the Dreamland Circus Side Show several years later, an amalgam of unusual people that included giants, dwarves, albinos, dog-faced boys, fat ladies, sword swallowers, microcephalics ("pinheads"), strongmen, and conjoined twins, among others. The success of Dreamland's freak show provided the impetus for other Coney Island institutions to create their own, including one at the Steeplechase Circus Big Show as well as several rival independent shows (Sam Wagner's World Circus Side Show, David Rosen's Wonderland Circus Side Show, and Fred Sindell's Palace of Wonders Freak Show) in addition to several animal freak shows (including Charlie Dooen's Freak Animal Show).

The birds and the bees: sex off the beach at Coney Island

Although its relationship to the trope of "external nature" is less obvious than that of animals, disasters and freaks, the clever manipulation of human sexuality played an enormously important role in Coney's formative history, as several commentators have noted. "It is clear from the descriptions of the time that one of the most important aspects of Coney Island was the way its entrepreneurs commodified sex as mass entertainment", notes Carlin (1989: 10), while Snow remarks that "Coney's clang and glitter always had a current of sexuality running through it" (1984: 15).

The overcrowded beaches and the debauchery of the Bowery were not the only factors which helped to earn Coney Island the scandalous moniker "Sodom by the Sea". Many of the rides themselves – particularly those commissioned by Tilyou at Steeplechase – were specifically designed to heighten heterosexual tension and create a space for the public transgression of sexual mores. The ride for which Steeplechase was named allowed a man and woman to snuggle together in double saddles astride wooden horses that "raced" round a track in imitation of the real thing. The exit from the Steeplechase ride led patrons through the Insanitorium and Blowhole Theater, a stage where couples would be poked, prodded, and generally humiliated slap-stick style by park actors while compressed air shooting up from holes in the floor would send a woman's skirts sailing over her head – a near-scandal in the first decade of the twentieth century – to the raucous amusement of the audience members, who had just endured the same ordeal themselves (Adams 1991: 45; Stanton 1999). Moreover, many of the Tilyou's rides, including Barrel of Fun (which was the only entrance to the park itself), the Earthquake Stairway, the Dew Drop, the Whichway, the Human Roulette Wheel, and the Wedding Ring, were designed to throw riders off balance and into each other's arms. Human sexuality was an integral part of Tilyou's rides, shows, and spatial forms.

NATURE'S CARNIVAL: PARADOX AND PLAY

Why, particularly during an historical period in which American society was undergoing intense urbanization and industrialization, would the tropes of nature continue to haunt – even dominate – an urban space whose fortunes hinged increasingly upon the mechanization of play? Interrogating the ubiquitous presence of nature-totems at the carnival, the fair, the circus, the exposition or the amusement park is an exercise in scrutinizing the obvious. Such symbols have become so familiar in this context as to render their analysis almost absurd. If a carousel didn't contain wooden animals, what would it contain? The question of what purpose a spinning platform of faunal effigies serves in the first place springs less readily to mind.

On the face of it, the question seems to answer itself: people go to carnivals to experience pleasure, and nature can be immensely pleasurable – amusing, absurd, surprising, strange, exciting, frightening, unexpected, familiar, dramatic, shocking, noble, sensual, comforting, edgy, weird, silly, awe-inspiring, *entertaining*. External nature is quintessentially carnivalesque. But I believe there is more to nature at the carnival than mere pleasure, and that pleasure, like the carnival, is itself far more complex than it appears on the surface. The pervasiveness of nature totems suggests that carnivals constitute one of a plethora of spaces where human beings congregate not only to gaze upon nature as spectacle, but to struggle with and work out their often contradictory and conflicting relationships with it. I believe that the key to understanding the significance of nature at Coney Island lies both in the essence of the carnival itself as well as the shifting material and ideological relationship between an increasingly urbanized population and the American landscape at the close of the nineteenth century.

The carnival has been described by many social critics as a sort of "safety-valve", an edge-space containing all the tensions of shifting modernity that allows people to challenge constricting social traditions:

> Carnivals are gross celebrations of foolishness in which proper, hard-working people participate in ritual transformations that call into question the certainty of their conscious perceptions and prove the reversibility of the social codes (legal, religious, financial) upon which their lives are based. The transgression of these codes, coupled with a basic human nostalgia for childlike symbolism and play, act as a sort of release valve, or social vaccine. A taste of irrationality is introduced so that society can shore up its defenses and then continue its normal life undisturbed. The transgressions of the carnival or the amusement park are not immoral. On the contrary, they service to reconcile social decorum with the primal human urges that threaten disruption on a more dangerous level. In other words, transgression becomes a ritualized game rather than true rebellion.

(Carlin 1989: 4–5)

Similarly, Foucault describes fairgrounds or festivals as *heterotopias*, or material "counter-sites" common to all societies: "a kind of effectively enacted utopia in which all the other real sites that can be found within the culture are simultaneously represented, contested, and inverted". Carlin's description of carnivals as "celebrations of foolishness" is redolent of what Foucault calls "heterotopias of deviation: those in which individuals whose behavior is deviant in relation to the required mean or norm are placed" (Foucault 1967: 4). Heterotopias of deviation, according to Foucault, might include such spaces as prisons, rest homes and psychiatric wards, but carnivals, as Carlin intimates, are similar in that they permit, legitimize, and even sanctify the transgression of social norms – and therefore isolate and contain that transgression from mainstream society, where it might inflict damage. According to Foucault, heterotopias "have a function in relation to all the space that remains", one of which is to "create a space of illusion that exposes every real space, all the sites inside of which human life is partitioned, as still more illusory" (Foucault 1967: 7).

Foucault describes fairgrounds themselves as "temporal heterotopias", cyclical ephemeral spaces where time itself is accelerated, fleeting, shifting:

> Opposite these heterotopias that are linked to the accumulation of time [museums, libraries], there are those linked, on the contrary, to time in its most flowing, transitory, precarious aspect, to time in the mode of the festival. These heterotopias are not oriented toward the eternal, they are rather absolutely temporal. Such, for example, are the fairgrounds, these marvelous empty sites on the outskirts of cities that teem once or twice a year with stands, displays, heteroclite objects, wrestlers, snakewomen, fortune-tellers, and so forth.
>
> (Foucault 1967: 6)

There is a sense of fleeting unreality to the carnival, as if it allows its participants to indulge in a brief moment of irrational madness which allows them to better tolerate the rationalized madness of normative time upon their return. Despite Foucault's focus on temporality, implicit in this description is a spatiality to the fairground – significantly, anchored "on the outskirts of cities". Abandoned to forlorn emptiness for most of the year, they intermittently swell with momentary and accelerated absurdity, a fun-house mirror reflecting the distorted image of modernity back on itself:

> The amusement park was a "temporary world within the ordinary world," where "special rules" obtained, and visitors literally stepped out of their "real" lives into a world of play and make-believe.
>
> (Nasaw 1993: 86, quoting Johan Huizinga's *Homo Ludens* (Boston 1955))

It is precisely this combination of madness and make-believe which makes the carnival a close cousin of "external nature" in the specific sense of the word – that which is *outside of*, yet *vital to*, the realm of normative human experience.

Environmental historian Richard White draws the connection between nature and play in an article wryly entitled "Are you an environmentalist or do you work for a living? work and nature" (White 1996). The title, he explains, originated on a bumper sticker that circulated around the logging town of Forks, Washington during the height of the spotted owl flak, and neatly sums up one of the fundamental critiques of the modern environmental movement. Mainstream environmentalists, argues White, tend to vilify all manual work as an injurious, degrading, even violent assault upon the passive victim of pristine nature, extending their disdain not only to work itself but to the labourers who perform that work. At the same time, those environmentalists who eschew work in nature tend to simultaneously celebrate play, and indeed posit leisure as the only legitimate modern role of humans in the wild:

> Environmentalists . . . readily consent to identifying nature with play and mak[e] it by definition a place where leisured humans come only to visit and not to work, stay, or live. Thus environmentalists have much to say about nature and play and little to say about humans and work . . . But the dualisms fail to hold; the boundaries are not so clear.
>
> (White 1996: 173)

The dualism collapses in part because the very "play" celebrated by environmentalists is in fact an attempt by those who do not make a living via manual labour in nature to mimic that labour through play in an effort to "know" nature the way workers know it: to know it as though their lives depended upon it (p. 174). While White describes the cordoning off of nature as a space fit solely for play as part of the modern mainstream environmental ethos, its material and ideological roots may lay farther back in American history, in the dawn of the wilderness movement. Roderick Nash (1982) traces the American obsession with wilderness to the beginning of the twentieth century, which commenced with an increasing tendency to treat the American landscape less as a dangerous threat to civility and more as a stalwart guardian of national character, upper-class gentility and rugged masculinity (see Nash 1982: chs 3–4). In his classic book *Wilderness and the American Mind*, Nash describes this reformulation of nature in terms of the development of a full-blown "wilderness cult" which emerged in the late nineteenth century out of a variety of material and cultural changes taking place on the American scene. Nash notes, importantly, that attitudes toward wild places were only able to shift toward the positive once the formidable wilderness no longer posed a material barrier to the white settlers of the American landscape. Only after wilderness (and its inhabitants, American Indians) had been subdued on the American continent by industrialization, colonization, and frontier settlement could it be conceived in a new, less menacing light. Fear would be replaced by celebration only gradually through the material subjugation of both the continent and its inhabitants; in a sense the sanctity of wilderness could only be established once the wilderness was no longer wild. The implication, of course, is that consciousness has something to do with a subject's relationship to the material conditions of production: nature became a potential place for play when it ceased to be a necessary place to work for large portions of the population.

But nature at the carnival was not about fostering a sense of either ruggedness or gentility. If other bits of nature in the city – such as the greenswards created by the nineteenth-century urban parks movement – were intended to refine, sooth, and improve (not to mention pacify) the labouring immigrant masses by exposing them to the cultivating effects of the pastoral order (Kasson 1978: 12), then the more chaotic, edgy and dangerous nature tropes that pervaded Coney Island served a rather different (and less studiously engineered) role. In modern environmental discourse, we often speak of external nature (and wilderness in particular) as object rather than subject, as something which requires protection from the contaminating influence of human beings, a vulnerable "other" with the capacity to be changed, damaged, even destroyed by the actions of its human overlords (Cronon 1996: 69). As White puts it, "Nature seems safest when shielded from human labor" (White 1996: 172). But at one time in Western history the tables were turned, and human beings were understood to be susceptible to the potentially harmful influence of a far more powerful force than themselves: nature as a dark, chaotic, savage-haunted wilderness, a potentially *be*wildering place where even the upstanding citizen could be led to his death through the maddening song of the Siren or the illusory notes of the pan-pipe (in this trope, as in all nature tropes, nature is heavily imbued with ideas about race and gender). Andrew Light has recently described these two views as the *romantic* and the *classical* views of wilderness (Light 1995: 195–196), the former characterized primarily by a sense of reverence, appreciation, and paternalism toward an externalized nature that can be controlled and dominated by humans; the latter by a sense of mistrust, suspicion, and fear of an externalized nature that can control and dominate us – "us", of course, understood as circumscribed and contingent in terms of both race and gender, for women and ethnic or racial "others" can slip easily into the category of nature-as-not-human, whether comprehended as threat or ward. In the classical view, nature is the "threat", in the romantic view, humans are (or might be), but the underlying difference is who occupies the driver's seat: are we object or subject?

Yet between these two views, there is a shadowy middle ground that brings together elements of both tropes in the same sort of "play" that is characteristic of carnivals themselves: the contradictory pleasure that derives from fear, wonder, chaos, and the potential danger of the unknown. The key lies in the incongruous sense of exhilaration that accompanies the uncertainty of giving oneself over to the power of the roller coaster – or any other experience characterized by controlled risk, including that which involves the extremes of external nature. The dialectical and irresolvable contradiction between the "classical" view of nature-as-adversary and the "romantic" view of nature-as-protectorate gives rise to a paradox which springs from the continued and inevitable fact of both human domination of and submission to the forces of external nature. The carnival itself signifies the same sort of spatially and temporally delimited danger as the tamed wilderness, a strategically contrived threat designed to awaken latent fears and desires, to tantalize with the possibility of danger, disorder, bewilderment, loss of control. The pleasure derived from Coney Island's nature, much like that derived from Coney Island's machines, represents a paradoxical shift from the domination *of* human beings to a domination *by* human beings, a turning of the tables which yet requires a contrived subjugation of people by both nature and machine in order for the fantasy to hold water for the believer – a nature so thoroughly externalized that is has been re-internalized on more favourable terms for the newly realized and lonely subject. White's environmentalists seek out much the same thrill in the Rockies as the carnival patron seeks in the Cyclone: the illusion of the reestablishment of an older power structure: fleeting, circumscribed, and entirely reversible.

External nature, whether in effigy or in the flesh, blends so seamlessly into the fabric of the carnival because it constitutes its own heterotopia, a counter-site where all the anxieties of modernity – and particularly those based upon our uncertain place as part-subject, part-object in the natural world – are projected, distorted, and reflected back: "As we gaze into the mirror it holds up for us, we too easily imagine that what we behold is Nature when in fact we see the reflection of our own unexamined longings and desires" (Cronon 1996: 69–70). Nature *belongs* at the carnival in the same way that risky mechanical contraptions, games of daring and chance, and sexual tension between strangers *belong* there – because they all stimulate the complex sense of pleasure that flows uneasily from the juxtaposition of control and abandon, certainty and doubt, order and chaos. We seldom bother to question the pervasive naturalistic symbolism of the carnival because it resonates with meaning at such a fundamental level.

CONCLUSION

Plus ça change, or so the saying goes. While an elephant execution is unlikely to draw anything but public outrage today in New York City, nature still constitutes spectacle in Gotham. At the time of this writing, the brightest stars on the stage of New York City wildlife are not elephants, but birds: namely, a pair of red-tailed hawks called Pale Male and Lola, residents of a tony Fifth Avenue perch above Central Park for the last eleven years and subject of both their own PBS documentary and their own biography (Winn 1997). On 7 December 2004, workers at the Fifth Avenue co-op where the birds had nested for over a decade on a twelfth floor cornice unceremoniously removed the nest and the pigeon-deflecting spikes which inadvertently anchored it in place, promptly sending the city into a tizzy. In the ensuing days, as word spread about the death of the twiggy homestead, a swift storm gathered over East 74th Street. Feather-bedecked bird-watchers protested in front of the co-op carrying placards saying "Honk 4 Hawks", and were answered raucously by the horns of passing taxi drivers and city busses. Pro-hawk celebrity shareholders in the co-op such as Mary Tyler Moore[2] squared off against anti-hawk celebrity shareholders like Paula Zahn and husband Richard Cohen, the co-op board president who was reputed to have spearheaded the effort to rid the building of its famous fowl. The *New York Times* ran a pro-hawk editorial under the headline "Squatting Rights" admonishing the wealthy residents at 927 Fifth Avenue to "learn to live with the hawks" (*New York Times* 2004b). The New York State Department of Environmental Conservation scrambled to sort out the legal mess as bird-lovers demanded reinstatement for the pair under an international treaty protecting migratory birds. The hawks circled nervously, reportedly making several unsuccessful attempts to rebuild their nest in brazen defiance of forcible eviction by their erstwhile landlords, thereby endearing themselves further to pro-hawk urbanites who quickly took them to heart as noble comrades in the endless struggle for affordable housing in New York City.

Pale Male and Lola, like Topsy, are mere chapters in the age-old story of human beings struggling for urban elbow-room with unpredictable non-human neighbours: some charismatic, some repugnant, most (liked hardened urbanites the world over) persistent in the face of encroachment. For some at the posh co-op, the hawks had become a pest (much as the rebellious Topsy had become a nuisance to her owners), sending bird droppings and the dead carcasses of small prey spiraling down onto the heads of their well-coiffed human neighbours. For others, such as hawk defenders at the *Times* and across the boroughs (much like Topsy's trainer, who refused to aid or even attend her execution), the hawks were a forcible and welcome reminder that New York City continues to be a part of nature (whether it likes it or not) and the rightful habitat of more than just humans. Nor was this the first time the city was forced to confront the less appealing side of such impressive birds: a few years ago, an innovative City Parks programme intended to rid Bryant Park of the never-ending scourge of rats and pigeons by unleashing predatory hawks in the vicinity came to a screeching halt after one of the birds nearly made off with a park-goer's precious pet chihuahua (WNBC.com 2003). The struggle continues: co-op, or coop? No matter how far urban humans manage to bend nature to their will, the conflict between the romantic and classical views of wild nature are never wholly resolved, while the paradoxical lurks quietly beneath, betraying our simultaneous hope and fear that we are not, after all, entirely in control of external nature.

NOTES

1 Stephen Weinstein notes that this trend would be reversed toward the mid-twentieth century, with automation waning as Coney Island's primary source of entertainment given the increasing importance of the beach, particularly with the rise to power of Robert Moses (1984: 282–283).
2 Mary Tyler Moore has become something of a veteran of celebrity environmentalism; ten years ago she spearheaded a campaign to have a 65-year-old lobster released from his tank at a Malibu restaurant and dumped back into his native habitat off the coast of Maine (Smith 1996: 35).

BIBLIOGRAPHY

Adams, J.A. (1991) *The American Amusement Park Industry: A History of Technology and Thrills*. Boston, MA: Twayne Publishers

Armbruster, E.L. (1924) *Coney Island*. New York: Published by the Author

Berman, J.S. (2003) *Coney Island*. New York: Barnes and Noble Books

Bodgan, R. (1988) *Freak Show: Presenting Human Oddities for Amusement and Profit*. Chicago, IL: University of Chicago Press

Brooklyn Daily Eagle (1896) "The big elephant in a blaze", *Brooklyn Daily Eagle*, 28 September 1896

Butler, L.H. (1991) *Coney Island Kaleidoscope*, text by J. Manbeck. Wilsonville, OR: Beautiful American Publishing Co

Carlin, J. (1989) *Coney Island of the Mind: Images of Coney Island in Art and Popular Culture*. New York: Whitney Museum of American Art

Cronon, W. (1996) "The trouble with wilderness; or, getting back to the wrong nature", in W. Cronon (ed.) *Uncommon Ground: Rethinking the Human Place in Nature*. New York and London: W.W. Norton & Company

Cudahy, B.J. (2002) *How We Got to Coney Island: The Development of Mass Transportation in Brooklyn and Kings County*. New York: Fordham University Press

Denson, C. (2002) *Coney Island: Lost and Found*. Berkeley, CA: Ten Speed Press

Foucault, M. (1967) *Of Other Spaces: Heterotopias*, originally published as "Des espaces autres", *Architecture/Mouvement/Continuité*, October 1984

Gilden, B. (1986) *Coney Island*. London: Westerham Press

Gillman, L. (1955) "Coney Island", *New York History* 36

Ierardi, E.J. (1975) *Gravesend, the Home of Coney Island*. New York: Vantage Press

Immerso, M. (2002) *Coney Island: The People's Playground*. New Brunswick, NJ: Rutgers University Press

Kasson, J.F. (1978) *Amusing the Million: Coney Island at the Turn of the Century*. New York: Hill & Wang

Kilgannon, C. (2004) "A happy tale for the birds: wings wide, Pierre is free", *New York Times*, 24 December 2004

Lapow, H. (1978) *Coney Island Beach People*. New York: Dover

Lee, J. (2004) "As hawks circle, all sides seek compromise", *New York Times*, 12 December 2004

Light, A. (1995) 'Urban wilderness", in David Rottenberg (ed.) *Wild Ideas*. Minneapolis, MN: University of Minnesota

Lueck, T.J. and Lee, J. (2004) "No fighting the co-op board, even with talons", *New York Times*, 11 December 2004

Lueck, T.J. (2004a) "New aerie is readied for Fifth Avenue hawks", *New York Times*, 22 December 2004

Lueck, T.J. (2004b) "Co-op to help hawks rebuild, but the street is still restless", *New York Times*, 15 December 2004

Lueck, T.J. (2004c) "Newly homeless above 5th Ave., hawks have little to build on", *New York Times*, 9 December 2004

McCullough, E. (1957) *Good Old Coney Island, A Sentimental Journey into the Past: The Most Rambunctious, Scandalous . . . Island on Earth*. New York: Scribner

McGowan, S., Hoerder, D. and Barrow, L. (1994) *The Coney Island Experience: A Case Study in Popular Culture and Social Change*. Bremen: Universität Bremen

Moses, R. (1939) *The Improvement of Coney Island*. New York: Department of Parks

Nasaw, D. (1993) *Going Out: The Rise and Fall of Public Amusement*. New York: Basic Books

Nash, R. (1982) *Wilderness and the American Mind*, third edition. New York and London: Yale University Press

New York Times (1884) "A jumbo house for Coney Island", 21 February 1884

New York Times (1902) "Elephant terrorizes Coney Island police", *New York Times*, 6 December 1902

New York Times (1903) "Coney elephant killed", *New York Times*, 5 January 1903

New York Times (2004a) "New York city hawk fans intensify protests over eviction", *New York Times*, 11 December 2004

New York Times (2004b) "Editorial: squatting rights", *New York Times*, 9 December 2004

Office of Brooklyn Borough President (1921) *Assessment Work 1921: Proposal for Bids, bid, Bond, Agreement and Specifications for the Improvement and Protection of the Public Beach at Coney Island, Together with All the Work Incidental Thereto*. New York: M.B. Brown Printing & Binding Co

Onorato, M.P. (1988) *Another Time, Another World: Coney Island Memories*. Fullerton: M.P. Onorato and California State University, Oral History Program

Onorato, M.P. (1997) *Steeplechase Park: Coney Island 1928–1964: The Diary of James J. Onorato*. Bellingham, WA: Pacific Rim Books

Onorato, M.P. (1998a) *Steeplechase Park: Sale and Closure 1965–1966: Diary and Papers of James J. Onorato*

Onorato, M.P. (1998b) *Steeplechase Park: Demolition of Pavilion of Fun 1966: Diary and Papers of James J. Onorato*

Peiss, K. (1986) *Cheap Amusements: Working Women and Leisure in Turn-of-the-Century New York*. Philadelphia, PA: Temple University Press

Pilat, O. and Ranson, J. (1941) *Sodom by the Sea: An Affectionate History of Coney Island*. Garden City, NY: Doubleday

Register, W. (2001) *The Kid of Coney Island: Fred Thompson and the Rise of American Amusements*. New York: Oxford University Press

Smith, N. (1984) *Uneven Development: Nature, Capital, and the Production of Space*. Oxford: Basil Blackwell

Smith, N. (1996) "The production of nature", in George Robertson, Melinda Mash, Lisa Tiekner, Jon Bird, Barry Curtis and Tim Putnam (eds) *FutureNatural: Nature/Science/Culture*. New York: Routledge

Snow, R.F. (1984) *Coney Island: A Postcard Journey to the City of Fire*. New York: Brightwaters Press

Snow, R. and Wright, D. (1976) "Coney Island: a case study in popular culture and technical change", *Journal of Popular Culture*, 9(4): 960–975

Stanton, J. (1997a) "Coney Island Timeline: 1880s. Coney Island History Site". Online. Available HTTP: <http: //naid.sppsr.ucla.edu/coneyisland/articles/1880.htm.>

Stanton, J. (1997b) "Coney Island – Dreamland Fire. Coney Island History Site". Online. Available HTTP: <http: //naid.sppsr.ucla.edu/coneyisland/articles/dreamlandfire. htm>

Stanton, J. (1999) "Coney Island Timeline: Coney Island – Second Steeplechase". Online. Available HTTP: <http: //naid.sppsr.ucla.edu/coneyisland/articles/steeplechase2. htm.>

U.S. Works Progress Administration (U.S.P.W.A.) (1939) "Coney Island", in *U.S. Works Progress Administration New York City Guide*. New York: U.S. Works Progress Administration

Weinstein, S.F. (1984) "The Nickel Empire: Coney Island and the creation of urban seaside resorts in the United States", Ph.D. Thesis, Columbia University

Weinstein, R.M. (1992) "Disneyland and Coney Island: reflections on the evolution of the modern amusement park", *Journal of Popular Culture*, 26(1)

White, R. (1996) "Are you an environmentalist or do you work for a living?: Work and Nature", in William Cronon (ed.) *Uncommon Ground: Rethinking the Human Place in Nature*. New York and London: W.W. Norton & Company

Winn, M. (1999) *Red-Tails in Love: A Wildlife Drama in Central Park*. New York: Vintage

WNBC (2004) "Hawks grounded in Bryant Park after chihuahua attack: bird may have mistaken dog for rat". WNBC.com Online. Available HTTP: <http: //www.wnbc.com/ news/2384499/detail.html.>

Zukin, S. *et al.* (1998) "Coney Island and Las Vegas in the urban imaginary: discursive practices of growth and decline", *Urban Affairs Review*, 33: 625–653

6 The desire to metabolize nature

Edward Loveden Loveden, William Vanderstegen, and the disciplining of the river Thames

Stuart Oliver

INTRODUCTION

This chapter examines the "improvement" of the river Thames in the late eighteenth and early nineteenth centuries as an example of the city's quest to create and exploit uneven patterns of development. It looks at how, through the construction of a series of locks and associated engineering works, those in control of the Thames interrupted and channelled its flow to make it both a vital part of England's infrastructure and a means for the exploitation of the natural world. It argues that the disciplining of the Thames is emblematic of the way in which people, politics, and political ecology have come to be bound together during the era of modernity to produce and enforce the disciplined nature that is characteristic of the contemporary city.

Contemporary urban political ecology is based on what is usually a relatively orthodox reading of the labour theory of value and its subject therefore becomes, in effect, landscape as the work of the flow of "labour value" channelled by the regulating powers of capitalist society. Even so, a considerable body of literature suggests that cultural values are so tightly woven into the economic text produced by that flow that they ought to be taken into consideration as factors of considerable importance in urban ecologies.

Literature from the cultural turn suggests that spatiality and spatial process are the result of the interplay of real, imagined, and symbolic elements (Shurmer-Smith and Hannam 1994). The corollary of this is that these elements channel the flows of life and are expressed in the needs created by the condition of the landscape, the wants constructed by the treatment of those conditions by culture, and the desires they arouse. The typically modern control established over the flow of the Thames in the eighteenth and nineteenth centuries can be seen as manifested in the solid constructions of locks and weirs designed to discipline the unprofitable, unseemly, and disturbing flows of the river. In this respect that control reflected the logic not just of capital and its institutional infrastructure but also of its desires. Such multiple logics were the result of the development of a capacity for controlling the river that was by turns economic, administrative, and emotional in nature.

Evidence to test the theoretical perspective of this approach can be found in the behaviour of two key individuals involved in the early administration of the Thames, Edward Loveden Loveden and William Vanderstegen. Through an analysis of the work in which they were involved it is possible to examine the links between landscape, public values, and private lives. By doing so it is possible to demonstrate ways in which these causal elements were constructed for and by the turbulent interplay between the river, the community, and individuals.

flow and its metaphors

The investment in fixed capital works such as the locks and weirs (and the associated infrastructure of navigation ferries, towpaths, bridges, and other works installed along the Thames) was such as to begin to make of the river a work of metabolized "second nature" (Smith 1990). In the language of political economy, by "improving" the river this metabolization was intended to further what Harvey (1999) identified as a necessity for the faster circulation of value and the increase of the capital stock by the cheapening of raw materials, the expansion of markets, and the acceleration of the turnover of capital.

The value represented in capital has, according to Marx (1954), a "purely social reality" (vol. I, p. 54) and measuring such a social conceit has proved problematic. A particularly helpful discussion is the analysis of labour value given by Spivak (1996) who argued that the textual nature of value makes a segregation between economic and cultural determinants of its nature in works of art impossible as the two intertwine in a way that makes them "irreducibly complicitous" (p. 120). The significance of this analysis of the cultural creation of value for understanding the metabolization of the Thames is that it provides a model for the questioning of the causality of the dominant economism of political ecology and also enables the questioning of culturalist interpretations of nature.

Spivak's argument about the nature of value is particularly important because of its relevance to the questioning of the prominence of the economic in explanation of places produced by the cultural turn in geography. It implies both that the logic of causality must be heterogeneous and that the deployment of culturalist values must be subject to intense scrutiny. Consequently, while the notion of there being anything other than relativity in the "value" of human labour is both impossible and absolutely necessary to affirm, it is also necessary to state that, built on the myth of a non-ideological culturalism, cultural "value" will be influenced by political economy.

The constraint of flow

The river, like any other landscape, acts as a topologue of human life, a texture imprinted by or interwoven with the meanings of human need. In a loose reading, therefore, it is still possible to say that power relations force the river to take on the contingent form of a "cultural landscape" (Sauer 1963) for which "Culture is the agent, the natural area is the medium, the cultural landscape the result" (p. 343).

However, the cultural landscape of the Thames in the late eighteenth and early nineteenth century can be seen as a more problematic construction. It was a landscape of discipline, formed by the holding back and channelling of the river's water in response to the manifestation of the "real" needs generated by the material demands of living, the "imagined" wants generated by culture, and the "symbolic" desires created by people's feelings. It was these forces of generation that gave the meaning of value to the metabolization of the river, to its construction in the period under examination as a built environment.

The *real* landscape of the Thames was subjected to an engineered discipline so that capital could flow freely. The locks, as solid works of fixed capital, enabled the holding back of water to make navigation, and the circulation of capital, easier. They therefore acted as a means for legitimizing the existing social structure because they created the necessary conditions for the reproduction of capital and also acted, in the way suggested by Harvey (1996: 183), as "manifestations and instanciations 'in nature'" of contemporary social relations.

The economic desirability of using waterways to benefit trade in England in the early-modern era led to the development of a movement for the improvement of rivers and the construction of canals (Jackman 1966) and a concern with the desirability of usable waterways came to motivate a number of thinkers' opinions about the Thames (Willan 1964). Through the installation of an engineered infra-structure the river was slowly but unavoidably changed into a partially metabolized work of culture. This process was carried out principally by replacing or supplementing the old weirs with new locks that were designed to ensure what Thacker (1968: I, 2) described as a regime of "easy and inexpensive facility". Their effect was to provide a faster circulation of value, and allow what Naruhito (1989: 24) listed as a "fall in freight rates . . . extra traffic generated . . . greater intensity in the use of barges, swifter speeds of passage, larger boats and lower costs".

Such improvements need to be seen in the context of the cultural *imaginary* of "circulation". The hegemonic discourse of circulation in the West went through a profound change at the beginning of the capitalist era, as Rublack (2002) showed, when medieval notions of the circulation of abundance gave way to notions of privately managed scarcity that were much better fitted for a new age of accumulation. Even so, according to Tuan's account of the hydrological cycle (1968), the influence of natural theology was such that as late as about 1850 the flow of the cycle was interpreted as a divine gift and the river's flow therefore to represent to some extent a divine bounty for human use.

According to Landreth and Colander (1994), until the late eighteenth century circulation was held to be of major importance in explaining the accumulation of wealth. In mercantilist theory, because the flow of goods was related proportionately to the level of wealth in the country, easy circulation was vital for the economy. The development of physiocratic thought in the mid-eighteenth century promoted the idea that transport was particularly important because it determined attrition of the value of the economy's productive surplus.

These ideas were reflected in the literature on the Thames. For Vallancey (1763) commerce was "the only means to render a State flourishing and formidable to its

Neighbours" (p. iii) and Burton (A Commissioner 1767) claimed the principal benefit of "improving" the river would be to enable the "Encrease of Trade" (p. 13). By the 1770s the mercantilist concern with circulation had, though, begun to be replaced by a physiocratic concern with the "surplus" represented by the value of water. Increasingly, therefore, what became most important was not the circulation of traffic on the river but the conservation of the river's water – a change that led A Commissioner (1772) to criticize the "Spirit of Benevolence" (iii) formerly shown to the millers and their use of the river water, and praise the thrifty regime regulating the use of the improved Thames.

The disciplining of the Thames also had significant *symbolic* resonance for repressed individuals fighting to establish boundaries over the unruliness of their own feelings. The significance of boundaries can be seen in those psychodynamic interpretations of desire which, influenced by Freud's mechanical model, stressed the significance of the flow of psychic energy. Marcuse (1956) put forward the interpretation that the repressive binding of energy by power is an inevitable function of the reality principle in any class-based society, and Reich (1970) presented the damming of the flows of desire as a means to engrave oppression into the inner self. Poststructuralist thought, with its emphasis more on becoming rather than on being, has likewise tended to represent the disciplining of any flow as inherently authoritarian. For Lyotard (1984) it is the "damming" up the flow of libidinal desire by the investment of libido in a "device" that polices the flow of desire, stabilizing it to make "locks, canals, regulators of desire" (p. 98).

Humanist writers on psychotherapy have suggested that cruel societies create individuals capable of destroying the alienated other of nature. Fromm believed that destructiveness is generated by conditions of emotional impotence: "I can escape the feeling of my own powerlessness in comparison with the world outside myself by destroying it" (1984: 158). For Miller the "repression of [emotional] injuries endured during childhood" acts as "the root cause of psychic disorders and criminality" (1987: 4). Influenced by Miller, Maguire (1996: 170) went further to suggest that "The unrecognized fear and hurt which fuel our absurd social and political process fuel at the same time both our aggression and our indifference towards the world we inhabit." These emotional pressures developed in modernity led to the instrumental relationship with nature characterized by Gutkind (1956: 21) as "I–it", as without "intimate and personal contact".

THE THAMES, AND THE CONSTRAINT OF ITS FLOW

The discursive currents of flow were lived out in the context of the specific "natural area" and historical conditions of the Thames.

Geoarchaeological reports indicate that the condition of the Thames has changed considerably in the period of human occupation (Gibbard 1985). This change meant a move away from a Thames of irregular shallows and pools to a more ordered river with many of its most awkward meanders cut through, its shallows dredged, its descent regulated by weirs. This change can be seen as the result of a long process by which a discourse of improvement was constructed in such a way that the river was represented as in need of engineering.

While at the beginning of the early modern era the condition of the Thames was seen as a manifestation of the harmonies of creation, by the eighteenth century it had generally come to be seen as a work of fallen nature. For Leland (written 1535–1543, published 1744) the condition of the river even at its source gave evidence of divine benevolence where "is the stream servid with many ofspringes" (vol. III, 100). Likewise, for Camden (1695) the river's upper course was "pleasant and gentle" and in harmony with the landscape (col. 137). Yet by the end of the sixteenth century, drawing on the direct experience of users of the river, there was a growing belief that the condition of the river was unsatisfactory. Most important in the development of this discourse was Bishop's petition of 1585 which was principally directed at the presence of watermills on the river that "stoppe the course of . . . Ryver [sic]" (in Furnivall 1908: IV, 418). Similarly Taylor's dyspeptic catalogue of the wrongs of the river (1632) linked the damage caused by private interests that had "barr'd its course with stops and locks" (unpaginated).

The desired improvement of the Thames was enabled by an administrative infrastructure established to control the river (Fishbourne 1882). The Thames was one of England's four Royal Rivers and from Saxon times had been administered by the Crown (Thacker 1968). In 1197 conservancy of the Thames was transferred by the Crown to the Corporation of London, which came to exercise a haphazard administrative control over the river below Staines. The Crown continued to intervene sporadically in the regulation of the river but the first time that an administrative apparatus for the river was made permanent was with the establishment of the Oxford-Burcot Commission in 1605 to improve the Thames between Cricklade and Burcot. Comprehensive and permanent regulation of the river was only established in 1751 with a commission given power over the whole river above Staines, but the 1751 Commissioners proved relatively ineffective and were re-established in 1771 with greater powers and an intended aim of re-engineering the river. This attempted solution to the problem of navigation on the Thames was initially met with opposition from the Corporation of London. It was only at the beginning of the nineteenth century that the Corporation, at first tentatively, accepted the Commissioners' case that building locks could improve the river and also protect the rights of property and trade associated with it.

The principal means of implementing improvement on the Thames was the installation of locks and associated works. Weirs existed on the Thames by as early as the end of the eighth century (Thacker 1968) and by 1771 there were perhaps 33 of them (vol. I, 171–173). They had mainly been installed to create a head of water for milling but also helped produce an increased depth of water for navigation. They were often known as "flashlocks" from the "flash" of water they created when opened to allow navigation and, as millers and navigators alike relied on an adequate head of water, this loss was usually regretted by both parties. From the seventeenth century flashlocks were replaced or supplemented by modern locks, usually then known as "poundlocks", that increased the head of water and economized the amount of water necessary to float a boat up or down the river. While until 1771 there were only 3 locks, such was the rapidity of their adoption that by 1793 this had increased to 25 (vol. I, 156).

To see these works in context it is, however, necessary not only to assess the real pressures of political economy and the imaginary of institutional affairs but also the symbolic realm of the personal feeling of the individuals involved in deciding upon them, so it requires now to turn to Edward Loveden Loveden and William Vanderstegen.

UNDERSTANDING LOVEDEN AND VANDERSTEGEN

It is possible to see the craving for improving the Thames as linked in part to the identities of the men who administered the river. Modernity has demanded more control of the inner selves of its inhabitants, and in eighteenth-century Europe new types of masculinity developed that were more suited to the new demands being made on men (Connell 1993). The principal change was to a more rigid control of the self, a change represented by outraged contemporaries as a "triumph of the sexless" over the blunt physicality of traditional manliness (Brunstrom 2001: 47).

This manliness manifested itself in individuals with a well-developed determination to control both self and alienated other. The ways in which this control was projected onto the river may be judged by examining the case of two men influenced by such pressures, Edward Loveden Loveden and William Vanderstegen, both Commissioners of the Thames and key protagonists in the "improvement" of the river. The relationship between the two is the subject of a brief examination by Hadfield (1969) and alluded to by Thacker (1968). For Hadfield (1969) they were the "two most influential" Commissioners, held "widely divergent views" (p. 24), and were engaged in active "rivalry" (p. 68) over the control of the river.

Edward Loveden Loveden

Edward Loveden Loveden was born Edward Loveden Townsend, probably in 1750, and attended Winchester School then Trinity College, Oxford (W. M. 1822; Foster 1888). The manor of Buscot was bequeathed to him by his uncle, Edward Loveden, on the condition that he took the Loveden surname – which he did in 1772 (Fisher 1986). He died at his country house, Buscot Park, Berkshire (now Oxfordshire) in 1822.

As a prominent public figure, and protagonist in a spectacularly unpleasant divorce case, there is considerable evidence concerning Loveden and his life. Loveden married three times: first in 1773 to Margaret Pryse; widowed in 1784, he married Elizabeth Nash in 1785; widowed again in 1785, he married Anne Lintall in 1794 and separated from her in 1808 (Fisher 1986). Loveden's public interests were many, and Mavor's obituary of him concluded with typical sympathy that "Few country gentlemen have performed a more honourable part in life than the deceased" (W. M. 1822: 89). He was a member of the Board of Agriculture from 1793, Sheriff of Berkshire from 1781 to 1782, Sheriff of Brecon from 1799 to 1800, and Lieutenant Colonel of the Berkshire militia from 1794 to 1796, he was also

Member of Parliament for Abingdon from 1783 to 1796 and for Shaftesbury from 1802 to 1812 (Fisher 1986).

Loveden seems to have been perceived by his contemporaries as a generous and cultured man. According to W. M. (1822: 89) "Mr Loveden was hospitable to a great degree, and his establishment at Buscot Park was on a scale of considerable expense." From school he "always delighted in Classical literature", and after his father's death he attended Trinity College as a Gentlemen Commoner. Nathaniel Wraxall concluded that "His figure, manners, and dress all bespoke a substantial yeoman rather than a person of education and condition; but he did not want plain common sense, nor language in which to clothe his ideas" (Wheatley 1884: 251). And W. M. added that "to the last, his appearance, his manners, and useful knowledge, always devoted to the best interests of society, caused him to be regarded as no common man" (1822: 89).

Loveden was a man of ambiguous political allegiance (Fisher 1986) which W. M. (1822: 89) interpreted as acting with "independence characteristic of his fortune and his principles". Originally elected an opponent of the Coalition, he declared in Parliament that he "considered himself as a free agent" (*Parliamentary History of England*: vol. 27, col. 908). A particularly interesting insight into his motivation is given in that same speech by his claim that his maxim was "*Nullis addictus jurare in verba magistri*". That quotation, from Horace's *Epistles* (vol. I, part i, l. 14), translated as "I am not bound over to swear as any master dictates" (Horace 1926: 251–253). Loveden, who "always delighted" in Classical literature, would assuredly have been familiar with the continuation of the paragraph which propounds an eclectic mixture of Stoic participation in public life and the Cyrenian teaching that individuals should control the world around themselves: "wherever the storm drives me I turn in for comfort. Now I become all action, and plunge into the tide of civil life, stern champion and follower of true Virtue; now I slip back stealthily into the rules of Aristippus, and would bend the world to myself, not myself to the world" (p. 253).

Despite the ambiguity of Loveden going "wherever the storm drives", his public claim was to be led by "an honest zeal for the promotion of the public welfare". This "honest zeal" he described as "a better principle, and a more becoming motive than either self-interest or ambition" (*Parliamentary History of England*: vol. 27, col. 908). The pride that Loveden felt in the comeliness of his motivation did not, however, protect him from political danger. Many of his constituents found him anything but W. M.'s generous host, with them "resentful of his parsimony" (Fisher 1986: 457) and there was strong opposition to his re-election at Abingdon in 1790.

Turning from Loveden's public to his private self, there is evidence that Loveden's sense of manliness was important to his character. Writing to Samuel Selwood he passed the then-common slur on William Pitt's unsexual (therefore unmanly) demeanour that he "appears to me of the doubtful gender" (BRO, A/AET 11). Significantly, Loveden also exerted or attempted to exert quite considerable manly control over the lives of the women in his household, as his relations with his daughters Margaret and Jane show. Loveden's elder daughter Margaret was married at the age of 21 against her father's wishes (BRO, D/ELV, catalogue entry) to the

Reverend Samuel Wilson (*The Times* 1796). Loveden attempted to disinherit her, and a marriage settlement was only negotiated in 1803 (D/ELV L23/1 to 8). More suggestive still is the forbidden engagement of Loveden's handicapped daughter Jane in 1809 to Mr R. Weeks, his protégé. On finding out from Jane about the proposal, Loveden wrote a curt one-sentence note requiring Weeks to leave Buscot Park immediately (D/ELV F 33/4). There is no evidence the rift was ever healed.

The most important evidence on Loveden's character is provided by the details on his divorce, a case that Stone (1993: 248) described as "a somewhat banal story of a bored, neglected, and childless young wife falling in love with a lively and attractive young man". Divorce then required, first, a civil suit for "criminal conversation" at the Court of King's Bench, then a suit of separation at the London Consistory Court, then finally a bill for divorce in the House of Lords. Loveden's case for criminal conversation failed, but he won at the London Consistory Court (Gurney 1811). Loveden had married Anne when he was 43 and she was 21. According to Stone (1993), Anne was an affectionate woman, but neglected and lonely; at the trial there was the hint that sexual relations between the couple were unsatisfactory. Although the jury at the Court of King's Bench had been unable to find against Barker, Scott's judgement was that there was "fair inference" of adultery between Anne and Barker (p. 2). In the end, the divorce was never finalized, apparently because Loveden objected to paying an annuity to Anne of £400 (Fisher 1986).

Loveden's responses to Margaret's and Jane's independence, and above all the events surrounding his divorce indicate he was a man who was prone to great anger, and even litigious revenge in response to hurt. In acting in this way he was, though, within his legal rights – and might be said to have been responding to his moral duty. Right and duty, as the example of Buscot Park shows, were issues that occasioned Loveden to act in a controlling way.

Loveden's desire for control was clear in his running of his estate. Buscot Park house was probably designed by Loveden himself and James Darley, and begun in 1779 (Hollings and Alexander 1924: 512). Some aspects of Loveden's management of the estate, particularly some of its innovations, hint further at a controlling desire. Witness the design of his weirkeeper's cottage which contained a fish house that "when locked, even the person who inhabits the cottage connected with them, could not open . . . without the certainty of detection" (Mavor 1808: 46).

Water was a necessary feature of Buscot Park, and something of Loveden's ostentatious experimentalism can be seen in his relationship with it. The estate also included Buscot Lock and a short canal from the Thames (called Buscot Pill) with a wharf, as well as rights over Eaton Weir. In addition to the Thames and Buscot Pill, the grounds had what Mavor described as "two fine pieces of water . . . in a pleasing natural style" (p. 44). The "pleasing" naturalness of the water features in Buscot Park were reinforced by Loveden's attitude to the control and display of water. Finding Buscot Park's wellwater brackish, he described himself as "determined to make a spring" (p. 45) – and pumped the house's water from subsurface drains.

Loveden's interest in the river was considerable in extent and proprietary in nature. Famously, the first barge that passed through the Thames and Severn Canal was greeted by a twelve-gun salute from Buscot Park (*Gentleman's Magazine* 1789). Loveden was a Commissioner from 1783 and most active in the years 1789 to 1792 (BRO, Commissioners of the Thames, Minutes). According to W. M. (1822: 89) he was:

> a principal promoter of the junction of the Thames and Severn; and the Thames Navigation was indebted to him for almost every real improvement in the upper districts; which has been made within a period of fifty years. So much was he attached to the prince of British streams, on whose banks a large portion of his estate lay, that he used to be called, jocularly by his friends, "Old Father Thames," an application which he did not dislike on suitable occasions.

Loveden ensured his revenue from Buscot Lock was that of a rack-renter, for which he was subjected to a thinly-veiled attack by Vanderstegen (1794a: 5) and Thacker (1968) gave Buscot as consistently the most expensive weir on the Thames above Oxford. Loveden's actions promoting navigation are likewise indicative of his enthusiasm for remunerative improvements – every increase in trade on the river increased his considerable revenue. In his essay on the threats posed to the Thames by canals Loveden claimed they would be "a robbery of the Thames" (1811: 3–4). More personal, indeed personalized, was his conclusion that urged the Commissioners to "unite in resisting the confederacy and conspiracy against old Father Thames" (p. 9).

Loveden chaired the meeting in 1781 to set up the Thames and Severn Canal, and was what Hadfield (1969) called a "very active proprietor" of the company (p. 34). But, Hadfield believed, he was also an opinionated partisan – most egregiously as chair of the Parliamentary Enquiry of 1793 into the Thames which Loveden made "a validation of his policy" (p. 25). When there was a proposal for the Wilts and Berks Canal to bypass the Thames (including Buscot) above Abingdon, Loveden argued for an extension of the Grand Junction Canal to Pinkhill – obliging barges to continue to use his own lock. In this counterproposal, according to Hadfield "he was in fact moved entirely by self-interest" (p. 124) and acted in a way that was "utterly unscrupulous" (p. 127). In order to win the vote he untruthfully claimed that the Commissioners were about to improve the Thames above Abingdon, but he was later defeated and the breathtaking consequences are worth quoting:

> Unable to conceal his chagrin or restrain his anger he defaced the minute book, the company's clerk recording in the margin: "After this Meeting was over Mr Loveden came to the Table, took the Book out of my hands, and struck his name out, saying he would not have his name appear when he did not approve the Resolutions."
>
> (p. 129)

In conclusion, Loveden was clearly "no common man" – he was able to depict himself plausibly as a man of especial talents and especial value. Fundamental to understanding him is his inference that he was a man who would turn for comfort "wherever the storm drives".

Loveden may have been a "substantial yeoman" but he was also a man of some intellectual sophistication – the member of learned societies, the active reformer of agriculture – and "utterly unscrupulous" in his pursuit of control and profit. In his politics, Loveden may well have claimed "disinterested independence", but this seems to have been at least at times, disingenuous. His ambition for power seems to have been betrayed by the egregiousness of his purported disinterest, for it is hard to imagine Loveden would have involved himself in the world of power had he felt no temptation for it. In his relationships with women, Loveden seems to have been haunted by his failures of control. When Anne found her own amusements Loveden seems to have been involved in the precipitation of that infidelity, absent about his public duties while she was asserting her desire for criminal abandon at home.

At Buscot Park Loveden displayed his taste and his control over landscaped nature in a very direct fashion. In the opulence of the house, the control of the landscaping, the barred fish house, and the use of water in the mimicked freedom, Buscot Park demonstrated the labour Loveden expended to subjugate nature. Each demonstrated his identity as an enlightened gentleman, each was a fiction of enlightened self-interest. But when Loveden was checked in his desires for prosperous, self-interested control, his uncontrolled rage was considerable – as the divorce proceedings and the Thames and Severn Canal's defaced minute book testify. Loveden may have enjoyed and been flattered by the nickname Old Father Thames, but his paternal authority seems to have been less than disinterested and to have been desperate for the satisfaction of his own wants, to prevent the "robbery" of . . . his own self.

A principled man who was unprincipled. A man sometimes a Stoic actively and self-sacrificingly engaged in public life, at other times a Cyrenian seeking to control life and others for his own purposes. As the storm blew him. A man who sought control, according to the needs of his uncontrolled controlling passions. A man, therefore, both extraordinary in public and ordinary inside.

William Vanderstegen

William Vanderstegen was probably born in 1737 (Foster 1888) and died in 1797 (Burke 1863). In the minimal details of his obituary he was "a very active magistrate" (*Gentleman's Magazine* 1797: 624) and his list of public duties was considerable: "a J.P. and a D.L. for Oxfordshire, Chairman of Quarter Sessions, High Sheriff in 1761, and one of the first Commissioners of the Thames" (Smith-Masters 1933: 32).

Vanderstegen's father was a Dutch Protestant immigrant who probably came to England in 1689 (Smith-Masters 1933) and at his death was described as "an eminent merchant" (*Gentleman's Magazine* 1754: 95). The Vanderstegen family had a considerable amount of money and William was left £18,000 at his majority (PRO, PROB 11/807, 90v–93r). They moved in high circles: his sister Elizabeth married Sir Charles Asgill and William married Elizabeth Brigham in 1759 which brought him the Brigham family estate – including Cane End House, Oxfordshire (Burke 1863).

Vanderstegen took an active part in county affairs, including the administration of the Thames as a Commissioner between 1783 and 1796 (BRO, Commissioners of the Thames, Minutes). The sources of his interest in the Thames were numerous: he felt himself one of those "possessing property in the center of navigation" (1794a: 76); the Cane End estate included part of the land on which Caversham Bridge was built (Pearman 1894) and Vanderstegen also became a shareholder in Whitchurch Bridge (E. M. F. 1988). His interest in the river may well also have been influenced by his brother-in-law, who served as Lord Mayor of London from 1757 to 1758 and was directly engaged in the administration of the Thames (*Gentleman's Magazine* 1788) as well as being "drawn to the riverside by memories of its fashionable reputation in his youth" (Hussey 1944: 992).

There is little information on Vanderstegen's life, though a few fragments can be gleaned. Elizabeth Brigham's wealth means he found a woman it was clearly advantageous for him to have married, but this by no means precludes a love match. Indeed, there is a small amount of evidence to suggest the family was close – perhaps very close. In 1794 Vanderstegen complained of his absences from the family, noting he had "attended many meetings, at all distances from home, and . . . met with no small difficulties of accommodation, some attention of mind, absence from my family and affairs" (1794a: 39–40). Vanderstegen and Elizabeth had one son (also called William) and one daughter. William junior was born in 1779 or 1780 (Foster 1888), eighteen to twenty years after his parents' marriage. Perhaps he too seems to have felt the need to marry a woman who was emotionally demanding: in Lybbe Powys' poem (1869) she comes across as a figure of mock dread for her smothering behaviour.

There are some indications other than the time into the marriage after which his son was born about the life that Vanderstegen seems to have pursued. As befitted a man of his status, Cane End House was a comfortable home, with its complement of servants, oak-panelled reception rooms, crimson damask Chippendale furniture, and a Chippendale doll's house commissioned by Vanderstegen (Dils 1994; Smith-Masters 1933). Vanderstegen kept a large pack of foxhounds and even in 1843 his grandson William Henry was still drinking the Cane End currant wine he had laid down – described as at its best "quite tip-top" (Lybbe Powys 1869: 15).

Yet Vanderstegen's life must also be seen alongside and against the intellectual constructions he placed on it. The key characteristics in his few public pronouncements about himself were of a man driven by duty and a sense of rectitude, even to the experience of "considerable expence" and "no small difficulties" with which he depicted his work (1794a: 40). The function of Vanderstegen's actions were plain: to discipline the Thames as a response to obligation and enlightened self-interest. In financial terms, the Thames might "at a very small expence, compared to that of making a canal, be made a navigation by far more beneficial to the public than any canal" (p. 2). In addition this would prevent the "injury" that "numberless individuals will sustain by the desertion of the River" (p. 2).

As well as being a "true lover of his country" (1794b: 76), Vanderstegen made "active able" efforts to make of the Thames a "safe and certain navigation" below Mapledurham and for it to become "compleat" as a "safe, easy, cheap and expeditious inland navigation" above that point (1794a: 8 and 11). During his time as a Commissioner he was present at the majority of occasions that proved to be significant in determining the policy towards locks on the river. It was he who presented the report to construct locks below Maidenhead at which the Commissioners, in gratitude "Ordered that the Thanks of this Meeting be given to M^r. Vanderstegen for his very active able and disinterested Conduct and attention to the business of the Navigation" (BRO, Commissioners of the Thames, Minutes, vol. 3, p. 305).

Rationality provided the key justification for Vanderstegen's interventions in public life. He was "a plain man, dealing only in the statement of plain facts" (1794b: 71). It served to differentiate him from those he opposed, being absent, for example, from the promoters of canal navigation who had behaved deplorably because "they have not given themselves time, cooly and impartially, to consider on which side the preference should be" (1794a: 1).

Vanderstegen presented probity as the motivation for his involvement in a public feud over the alleged evasion of salt tax. His own explanation of these events was that an un-named relative inherited a saltworks and through her he became aware of a widespread evasion of taxation. His involvement was because, he claimed, he was "a lover of justice and good order" (1794b: 1). Not so, according to the rebuttal produced by Thomas Weston and Co. (1794) who accused him of being a man of "malice", his actions "pointed, in a very envenomed manner" (p. 3), his "emnity" (p. 4) indicative of an "indefatigable energy" (p. 27) to ruin the company after it had ended a commercial relationship with his relative. Vanderstegen's reply to their pamphlet (1794c) angrily rebutted these claims as "wilfully mistaken" (p. 3).

Vanderstegen strove with tenacious energy to protect his honour. In his anonymous reiteration of his attack on Weston (Anonymous 1794) he claimed his only object was "to obtain greater justice to the community in general, and individuals in particular" (p. iii). Relinquishing his duty, he claimed, "will tend to greater injury to the State, than a manly perseverance; and would certainly reduce him to a disgraceful and contemptible situation" (p. vii). This striving to be above contempt reflected his family's public myths as depicted in its coat of arms: a lion between two fish (see Burke 1863) – the lion being the beast of grandeur, of honour (Room 1999).

In Vanderstegen's account he was a man from whose honour flowed the obligation of duty: "a true lover of his country", he wrote of himself in the third person (1794b: 76), "can never think himself employed so agreeably as in promoting its welfare". This duty was not all encompassing (it allowed for enjoyments) but it seems to have nevertheless represented a burden for him. It led him to what his great-grandson Douglas recalled as the "considerable" expense of his official duties as High Sheriff (D. Vanderstegen 1935: 10).

Even so, duty involved Vanderstegen in inflicting punishment: "it will appear to the candid Reader that, throughout the whole of the Author's proceeding, his object has only been to obtain justice" (Anonymous 1794: iii). Justice, as in the courts over which he presided, necessitated him in the punishment of offenders. For transgressors he was direct in his condemnations – over the actions of the "ungovernable" bargemasters (1794a: 6), the impertinences of "persons who call themselves engineers" (p. 26), and the prevarications of the Corporation of London that were the result of its "unwillingness to expend the money necessary" (p. 41). Yet while Vanderstegen may have addressed the "candid Reader" he was not always a candid man himself. When his duty compelled him to mount a sustained attack on the integrity of Loveden and on the opponents of improvement at Windsor (1794a) none were mentioned by name but all would have been readily identifiable to his audience. Such an attack allowed him the luxury of punishment with the benefits of plausible deniability.

In conclusion, Vanderstegen's adherence to a code of rational obligation reflected in his notion of "manly perseverance" the converse of what he alluded to as a "disgraceful and contemptible situation". In his attitude to the outlawed "contemptible" may perhaps be seen his attitude to the outlawed natural of the foxes he hunted as much as those who broke the laws relating to salt duties. A similar approach can be seen in his attitude to the river. In Vanderstegen's discourse "Plain" common sense, allied to "indefatigable energy" was the solution to improving the irrational out-of-placeness of a fallen nature in need of "good order".

Both the Thames and its navigation were the source of considerable interest to Vanderstegen, and as an active member of the riparian gentry and brother-in-law of Charles Asgill this is unsurprising. As he demonstrated in his pamphlet on the Thames (1794a), this interest was not merely important because it took him away on his "active able" business, but also because it provided a focus for his "indefatigable energy", the very imperfectness of the river's nature seems to have been a provocation for him and the aim of his work seems to have been to project his own values over the ungovernable Thames, to make of it a river subject to the "justice and good order" of which he was a lover.

For Vanderstegen, "good order" was a matter both of honour and of property. It is plain that the notion of "deserting the navigation" must have had considerable resonance for a man of honour, let alone a man of agrarian property, but there is more than that. The propertied order of the "center of the navigation" was threatened with decentring by the proponents of new canals, its value by devaluation. The fairness for which Vanderstegen strove was under real and metaphorical attack by these new routes that would bypass Cane End and reject the nature represented by the river and property relations.

Vanderstegen saw himself as a needed gentleman at the centre of power. The respect he earned and the values he ascribed to, as part of an immigrant family, was threatened by change. As he showed on a number of occasions when angered, although his rational self always seems to have justified his anger against those who transgressed good order, Vanderstegen was capable of responding with zeal.

GENERAL DISCUSSION

In order to provide a constructive critique of political ecology, this chapter has examined the metabolization of the Thames from a culturalist perspective. This should not be taken as any sort of dismissal of the dominant approach in political ecology; that there were "real" needs for the improvement of the Thames is not contested – the state of the river ensured that. Nor is it contested that there were "imaginary" wants for the improvement of the Thames – the political discourse of the period ensured that. But it suggests that these needs and wants were only partial things, that they cannot be fully understood without a recognition that they provoked and were themselves changed in a dialectic with the very feelings of desire they helped to construct. More specifically, the evidence produced here indicates that feelings, ideology, and economic calculation were all "irreducibly complicitous" in the improvement of the Thames.

In particular, this chapter has indicated how, within the constraints of the economic and cultural pressures of the time in which they found themselves, Loveden and Vanderstegen were driven to impose their emotional identity on the river. In part their involvement was one of economic calculation, to prevent the "confederacy and conspiring against old Father Thames" (Loveden 1811: 9) and substitute for it a "safe, easy, cheap and expeditious inland navigation" (Vanderstegen 1794a: 11) along the improved river. But Loveden and Vanderstegen also demonstrated a desire for the ordered flow of control in their lives, a desire that enabled them to fit in to the very centre of a culture requiring control over nature. Occasionally their own feelings surged out in a flash of rage, in pamphlets or in the pages of a minute book, but they acted to subject the flow of the Thames to an order that their inner selves seem at times powerless to accept.

The metabolization of the Thames was part of a process of inscribing meaning on the river, of creating a disciplining of its flow and order over its disorder, to make value. The cases of Loveden and Vanderstegen give some evidence of the concrete manifestations of these processes in everyday living and everyday landscapes. The shaping of their desires by the broader needs and wants of their time indicates something of the dialectical flows involved in the construction of value. The evidence that members of the gentry, free to live lives relatively unconstrained by unfulfilled needs and wants, were even so men of their time whose actions can be seen as works of desire connected to a particular historical time and place is significant. The feelings of Loveden and Vanderstegen may ultimately be seen as profoundly influenced by the material conditions of the surrounding world. That these feelings had meaning for both Loveden and Vanderstegen is undeniably true, but these meanings can only be understood in the context of the needs generated by the logic of the flow of capital and the wants generated by the logic of the political imaginary.

Given this analytical perspective it remains here to return to the type of "improvement" the resultant dialectic of change created to make an ordered Thames through the construction of locks, weirs, and their associated works. The arguments presented here demonstrate that while this built environment, in part merely a network of capital investment, was an attempt to fix and maintain the value maintained and represented in the channel of the river, it was also the manifestation of imaginary wants and symbolic desires. In their solid rigidity standing against the flow of water these works were, in both senses, an engineering of powerful control.

The infrastructure of locks acted in a real sense to control the river and its flow by holding back the river to ease navigation. They acted in the imaginary of public discourse to control the wastefulness of the river by economizing the use of the water. They acted in a symbolic sense to control the disturbing fluxes of the moving river's flow as a rational man might control his feelings.

BIBLIOGRAPHY

A Commissioner [John Burton] (1767) *The Present State of Navigation on the Thames Considered; and Certain Regulations Proposed*. London: Daniel Price

A Commissioner (1772) *Extracts from the Navigation Rolls of the Rivers Thames and Isis: With Remarks Pointing out the Proper Methods of Reducing the Price of Freight*. London: C. Bathurst

Anonymous [William Vanderstegen] (1794) *An Address to the Public, in Justification of the Conduct of the Author of the Pamphlet Entitled Observations on Frauds Practised in the Collection of the Salt Duties, and the Misconduct of Officers Fairly Stated*. Reading: Smart and Cowslade

Berkshire Records Office, BRO: A/AET 11; D/ELV, catalogue entry; D/ELV F33/4; D/ELV L23/1 to 8; Commissioners of the Thames, Minutes, 3, 305

Brunstrom, C. (2001) "'Be male and female still': an ABC of hyperbolic masculinity in the eighteenth century", in C. Mounsey (ed.) *Presenting Gender: Changing Sex in Early-modern Culture*. Lewisburg, PA: Bucknell University Press

Burke, B. (1863) *A Genealogical and Heraldic Dictionary of the Landed Gentry of Great Britain and Ireland*. London: Harrison

Camden, W. (1695; new edn 1971) *Camden's Britannia 1695: A Facsimile of the 1695 Edition, Published by Edmund Gibson*. Newton Abbot: David and Charles

Connell, R.W. (1993) "The big picture: masculinities in recent world history", *Theory and Society*, 22: 597–623

Dils, J. (ed.) (1994) *Rural Life in South Oxfordshire 1841–1891: Cane End, Kidmore End, Gallows Tree Common: With an Appendix on Emmer Green*. Oxford: Sonning Common W.E.A. Local History Group

E.M.F. [E. Fitzeustace] (1988) "The Vanderstegens [sic] activity in St Peter's", *Caversham Bridge*, April 1988

Fishbourne, E.H. (1882) *The Thames Conservancy*. London: Davis and Son

Fisher, D.R. (1986) "Loveden, Edward Loveden", in R.G. Thorne (ed.) *The House of Commons 1790–1820*, vol. IV. London: Secker and Warburg

Foster, J. (1888) *Alumni Oxonienses: The Members of the University of Oxford, 1715–1886: Their Parentage, Birthplace, and Years of Birth, with a Record of their Degrees*. Oxford: Parker and Co

Fromm, E. (1984) *The Fear of Freedom*. London: Ark

Furnivall, F.J. (1908) *Harrison's Description of England in Shakespeare's Youth: Being the Second and Third Books of his* Description of Britaine and England: *Edited from the First Two Editions of Holinshed's* Chronicle, *A.D. 1577, 1587*. London: Chatto and Windus

Gentleman's Magazine (1754) "Deaths", *Gentleman's Magazine*, 24: 95

Gentleman's Magazine (1788) "At Richmond, Sir Charles Asgill, bart. banker", *Gentleman's Magazine*, 58: 841

Gentleman's Magazine (1789) "Gloucester, Nov. 19", *Gentleman's Magazine*, 59: 1139

Gentleman's Magazine (1797) "Deaths", *Gentleman's Magazine*, 68: 624

Gibbard, P.L. (1985) *The Pleistocene History of the Middle Thames Valley*. Cambridge: Cambridge University Press

Gurney, Mr. (amanuensis) (1811) *Loveden and Loveden: The Judgement Pronounced by Sir William Scott in the Consistory Court of London, on the 13th July 1810, in a Suit Instituted by Edward Loveden Loveden, Esq. M.P. for a Divorce from Ann* [sic] *Loveden his Wife*. London: John Stockdale

Gutkind, E.A. (1956) "Our world from the air: conflict and adaption", in W.L. Thomas Jnr (ed.) *Man's Role in Changing the Face of the Earth*. Chicago, IL: University of Chicago Press

Hadfield, C. (1969) *The Canals of South and Southeast England*. Newton Abbot: David and Charles

Harvey, D. (1996) *Justice, Nature and the Geography of Difference*. London: Blackwell

Harvey, D. (1999) *The Limits to Capital*. London: Verso

Hollings, M. and Alexander, N. (1924) "Buscot", *The Victoria History of the Counties of England: A History of Berkshire*, vol. IV, St Catherine Press

Horace (1926) *Satires, Epistles and Ars Poetica*. Cambridge, MA: William Heinemann

Hussey, C. (1944) "Asgill House, Richmond, Surrey: the home of Mr. H. Ward and Mr. B.A. Stirling Webb", *Country Life*, 9 June 1944: 992–995

Jackman, W.T. (1966) *The Development of Transportation in Modern England*. London: Frank Cass

Landreth, H. and Colander, D.C. (1994) *History of Economic Thought*. London: Houghton Mifflin

Leland, J. (1744; new edn 1964) *The Itinerary of John Leland in or about the Years 1535–1543*. London: Centaur Press

Loveden, E.L. (1811) Untitled essay in Commissioners of the Thames Navigation (eds) *Two Reports of the Commissioners of the Thames Navigation, on the Objects and Consequences of the Several Projected Canals, which Interfere with the Interests of that River; and on the Present Sufficient and Still Improving State of its Navigation*. Oxford: Commissioners of the Thames Navigation

Lybbe Powys, P. (1869) *The Lay of the Sheriff*. London: privately published

Lyotard, J.-F. (1984) "Several silences", *Driftworks*, New York: Semiotext(e)

Maguire, J. (1996) "The tears inside the stone: reflections on the ecology of fear", in S. Lash, B. Szersznski and B. Wynne (eds) *Risk, Environment and Modernity: Towards a New Ecology*. London: Sage Publications

Marcuse, H. (1956) *Eros and Civilization: A Philosophical Inquiry into Freud*. London: Routledge and Kegan Paul

Marx, K. (1954) *Capital: A Critique of Political Economy*. London: Lawrence and Wishart

Mavor, W. (1808) *General View of the Agriculture of Berkshire*. London: Richard Phillips

Miller, A. (1987) *For Your Own Good: The Roots of Violence in Child-rearing*. London: Verso

Naruhito, Prince (1989) *The Thames as Highway: A Study of Navigation and Traffic on the Upper Thames in the Eighteenth Century*. Privately published

Pearman, M.T. (1894) "Historical notices of Caversham", *Transactions of the Oxfordshire Archaeological Society*, 34

Public Record Office, PRO: PROB 11/807, 90v–93r

Reich, W. (1970) *The Mass Psychology of Fascism*. London: Souvenir Press

Room, A. (1999) *Brewer's Dictionary of Phrase & Fable*. London: Cassell

Rublack, U. (2002) "Fluxes: the early modern body and the emotions", *History Workshop Journal*, 53: 1–16

Sauer, C.O. (1963) "The morphology of landscape", in J. Leighly (ed.) *Land and Life: A Selection from the Writings of Carl Ortwin Sauer*. Berkeley, CA: University of Berkeley Press

Shurmer-Smith, P. and Hannam, K. (1994) *Worlds of Desire, Realms of Power: A Cultural Geography*. London: Edward Arnold

Smith, N. (1990) *Uneven Development: Nature, Capital and the Production of Space*. Oxford: Basil Blackwell

Smith-Masters, J.E. (1933) *The History of Kidmore End, Oxfordshire: With Notes of Sonning, Eye and Dunsden, Caversham, and Mapledurham, from the Earliest Times*. Leighton Buzzard: Faith Press

Spivak, G.C. (1996) "Scattered speculations on the question of value", in D. Landry and G. Maclean (eds) *The Spivak Reader: Selected Works of Gayatri Chakravorty Spivak*. London: Routledge

Stone, L. (1993) *Broken Lives: Separation and Divorce in England 1660–1857*. Oxford: Oxford University Press

Taylor, J. (1632) *Taylor on Thame Isis: Or the Description of the Two Famous Riuers of Thame and Isis, who Being Conioyned or Combined Together, are Called Thamisis, or Thames*. London

Thacker, F.S. (1968) *The Thames Highway*. Newton Abbot: David and Charles

The Parliamentary History of England, from the Earliest Period to the Year 1803 (1816). London: Longman, Hurst, Rees, Orne and Brown

Thomas Weston and Co. (1794) *Refutation of the Charges Brought by Wm. Vanderstegen, Esq. against Mr. Thomas Weston, and Other Merchants Concerned in the Salt Trade, as Far as Those Charges Respect the Thames Street Company of Salt Importers*. London: G.G.J. and J. Robinson

The Times (1796) "Married", *The Times*, 30 September, 1796

Tuan, Y.-F. (1968) "The hydrologic cycle and the wisdom of God: a theme in geoteleology", *University of Toronto Department of Geography Research Publications*, 1

Vallancey, C. (1763) *A Treatise on Inland Navigation, or, the Art of Making Rivers Navigable, of Making Canals in all Sorts of Soils, and of Controlling Locks and Sluices*. Dublin: George and Alexander Ewing

Vanderstegen, D. (1935) "High Sheriffs", *The Times*, 11 April 1935: 10

Vanderstegen, W. (1794a) *The Present State of the Thames Considered; and a Comparative View of Canal and River Navigation*. London: G.G. and J. Robinson

Vanderstegen, W. (1794b) *Observations on Frauds Practised in the Collection of the Salt Duties, and the Misconduct of Officers Fairly Stated*. Reading: Smart and Cowslade

Vanderstegen, W. (1794c) *A Reply to a Pamphlet, Entitled Refutation of Charges, &c. Respecting Frauds Committed in the Collection of the Salt Duties*. Reading: Smart and Cowslade

Wheatley, H.B. (ed.) (1884) *The Historical and the Posthumous Memoirs of Sir Nathaniel Wraxall 1772–1764*. London: Bickers and Son

Willan, T.S. (1964) *River Navigation in England 1600–1750*. London: Frank Cass

W.M. [William Mavor] (1822) "Edward L Loveden, LL.D.", *Gentleman's Magazine: And Historical Chronicle*, new series, 15: 88–89

7 Turfgrass subjects

The political economy of urban monoculture

Paul Robbins and Julie Sharp

> Suburbanites – advised by nurserymen who in turn have been advised by the chemical manufacturers – continue to apply truly astonishing amounts of crabgrass killers to their lawns each year. Marketed under trade names, which give no hint to their nature, many of these preparations contain such poisons as mercury, arsenic, and chlordane. Application at recommended rates leaves tremendous amounts of these chemicals on the lawn.
>
> (Carson 1962: 80)

> One of our dogs was very allergic to the [lawn chemical] treatment. In the spring when they would start to fertilize, his paws would just get raw and bleed. We would have to take him to the vet two or three times a week and they would do these whirlpool treatments and finally we realized it was the lawn chemicals. So, for a couple of days after we had the grass done we would put these little booties on the dog. Otherwise it would really hurt him, and he would just bite and chew at his paws and they would bleed all over the place. We felt so badly for him.
>
> (Suzanne, Ohio homeowner)

Nothing more captures such anxieties associated with urban living than the strange case of the American lawn. The momentum of the nascent lawn chemical industry, viewed at its inception by Rachel Carson in the 1950s (above), was fully realized by the late 1990s when the lawn chemical economy at last came of age. At the dawn of the twenty-first century, more people in the United States apply chemicals to their lawn than do not. In an analysis of national water quality, the United States Geological Survey reveals that 99 percent of urban stream samples contain one or more pesticides and that in urban watersheds insecticides were detected more often and at higher concentrations than in non-urban systems (United States Geological Survey 1999). Though these chemicals are coming from a range of urban sources, lawn care is an important contributor. But what does this mean for the people who actually use them with such hesitation?

Mostly, it means contradiction and ambivalence. At the same time that homeowners like Suzanne (above), who told us her story in an interview during the Spring of 2003, live in an internally warring state of anxiety and responsibility over chemical use, their applications continue to increase annually. US households spend

$222 each on lawn care equipment and chemicals with consumer lawn care input purchases recently reaching an all-time high of $8.9 billion; 55 percent of households apply insect controls and 74 percent apply fertilizer (National Gardening Association 2000). So even while national pesticide consumption has decreased, and fewer chemicals are being used in industrial and commercial sectors, pesticide use on private lawns remains high and continues to climb (United States Geological Survey 1999). With private turfgrass estimated to exceed 23 percent of urban land cover, and lawn grass coverage increasing by well more than one hundred thousand hectares annually (Robbins and Birkenholtz 2003), such applications cover more ground every year, no matter what unease may pervade the imagination of the average homeowner.

Such conditions reflect the enigma of urban political ecologies, which are complicated, in part because they are so remarkably unspectacular, and yet so dangerously far-reaching. Unlike the great and exotic catastrophes of large-scale tropical deforestation or mass extinction, both worthy topics of critical environmental analysis, urban political ecologies expand and insinuate themselves largely "below the radar", through the daily disaggregated practices of hundreds of millions of people, who consume and produce the world around them in the conversion of land, the puddling of wastewater, and the puffing of emission.

That such daily ecologies aggregate to vast effects is, of course, well known. The average automobile, for example, will produce its own weight in carbon over a year's time, making the daily commute of the average Los Angelean a matter of clearly global significance. Similarly, the linkage of such local urban ecologies to large scale economic interests and power is increasingly well understood. The relationship between land developers and speculators has as much to do with urban sprawl as any consumer choice.

Even so, the fragmented character of urban ecology causes the political economy of nature in the city to be *experienced* by its participants (consumers, workers, managers) with a degree of ambivalence and contradiction. People living in systems that produce unevenly distributed costs, environmentally unsound externalities, and shifting risk ecologies are driven to feel simultaneously distant from ecological process, while acknowledging at some level their intimate relationships to nature all around them. The remarkable success of household recycling in the United States and Canada, despite its mixed record of effectiveness (Ackerman 1997), is a tribute to the unease experienced by normal people living in a dawning "risk society" (following Beck 1992).

The lawn is emblematic in this sense. In the time between Carson's prognosis and Suzanne's dawning concern, the lawnscape itself expanded across North America to become ubiquitous. Estimates of lawn coverage are difficult to make, but turfgrass cover is estimated to be around 16 million hectares nationally. For Franklin County Ohio, a typical urban/suburban region encompassing Columbus Ohio and its satellite communities, roughly a quarter of total urban land is under lawn. With urban land in the United States expanding by 675 thousand hectares per year between 1982 and 1997 (Natural Resources Conservation Service 2000), this means ongoing expansion of turfgrass cover, especially in residential areas. The proportion of private land given over to lawn coverage – as opposed to the footprint of the residence, shrub/tree cover, sidewalks, and driveways – also increases with every housing start, and in increasing proportion of total lot size (Robbins and Birkenholtz 2003).

These monocultural turf landscapes of the United States have their aesthetic roots in the gardens of English manor houses following the landscape fads and Italian landscape paintings of the eighteenth century (Jenkins 1994). The lawn in its modern form, however, no matter how common in contemporary cities, is a relatively recent phenomenon. The key species of the monocultural lawn came to North America in the last century and the high-input chemical management system is even more recent. As late as the 1930s, lawn maintenance practices were largely weed-tolerant and involved hand-pulling and keeping of chickens for weeds and grubs. The use of chemicals was in fact discouraged, since it retarded the growth of edible greens (Barron 1923; Dickinson 1931). It was only in the post-World War II era that the quantity of lawn coverage and the intensity of its management began to accelerate (Bormann *et al.* 1993), and with it a broader political economy that produces and is reproduced by a vast community of homeowners, who together maintain and service the lawn monoculture.

Yet the lawn and the lawn owner are rarely addressed when considering municipal health risks, ecosystem degradation, or ecosystem function more generally. The reasons for this are twofold. Firstly, the cultural landscapes associated with daily life (homes, gardens, offices, stores) tend to vanish because of their normalness, ordinariness, and ubiquity. More pointedly, the lawn rarely receives critical scrutiny because it is largely viewed as a cultural artifact, rather than a political or economic one. People have lawns because they like lawns, the intuitive line of thinking suggests, and do what is required to maintain them (Schroeder 1993). Thus the lawn, a significant chemical input and a massive multi-national economy, remains in hiding, if directly in plain sight, an artifact of personal choice, nested in a vast economy, driving an inchoate sense of consumer anxiety.

Launching our investigation from exactly this point, we here summarize research that surveys the political and economic character of the American lawnscape. The research, part of a three-year project utilizing national surveys, interviews, aerial photo assessment, and industry analysis, sets out to answer the simple question: what determines the extent and management of the lawn, and what perpetuates its existence when those who maintain it do so with such profound hesitation?

Examining the linkages of the turfgrass yard to ecosystems, chemical production economies, and community values and priorities to answer these questions, we conclude here that direct and aggressive sales of chemicals to consumers are spurred in part by crises in the chemical formulator industry and by declining margins in the worldwide chemical trade. The evidence supports a broader understanding of the question, however: the lawn is a capitalized system that produces a certain kind of person, one who answers to the needs of landscape, despite an urge to the contrary. We conclude, therefore, that it is not that American communities that produce lawns or that global industry produces individual desires, but instead that the lawn itself, as a socio-technical system implicated in capitalized production, produces turfgrass subjects – that urban/suburban subject whose identity is inter-pellated (literally "hailed", following Althusser) by the purified lawn, and whose identity and life is disciplined by the material demands of the landscapes they inherit. So too, the urban community, which appears to create normative pressure to produce lawns, is itself formed by the specific material demands of turfgrass, and the cycles of daily life directed by cutting, watering, and tending, this ravenous shared ecology. For urban political ecology, our results suggest a serious and renewed engagement with human ideology, experience, and desire. For lawn owners, including the authors themselves, these results suggest a critical appraisal of the political and ecological economy of our own identities.

LAWN CHEMICAL USE, ECOLOGY, AND RISK

> Certainly I've been responsible over the years. When there were younger kids in the neighborhood, I made sure they weren't getting on the grass, and put those little flags up, keeping them off so that they don't walk through it and put it in their mouth.
>
> (Tom, Ohio homeowner)

The pronouncements of people like Tom, interviewed in his home in 2003, suggest a growing awareness that lawn chemicals have social and ecological effects. More generally, there is a growing acknowledgment that the demands of lawn care, when met by input-oriented control solutions, inevitably involve a certain degree of risk, sometimes made manifestly obvious in cases of acute exposure (as in the case of Suzanne above). Even the Scotts Company, the industry leader in lawn chemical retail sales with 52 percent of market share, explained to its investors in 2001, "We cannot assure that our products, particularly pesticide products, will not cause injury to the environment or to people under all circumstances" (United States Securities and Exchange Commission 2001: 16).

This is largely a result of the specific ecology of lawn species. Anglo-Americans originally introduced these landscapes from Eurasia, along with almost all of their constituent species, including Bahiagrass (*Paspalum notatum*), Bermudagrass (*Cynodon* spp.), Kikuyugrass (*Pennisetum clandestinum*), Annual ryegrass (*Lolium multiflorum*), Colonial bentgrass (*Agrostis tenuis*), Kentucky bluegrass (*Poa pratensis*), Perennial ryegrass (*Lolium perenne*), and Tall fescue (*Festuca arundinacea*). As a result, and despite its cultural significance throughout North America, the turfgrass lawn is an exogenous ecosystem in Canada and the United States, where it dominates, and the requirements for its propagation are high as a result. Though these species are robust, the climatic demands of many regions, including the humid south, the arid west, and the frigid north, all make tremendous demands on homeowners seeking to nourish exotic monoculture, specifically the use of pesticide and fertilizer inputs.

The most commonly used home pesticides (both insecticides and herbicides) in the US are shown in Table 7.1 (Extension Toxicology Network 2000; United States Environmental Protection Agency 2000). The deposition of these chemicals is largely unregulated and has been identified as a serious ecosystem risk in both the

United States and Canada (Fuller *et al.* 1995). Indeed, many of the same chemicals for which registration and training are required in the agricultural sector, are sold over-the-counter to lawn owners in unregulated quantities (Guerrero 1990).

Case reports of childhood tumours and leukemia associated with lawn chemical usage began to emerge in the late 1970s and 1980s, with a growing body of work substantiating these concerns. Though the effects of 2,4-D, the most common of all yard herbicides, on human health are generally debated, expert panels have concluded that the weight of evidence supports the possibility that exposure can cause human cancers (Ibrahim *et al.* 1991). So too, neurotoxins like chlorpyrifos appear far more significant than has been generally accepted to date (Zartarian *et al.* 2000), and yard treatments have been shown to have strong associations with soft tissue sarcomas (Leiss and Savitz 1995). This is especially true for children; pesticide usage, specifically including Diazanon as well as yard weed herbicides, has been shown to be associated with childhood brain cancer (Davis *et al.* 1993; Zahm and Ward 1998).

More troubling, lawn chemicals, rather than residing on lawns where they have a relatively short half-life, are commonly tracked into homes and deposit on clothing, where they represent ongoing exposure risks, and where they become considerably more persistent than previously thought. These chemicals accumulate in house dust and on surfaces and carpets where small children – precisely the group whose risk levels are highest – are disproportionately exposed (Lewis *et al.* 1991; Leonas and Yu 1992; Lewis *et al.* 1994; Nishioka, Brinkman *et al.* 1996; Nishioka, Burkholder *et al.* 1996, 1999a and 1999b). Controlled studies demonstrate that measurable quantities of herbicides are absorbed by dogs and remain in their urinary system for several days after lawn exposure, underlining the possible exposure impacts on other mammals, including humans (Reynolds *et al.* 1994).

Chemicals and other inputs on lawns have been demonstrated to have severe and detrimental ecological effects, moreover. Beyond direct chemical deposition, with

Table 7.1 Pesticides used in US homes and gardens

Pesticide	Mt Active*	Type	Toxicity (EPA)	Environmental toxicity
2,4-D	3150-4050	Herbicide	Slight to High	Birds Fish Insects
Glyphosate	2250-3600	Herbicide	Moderate	Birds Fish Insects
Dicamba	1350-2250	Herbicide	Slight	Aquatic
MCPP	1350-2250	Herbicide	Slight	NA
Diazanon	900-1800	Insecticide	Moderate	Birds Fish Insects
Chlorpyrifos	900-1800	Insecticide	Moderate	Birds Fish
Carbaryl	900-1800	Insecticide	Moderate to High	Fish Insects
Dacthal (DCPA)	450-1350	Herbicide	Low	Birds Fish

*Millions of metric tonnes of active ingredient used in the US (United States Environmental Protection Agency 2000).

its serious implications for ambient insect, fish, and bird populations (see Table 7.1), lawn maintenance degrades air quality through the use of relatively dirty two-stroke engines (Priest *et al.* 2000; Sawyer *et al.* 2000; Christensen *et al.* 2001). The fragmentation of the landscape by lawns also adversely affects reproduction, survivorship, and dispersal of birds species (Marzluff and Ewing 2001). In sum, the lawnscape as an exogenous monoculture demands and receives increasing quantities of inputs per unit land.

THE POLITICAL ECONOMY OF LAWN CHEMICAL PRODUCTION

We wish to suggest here, moreover, that the expansion of the lawn and the increasing intensity of its ecology occur at appreciable expense and represents the end of an extensive commodity chain, with political economic pressures for its development exerted at multiple scales. Pressures for the development of the lawn monoculture are most evident at the local scale where the economy of urban development assures a steady supply of spaces for management and an enforced demand for normative lawn aesthetics.

Producing customers: the formulator industry

Much of the expansion of turf chemical use parallels changes in the industry of turf chemical production. In particular, the formulator industry – companies that purchase raw chemical inputs to combine them into consumer chemical products –have dramatically changed their relationship to consumers in recent years. Specifically, since the late 1980s, formulator firms have turned to "pull" marketing: direct advertising by mail, radio, and television. This departs radically from traditional ("push") marketing, where formulator companies fill seasonal bulk orders to independent retail hardware and garden stores, who interact with customers (Baker and Wruck 1991; Williams 1997). "Pull" marketing means the direct marketing of products by formulators, with familiar company names like Scotts and Bayer. This change in strategy is notably recent and has been received by the trade as revolutionary, innovative, and crucial for industry survival (Journal of Business Strategy 1989; Cleveland Plain Dealer 2000; Robbins and Sharp 2003a and b).

The strategy requires a shift of resources towards the production of image and brand recognition and the devotion of significant budgets towards market research and the investigation of household chemical habits, with the specific goal of changing them. The concomitant massive increase in advertising costs is directed not only to television, radio, and print advertising, but also to toll-free hotlines, in-store sales representatives, web pages, and email lists (Hagedorn 2001; US Securities and Exchange Commission 2001). Scotts, the industry leader, commonly spends twice as much as traditional firms to advertise its product, spending millions of dollars on television advertising, where traditionally such expenses were shouldered by retailers (Jaffe 1998; Scotts Company 2000). This marketing revolution has proven somewhat successful and consumer spending on lawn chemicals has increased in many markets, despite declines in other areas of herbicide and insecticide sales (National Gardening Association 2000).

Why increase a long-shunned and somewhat risky shift to direct marketing strategies and shoulder the considerable costs associated with it? The answer to this question centres on the narrowing margins in the industry, which have created an imperative to expand the number of chemical users and the intensity of chemical use per lawn. Firstly, the industry has become increasingly reliant on big box discount stores and home improvement warehouses, as small hardware stores and other traditional retailers shun the standing warehouse stock required for seasonal industries like lawn care, and as they disappear altogether (Bambarger 1987; Cook 1990; Williams 1997). Mass sales and bulk wholesaling reduces formulator industry receipts as a result (Scotts Company 2002).

Secondly, the formulator industry has undergone a series of aggressive and capital-intensive product acquisitions in recent years, which have reduced credit ratings and stock share prices, while increasing standing debt, aggravated by closed facilities, severance packages, and product recalls (Baker and Wruck 1991; Chemical Week 1998; Cleveland Plain Dealer 2000). The Scotts Company, in a prominent example, spent $94 million on interest payments in the fiscal year 2000, a figure inflated by the high interest rates of that year, placing tremendous pressure on cash flow (United States Securities and Exchange Commission 2001).

These kinds of increased expenses and reduced receipts have been coupled with the rising direct costs and opportunity costs associated with the difficult patenting systems, increased regulation of inputs, and environmental citations and fees from government agencies (Scotts Company 2001; United States Securities and Exchange Commission 2001; Scotts Company 2002a; Scotts Company 2002b).

The environment itself poses further barriers to accumulation, since lawn products sell most vigorously in Spring and Summer, while highest expenses and debt service payments tend to come in Fall and Winter. Wet years slow fertilizer sales, dry ones decrease pesticide sales, and cold seasons retard sales overall (United States Securities and Exchange Commission 2001; Scotts Company 2002a and b).

In sum, the formulator industry is in a production squeeze with tight and decreasing margins, ongoing consolidation, and debt. It is this business climate that drives the revolution in high-expense, high-risk, direct sales and its concomitant increase in chemical usage. At the same time, moreover, agrochemical firms, who supply raw chemical materials to the formulator and applicator industries, also face increasing pressure to find, produce, and exploit household chemical markets.

Producing markets: the global agrochemical industry

The term "agrochemical industry" specifically refers to those few, large, diversified chemical companies that globally manufacture and sell the active ingredients in pesticides and fertilizers. These firms and their respective markets have also undergone dramatic recent changes.

The establishment and expansion of large-scale pesticide and fertilizer development and manufacturing is a product of military technology and processing power developed during World War II. The fight against typhus and malaria on the front, coupled with the search by chemists for chemical warfare agents, led to the discovery and the development of DDT (in 1939), the discovery of the herbicidal properties of 2,4-D and MCPA (in 1944), and the first organic weed killers based on the regulation of plant hormones (Whitten 1966; Aldus 1976; Anderson *et al.* 2003). Under the direction of research and development by the Chemical Warfare Service of the US military, pesticide production blossomed and proved to be effective and economically efficient. As historian Edmund Russell notes (2001: 149): "the finances of making DDT could not be beat", especially since government tax amortization was dispensed to chemical producers as the war neared its end.

The profitability of the industry expanded during the post-war period, as demographic shifts increased food demands, agricultural land prices rose, farm labour became increasingly scarce, and growing affluence meant increased demands for unblemished food products with no signs of damage or disease (Stephens 1982). Inexpensive, easy-to-use farm chemicals continued to grow (Green *et al.* 1987), especially as the petrochemical industry entered pesticide manufacture as a way to market production by-products, and academic researchers and extension agents continued to sing their praises (Young *et al.* 1985).

However, contraction of farm chemical markets began following the energy crisis of the 1970s, the US farm crisis of the mid-1980s, and the concurrent economic recession. By the 1980s, there were fewer acres in crops, pesticides had reached saturation in most markets, and demand for farm chemicals began to drop, remaining low throughout the 1990s (United States Department of Commerce 1985; Eveleth 1990; British Medical Association 1992). Continued depression in commodity prices meant a continued reduction in the market for and drop in the price of agricultural chemicals (Reich 2000).

At the same time, the costs of raw materials, solvents, and other chemicals needed for the reactions and purification processes in chemical manufacture have risen (British Medical Association 1992). So too, research and development costs have skyrocketed, becoming higher in the pesticide industry than in manufacturing as a whole (Reich 2000). About 15,000 new compounds must be tested to yield one marketable pesticide and it takes 8–10 years to bring a pesticide from the stage of initial synthesis to the commercial market; the extent and cost of registration trials, required by the EPA under the Federal Insecticide, Fungicide, and Rodenticide Act (FIFRA) and the FDA under the Federal Food, Drug, and Cosmetic Act (FFDCA), have risen in recent years as regulation of pesticides has increased (Staggers 1976; Anderson 1996). With the passage of the Food Quality Protection Act (FQPA) in 1996, all pesticides previously declared safe were subject to review by EPA (Hanson 1998; Thayer 1999; Hess 2000). The cost to develop a single new pesticide, as a result, ranges from $20 million to $50 million (Zimdahl 1999; Rao 2000).

Because the research and application process takes so long, a new pesticide will not show profit until about 10 years after application, leaving only 10 years before patent expiration. Producers must constantly research new pesticides as older patents expire. This, coupled with increasing pest resistance to existing formulae, has led to consolidation. In the mid-1980s, the patents on several major herbicides expired, fueling a series of mergers and acquisitions by chemical companies. Pesticide manufacturing is currently dominated by only a few large firms with familiar names like DowElanco, du Pont, and Ciba-Geigy (United States Department of Commerce 1985; British Medical Association 1992).

In sum, saturated agricultural markets, rising costs of materials, expense and lengthy time requirements for research and development, extensive and retro-active regulatory requirements, patent expiration, growing problems of pest resistance, together set the competitive conditions of early twenty-first-century agrochemical production. As a result, agrochemical manufacturers are increasingly turning away from conventional agriculture and seeking new markets (Zimdahl 1999). So far, attempts to sell pesticides to the developing world have been less profitable than predicted (US Department of Commerce 1985). Despite the potential promise of biotechnological innovation, genetic innovation has only resulted in further consolidation; because biotechnology research is expensive, smaller firms are usually taken over by larger, better-capitalized firms, hastening the concentration of the industry, increasing the ferocity with which non-biotechnological manufacturers must compete (Thayer 1999).

Raw, non-agricultural pesticides, on the other hand, represent a worldwide market currently worth around $7 billion, which is growing at 4 percent per annum, far more rapidly than the agricultural sector. 40 percent of these sales represent US household consumption. The turf care market for raw chemicals is itself about a billion dollars, and is also increasing annually (Agrow Reports 2000). By way of illustration, over 500,000 pounds of lawn care chemicals are applied annually in New Jersey, as compared to 63,000 pounds for mosquito control and 200,000 pounds for golf courses (New Jersey Dept. of Environmental Protection 2002).

Agrochemical companies are therefore finding yard chemical formulators to be their most reliable customers and formulator companies have developed several agreements with chemical manufacturers to secure exclusive access to pesticide and fertilizer ingredients (United States Securities and Exchange Commission 2001). Contracting margins in the agrochemical industry mean that chemical manufacturers will continue to seek out relationships like these, which in turn strengthen the ability of formulators to develop new marketing plans and increase the ranks of chemical-using lawn managers. Thus, changes in the broader economy of agricultural chemical manufacturing have paved the way for increases in the sales of lawn chemicals. An increasingly constricted industry is the central engine for the expansion of chemical commodity markets and the invention of new arenas for the consumption of toxins. It is ultimately the supply of pesticides, herbicides, and fertilizers that directs the imperatives for chemical demand.

TURFGRASS SUBJECTS

> I would not let the house run down. I would not let it grow up to look unseemly. That's just out of common courtesy. You want to keep up what you paid a heck of a lot for. You start to let it go downhill and then the neighborhood changes. Not that I mean by kinds of people, because we have all kinds here, all . . . nationalities. I just mean things start to go downhill.
>
> (Walter, Ohio resident, interviewed Spring 2003)

The imperatives of a global industry, the complexity of worldwide markets, the ongoing decline in the rate of profit, together seem like inexorable engines for lawn pesticide use. It is tempting therefore, to leave discussion here, following the traditional chain of explanation in political ecology (Blaikie and Brookfield 1987). Chemicals create risks. Why? Because chemical users over-apply them. Why? Because direct marketing of aesthetics influences them? Why? Because formulators are in a production squeeze and producers are in a state of consolidation.

And yet, there is something unconvincing in this account. It depends on a model of human behaviour and volition that seems too crude. How do people become convinced that certain ways of being and doing are normal, especially ways that contradict their better judgment? In a national surveys of lawn owners, for example, it has previously been discovered that those people who acknowledge the environmental risks of chemicals are the ones most likely to use them (Robbins and Sharp 2003a and b).

To burrow to the heart of this question and link global capital to local desire, therefore, requires a more careful assessment of who lawn tenders are and how they become this way. What kinds of social communities, knowledge communities, and communities of practice emerge from lawn care? The missing factor for answering this question is the one we most commonly encountered in our interrogation of lawn chemical users – they do it for their neighbours.

The lawn as community ideology

> The neighbors have a lawn service and their guy comes out on Wednesdays. So, I try to cut my grass on Wednesdays also because our yards kind of flow together. And the neighbor behind us, if they see us out they will also cut their grass on the same day, to keep it all looking nice at the same time . . . So we kind of keep an eye on each other, thinking OK, this is grass cutting day.
>
> (Suzanne, Ohio resident, Spring 2003)

As Suzanne suggests and as echoed by myriad other informants, the needs of the lawn are the business of the community. Survey results indicate that lawn managers who use chemicals are statistically more likely to know their neighbours by name than those who do not use chemicals, they are more likely to claim that their neighbours use chemicals, and they are more likely to report being interested in knowing what is happening in their neighbourhoods (Robbins and Sharp 2003a and b). Lawn chemical users further are more likely to report that their neighbours' lawn care practices, no matter what they were, have a positive impact on "neighbourhood pride".

As a result, intensive lawn management tends to cluster. If one's neighbours use lawn chemicals, then one is more likely to engage in a number of intensive lawn care practices, including hiring a lawn care company, use of do-it-yourself chemicals, and doing so more often. In addition, lawn management in general is associated with positive neighbourhood relations. People who spend more hours each week working in the yard and report greater enjoyment of lawn work, feel more attached to their local community (Robbins and Sharp 2003).

This sense of community is supported by both positive and negative institutions, where individuals feel an "obligation" to manage the private lawn intensively, not only in defence of their neighbours' property values, but also in support of "positive neighbourhood cohesion", "participation", and "holding the neighbourhood together". Thus, yard management is not an individual activity but is rather carried out for social purposes: the production of community. As Patrick, an Ohio homeowner explained to us: "[There are neighbourhoods where] . . . if you don't cut twice a week you are a communist! It's like, oh man!"

Using lawn chemicals further confers social rewards on the user. Lawn chemical users stress the way in which a well-maintained lawn reflected good character and social responsibility. They also commonly report feelings of social anxiety, describing comments from their neighbours and shame when lawn maintenance falls behind. In some cases lawn owners have been sued by their neighbours or had their yards mowed, turf restored, and saplings destroyed by neighbours or subdivision managers without legal action or permission (Crumbley 2000a, b and c; Crumbley and Albrecht 2000; Van Sickler 2003). Again, these responses and behaviours are most common where people know their neighbours by name and are cognizant of things going on in the community around them. This social process, where the neighbourhood resembles a small village, governed by face-to-face norms, suggests the disciplinary character of turfgrass institutions.

The lawn, therefore, must be viewed as a techno-social system. This convergence of technologies (Diazanon) and social norms (collective community desire for green grass) become mutually reinforcing as lawn is forged out of human labour, normative ideologies and aesthetics, and capitalized inputs. Such an assessment complicates the clean distinction between production and consumption (is turf consumed or produced by homeowners?) as well as public and private (is the lawn a private or community good?).

Even this potent explanatory cocktail under-specifies the nature of the systemic linkages in the lawn system, and leaves unanswered questions. How, after all, could large firms influence the structure of community desire? In part, of course, one could imagine community desire to be a form of institutionalized instrumental capitalist logic; the preservation of housing values in upper-middle-class neighbourhoods creates a conspiracy of mutual poisoning. And yet here too, explanation rings hollow. What produces this complex and internally jagged ideology – the desire to tend, mow, clip, and apply toxins, specifically amongst people who recognize and acknowledge the associated risks? How does this system of ideas take hold, maintain, and reproduce itself in growing legions of chemical users? What mechanism is the conduit for such an ideology, connecting global agrochemistry, community norms, and personal aesthetics?

The lawn interpolates the subject

Ideology, according to Louis Althusser (1971), functions by appearing as non-ideological – indeed by denying and repelling its own ideological character. Social agents have ideas (e.g. turf aesthetics) that are actions (e.g. chemical application) inserted into material practices (e.g. lawn care). These practices, however, as Althusser further asserts, are themselves defined by the *material ideological apparatus*, a system of ideas through which the conditions of production and reproduction of the world (labour, chemicals, surplus, etc.) are represented back to the individual as a system of natural necessity and immediate practice (home, community, and nature).

As such, Althusser insists, ideology exists only by constituting individuals as *subjects*. The term subject has critical dual meaning here. It asserts a "free" subjectivity and an actor who acts freely – as in the subject of a sentence – while simultaneously implying a "subjected being, who submits to a higher authority". This dual identity is essential to the function of ideology to erase its own ideological character. The individual as subject must act "freely" while submitting fully. Subjects must "work by themselves" without evident coercion (Althusser 1971: 182). Only then can the system of production and flow of surplus from the economy remain stable.

Althusser's argument remains somewhat abstract, by his own admission, specifically because the mechanisms through which social participants are "trained" and "have their roles assigned" to them in a capitalist society require a process of recognition, where the subject comes to recognize herself as a subject, and respond accordingly. The subject must be "hailed", named, recognized, self-recognized, or in Althusser's term: interpellated. In explaining this concept, Althusser draws on the example of a policeman calling to an individual on the street; in the moment the individual turns in recognition of the call, guiltily, they are the subject.

Such an explanation seems compelling for social structures like law enforcement, or perhaps the church, but is says little about the daily interactions that actually dominate our human behaviours in nature, economy, and community. In the case of the vast chemical economy, what does the interpellating? Who calls to the lawn chemical user so that they consistently respond as lawn tenders? Who's voice does the lawn owner hear as they open the door and look out on the grass, checking the moisture to determine whether it is time to mow?

The answer is, of course, the lawn itself. Desire and Diazanon, it can be argued, are demanded by turfgrass. When the lawn needs cutting, when its constituent species are rivaled by wild mints or fungi, when it becomes dry, its signals are apparent to the individual, whose response is an act of subjection, not only to the lawn, but to the ideology of community, and the global economy of turf maintenance. And in gazing into their landscapes, responding to the demands of the grass, and answering these calls, individuals become new kinds of subjects. Thus, as the turf draws its demands from the culture and the community, it helps to mould the capitalist economy into specific forms, and helps to produce peculiar kinds of people – turfgrass subjects. It is only these sorts of subjects who can together constitute lawn communities and produce lawn chemical economies. And they do so, working by themselves, in an effort to purify, tend, and maintain an object whose essential ecology is high maintenance, fussy, and energy demanding. Following Donna Haraway, therefore, we can suggest that subjection is not simply a process in which institutions act on the subject. Non-humans, like the lawn, "mere" objects, help to construct the subject (Haraway 1997).

It would not be trivial to add that the benefits of such subjection are passed along to the lawn itself, both as an individual resource-hungry front yard, but also as one of the largest monocultures across the face of North America. Moreover, the specific rhythms of the lawn, the timing of its specific needs, and the material practices required to produce them, all become the rhythms of the neighbourhood, where the chorus of mowers can be heard in near unison during dry daylight weekend afternoons, and the clicking of the Rainbird sprinkler sounds in the early morning hours.

It cannot therefore be argued that the industry and the advertising system it promulgates (showing happy people mowing lawns) produce the desires of the subject in any simple way. Indeed in this context, advertising, including detailed flyers explaining when and how to apply fertilizers and pesticides, must be seen as largely informational. Industry is not producing desire, but is rather responding to the need for information required for the material practice of lawn care by the turfgrass subject. Neither does community pressure, a clear driver for individual behavior, emerge in some simple way through the demands of industry. Rather, it can far more easily be argued that community pressures suit most directly the demands of turfgrass. Neighbors respond to the needs of Poacea not shareholders, staring over their back fences at brown patches and dandelions.

The lawn speaks: objects, subjects, and political ecology

None of this is to say that the lawn isn't itself produced or that it somehow pre-exists the capitalist economy of its creation. Nevertheless, the lawn itself has independent power in the process of producing that economy, its constituent agents, and the ideas of those agents. It is not the prime mover of such a system, but it is an essential part. It has its own interests too, mediated by the structured flows of fertilizer, water, and pesticides in the urban infrastructure where it is resident. This role, previously only assigned to social actors and institutions (policemen, courts, offices, families), must be extended to the non-human if there is any hope of resisting and dismantling the political economies of nature in which we are so tightly bound.

This approach to the lawn, we would further argue therefore, holds epistemological implications for a far wider problem in political ecological explanation. How do objects matter in a world where culture mediates our experience of objects?

How can the lawn matter independently of its own construction? To date, soft answers have been given to this problem. Objects are part of political actor-networks, it has for example previously been suggested (Robbins 2004). The lawn, the chemical company, and the consumer, together stabilize a regime of reality that is mutually reinforcing, and mutually constitutive.

This answer, however satisfying and self-evidently true, under-specifies the character of object–subject interactions. As Timothy Mitchell asserts, such an approach does much to dismantle the simple and problematic subject/object and human/nature dualisms that proliferate in modern thinking. But, as he further insists,

> To put in question these distinctions, and the assumptions about agency and history they make possible, does not mean introducing a limitless number of actors and networks, all of which are somehow of equal significance and power. Rather it means making the issue of power and agency a question, instead of an answer known in advance.
>
> (Mitchell 2002: 53)

We have not argued here, therefore, that turfgrass exists outside of our categorical understanding or our social imagination of "lawn" per se. Nor have we argued that turfgrass has a monopoly on the power to call the subject into being. Nor have we argued that turfgrass acts prior to, or outside of, the capitalist forces of production we have so exhaustively detailed here. "Ideas and technology did not precede this mixture as pure forms of thought brought to bear on the messy world of reality. They emerged from the mixture and were manufactured in the processes themselves" (Mitchell 2002: 52). Instead we have asserted, also following Timothy Mitchell, that *nature speaks*. It does so we suggest by hailing into existence specific kinds of human subjects, whose system of material ideological practices have ecological consequences as well as causes. To examine this flow of power and chemicals is to begin to shed light on the active role of natural objects in capitalized ecosystems. This network of power, between capital, community, and turf creates a flow not only of value, but of chemical externalities that are quite simply bad for children, wildlife, and other living things. Only once we recognize the chemical user as subjected, therefore, can we unimagine the ideological formations that make it possible.

In this way the lawn is in no way unique. To examine this flow of power and chemicals is to shed light on the active role of all kinds of subject-producing objects. Light switches, birds, filing cabinets, bacteria, trees, swimming pools, coyotes, and ATMs, all act to produce the subject we recognize as ourselves, as we live out our "joint lives", things and people together, locked in mutually constituted identities (Haraway 2003: 16–17). And while human actors obey socioeconomic institutions and while they are in turn enmeshed in an economy driven by a massive industry, itself in a state of crisis and consolidation, as is so evident from ever-hungry turfgrass, it is the demands of objects that set the pace and character of subjected community lives. People's behaviours are tied to the exigencies of capitalist power, to be sure, but with the localized experience of neighbourhood objects holding independent, prior, and often ultimate authority.

An urban political ecology with any hope of political and explanatory efficacy must begin precisely from here, therefore, through the documentation of the reachings, "prehensions", or "graspings", following Haraway (2003: 6), of objects and subjects into one another, each created through its interaction with the other. Political ecology must acknowledge the agency of nature as well as its socially constructed character. It must recognize the consciousness of human subjects even while recognizing its constitution by the non-human. Unthinking the pervasive and expanding hazards of urban life, like the North American lawn, requires respecting the power of not only distant multi-national capital, therefore, but also of the intimate odds and ends of daily life. It means understanding ourselves in the myriad objects of the world around us.

ACKNOWLEDGEMENTS

This material is based upon work supported by the National Science Foundation under Grant No. 0095993, and by further support from the Ohio State University Center for Survey Research, and the Ohio State Environmental Policy Initiative. Our fundamental leap, from Althusser to organophosphates, is owed to Joel Wainwright.

BIBLIOGRAPHY

Ackerman, F. (1997) *Why Do We Recycle?* Washington, DC: Island Press
Agrow Reports (2000) *World Non-Agricultural Pesticide Markets*. London and New York: PJB publications
Aldus, L.J. (ed.) (1976) *Herbicides: Physiology, Biochemistry, Ecology*, Second Edition, Volume I. London: Academic Press
Althusser, L. (1971) *Lenin and Philosophy and Other Essays*. New York: Monthly Review Press
Anderson, G.L., E.S. Delfosse, *et al.* (2003) "Lessons in developing successful invasive weed control programs", *Journal of Range Science*, 56(1): 2–12
Anderson, W.P. (1996) *Weed Science: Principles and Applications*. St. Paul, MN: West Publishing Company
Baker, G.P. and K. Wruck (1991) "Lessons from a middle market LBO: the case of O.M. Scott", *The Continental Bank Journal of Applied Corporate Finance*, 4(1): 46–58
Bambarger, B. (1987) "O.M. Scott and Sons", *Lawn and Garden Marketing*, October: 24
Barron, L. (1923) *Lawn Making: Together with the Proper Keeping of Putting Greens*. New York: Doubleday, Page and Co.
Beck, U. (1992) *Risk Society: Towards a New Modernity*. London: Sage Publications
Blaikie, P. and H. Brookfield (1987) *Land Degradation and Society*. London and New York: Methuen and Co.
Bormann, F.H., D. Balmori, *et al.* (1993) *Redesigning the American Lawn: A Search for Environmental Harmony*. New Haven and London: Yale University Press
British Medical Association (1992) *The British Medical Association Guide to Pesticides, Chemicals, and Health*. London: Edward Arnold
Carson, R. (1962) *Silent Spring*. New York: Houghton Mifflin

Chemical Week (1998) "Monsanto completes pesticide sales; more divestments to come", *Chemical Week*, 160: 13

Christensen, A., R. Westerholm, *et al.* (2001) "Measurement of regulated and unregulated exhaust emissions from a lawn mower with and without an oxidizing catalyst: a comparison of two fuels", *Environmental Science and Technology*, 35(11): 2166–2170

Cleveland Plain Dealer (2000) "At Scotts they call it pull", *Cleveland Plain Dealer*, Cleveland, OH: 44

Cook, A. (1990) "Digging for Dollars", *American Demographics*: 40–41

Crumbley, R. (2000a) "Neighborhood dispute over unmanicured yard headed to court", *Columbus Dispatch*, Columbus, OH: D7

Crumbley, R. (2000b) "Neighbors sue over high grass", *Columbus Dispatch*, Columbus, OH: 4C

Crumbley, R. (2000c) "Reynoldsburg says resident can let back yard grow wild", *Columbus Dispatch*, Columbus, OH: B4

Crumbley, R. and R. Albrecht (2000) "It's mowing versus growing in area's turf war grass-height laws", *Columbus Dispatch*, Columbus, OH: 1B

Davis, J.R., R.C. Brownson, *et al.* (1993) "Family pesticide use and childhood brain cancer", *Archives of Environmental Contamination and Toxicology*, 24(1): 87–92

Dickinson, L.S. (1931) *The Lawn: The Culture of Turf in Park, Golfing, and Home Areas*. New York: Orange Judd Publishing

Eveleth, W.T. (ed.) (1990) *Kline Guide to the U.S. Chemical Industry*, Fifth Edition. Fairfield, NJ: Kline and Company, Inc

Extension Toxicology Network (2000) Ecotoxnet, Online. Available HTTP: <http://ace. ace.orst.edu/info/extoxnet/>

Fuller, K., H. Shear, *et al.* (eds) (1995) *The Great Lakes: An Environmental Atlas and Resource Book*. Chicago, IL and Toronto: Great Lakes National Program Office, US Environmental Protection Agency and Government of Canada

Green, M.B., G.S. Hartley, *et al.* (1987) *Chemicals for Crop Improvement and Pest Management*, Third Edition. Oxford: Pergamon Press

Guerrero, P.F. (1990) "Lawn care pesticides remain uncertain while prohibited safety claims continue". *Statement of Peter F. Guerrero before the Subcommittee on Toxic Substances, Environmental Oversight, Research and Development of the Senate Committee on Environment and Public Works*. Washington, DC: US General Accounting Office

Hagedorn, J. (2001) *Priorities for the Future: from James Hagedorn, President and Chief Executive Officer of the Scotts Company*. Marysville, OH: The Scotts Company

Hanson, D. J. (1998) "Pesticide law off to a rough start", *Chemical and Engineering News*: 20–22

Haraway, D. (1997) *Modest_Witness@Second_Millenium.FemaleMan_Meets_ Oncomouse™*. New York: Routledge

Haraway, D. (2003) *The Companion Species Manifesto: Dogs, People, and Significant Otherness*. Chicago, IL: Prickly Paradigm Press

Hess, G. (2000) "Pesticide manufacturers are unhappy with EPA's crackdown on chlorpyrifos", *Chemical Market Reporter*, 257: 1,13

Ibrahim, M.A., G.G. Bond, *et al.* (1991) "Weight of evidence on the human carcinogenicity of 2,4-D", *Environmental Health Perspectives*, 96: 213–222

Jaffe, T. (1998) "Lean green machine", *Forbes*, 162: 90

Jenkins, V.S. (1994) *The Lawn: A History of an American Obsession*. Washington and London: Smithsonian Institute Press

Journal of Business Strategy (1989) "Why I bought the company", *Journal of Business Strategy*, 10: 4–8

Leiss, J.K. and D.A. Savitz (1995) "Home pesticide use and childhood-cancer: a case control study", *American Journal of Public Health*, 85(2): 249–252

Leonas, K.K. and X.K. Yu (1992) "Deposition patterns on garments during application of lawn and garden chemicals – a comparison of six equipment types", *Archives of Environmental Contamination and Toxicology*, 23(2): 230–234

Lewis, R.G. *et al.* (1991) "Preliminary results of the EPA house dust infant pesticides exposure study (HIPES)", *Abstracts of the Papers of the American Chemical Society*, 201(89-Agro Part 1, April 14)

Lewis, R.G., R.C. Fortmann, *et al.* (1994) "Evaluation of methods for monitoring the potential exposure of small children to pesticides in the residential environment", *Archives of Environmental Contamination and Toxicology*, 26(1): 37–46

Marzluff, J.M. and K. Ewing (2001) "Restoration of fragmented landscapes for the conservation of birds: A general framework and specific recommendations for urbanizing landscapes", *Restoration Ecology*, 9(3): 280–292

Mitchell, T. (2002) *Rule of Experts: Egypt, Techno-Politics, Modernity*. Berkeley, CA: University of California Press

National Gardening Association (2000) *National Gardening Survey*. Burlington, VT: National Gardening Association

Natural Resources Conservation Service (2000) *Summary Report: 1997 National Resources Inventory (revised December 2000)*. Washington, DC: United States Department of Agriculture

New Jersey Dept. of Environmental Protection (2002) *New Jersey Comparative Risk Project, Draft Report*. Trenton, NJ

Nishioka, M.G., M.C. Brinkman, *et al.* (1996a) *Evaluation and Selection of Analytical Methods for Lawn-Applied Pesticides*. Research Triangle Park, NC: US Environmental Protection Agency, Research and Development

Nishioka, M.G., M.C. Brinkman, *et al.* (1996b) "Measuring transport of lawn-applied herbicide acids from turf to home: correlation of dislodgeable 2,4-D turf residues with carpet dust and carpet surface residues", *Environmental Science and Technology*, 30(11): 3,313–3,320

Nishioka, M.G., M.C. Brinkman, *et al.* (1999a) *Transport of Lawn-Applied 2,4-D from Turf to Home: Assessing the Relative Importance of Transport Mechanisms and Exposure Pathways*. Research Triangle Park, NC: National Exposure Research Laboratory

Nishioka, M.G., M.C. Brinkman, *et al.* (1999b) "Distribution of 2,4-dichlorophenoxyacetic acid in floor dust throughout homes following homeowner and commerical applications: quantitative effects on children, pets, and shoes", *Environmental Science and Technology*, 33(9): 1,359–1,365

Nishioka, M.G., H.M. Burkholder *et al.* (1996) "Measuring transport of lawn-applied herbicide acids from turf to home: Correlation of dislodgeable 2,4-D turf residues with carpet dust and carpet surface residues", *Environmental Science and Technology*, 30(11): 3,313–3,320

Nishioka, M.G., H.M. Burkholder *et al.* (1999a) *Transport of Lawn-Applied 2,4-D from Turf to Home: Assessing the Relative Importance of Transport Mechanisms and Exposure Pathways*. Research Triangle Park, NC: National Exposure Research Laboratory

Nishioka, M.G., H.M. Burkholder *et al.* (1999b) "Distribution of 2,4-dichlorophenoxyacetic acid in floor dust throughout homes following homeowner and commerical applications:

quantitative effects on children, pets, and shoes", *Environmental Science and Technology*, 33(9): 1,359–1,365

Priest, M.W., D.J. Williams, *et al.* (2000) "Emissions from in-use lawn-mowers in Australia", *Atmospheric Environment*, 34(4): 657–664

Rao, V.S. (2000) *Principles of Weed Science*, Second Edition. Enfield, NH: Science Publishers, Inc

Reich, M.S. (2000) "Seeing Green", *Chemical and Engineering News*, 78: 23–27

Reynolds, P.M., J.S. Reif, *et al.* (1994) "Canine exposure to herbicide-treated lawns and urinary excretion of 2,4-dichlorophenooxyacetic acid", *Cancer Epidemiology Biomarkers and Prevention*, 3(3): 233–237

Robbins, P. (2004) *Political Ecology: A Critical Introduction*. New York: Blackwell

Robbins, P. and T. Birkenholtz (2003) "Turfgrass revolution: measuring the expansion of the American lawn", *Land Use Policy*, 20: 181–194

Robbins, P. and J.T. Sharp (2003a) "Producing and consuming chemicals: the moral economy of the American lawn", *Economic Geography*, 79(4): 425–451

Robbins, P. and J.T. Sharp (2003b) "The lawn chemical economy and its discontents", *Antipode*, 35(5): 955–979

Russell, E. (2001) *War and Nature: Fighting Humans and Insects with Chemicals from World War I to Silent Spring*. Cambridge: Cambridge University Press

Sawyer, R.F., R.A. Harley *et al.* (2000) "Mobile sources critical review: 1998 NARSTO assessment", *Atmospheric Environment*, 34(12–14): 2,161–2,181

Schroeder, F.E.H. (1993) *Front Yard America: The Evolution and Meanings of a Vernacular Domestic Landscape*. Bowling Green, OH: Bowling Green State University Popular Press

Scotts Company (2000) "Business segments overview: North American consumer", Online. Available HTTP: <www.smgnyse.com/html/consumerlawn.cfm>

Scotts Company (2001) *The Scotts Company: 2000 Summary Annual Report and 2001 Financial Statements and Other Information*. Marysville, OH: The Scotts Company

Scotts Company (2002a) *The Scotts Company: 2001 Summary Annual Report and 2001 Financial Statements and Other Information*. Marysville, OH: The Scotts Company

Scotts Company (2002b) *Scotts, UK Government Reach Unique Agreement on Regeneration of Environmentally Sensitive Peatlands*

Staggers, D.T. (1976) "The Search for New Herbicides", *Herbicides: Physiology, Biochemistry, Ecology*, Second Edition, Volume II. London: Academic Press

Stephens, R.J. (1982) *Theory and Practice of Weed Control*. London: The Macmillan Press

Thayer, A.M. (1999) "Transforming agriculture", *Chemical and Engineering News*, April 19: 21–35

United States Department of Commerce (1985) *A Competitive Assessment of the U.S. Herbicide Industry*. Washington, DC: International Trade Administration, US Department of Commerce

United States Environmental Protection Agency (2000) *1998–1999 Pesticide Market Estimates*. United States Environmental Protection Agency

United States Geological Survey (1999) *The Quality of Our Nations Waters: Nutrients and Pesticides*. Washington DC: United States Geological Survey

United States Securities and Exchange Commission (2001) *The Scotts Company Annual Report*

United States Securities and Exchange Commission (2001) *The Scotts Company Quarterly Report*. US Securities and Exchange Commission

Van Sickler, M. (2003) "Lawsuit springs from lawn dispute", *St. Petersburg Times*: 4

Whitten, J.L. (1966) *That We May Live*. Princeton, NJ: Van Nostrand Company Inc

Williams, B. (1997) "Storms past, Scotts finds seeds of change yield a blooming success", *The Columbus Dispatch*, Columbus, OH: 27 July

Young, R.D., D.G. Westfall, *et al.* (1985) "Production, marketing, and use of phosphorous fertilizers", *Fertilizer Technology and Use*, Third Edition. Madison, WI: O.P. Englestad

Zahm, S.H. and M.H. Ward (1998) "Pesticides and childhood cancer", *Environmental Health Perspectives*, 106 (Suppl. 3): 893–908

Zartarian, V.G., H. Ozkaynak, *et al.* (2000) "A modeling framework for estimating children's residential exposure and dose to chlorpyrifos via dermal residue contact and nondietary ingestion", *Environmental Health Perspectives*, 108(6): 505–514

Zimdahl, R.L. (1999) *Fundamentals of Weed Science*, Second Edition. San Diego, CA: Academic Press

8 Justice of eating in the city

The political ecology of urban hunger

Nik Heynen

URBAN HUNGER, LIKE THE BITE OF CRABS

I can think of no configuration of socionatural relations more debilitating to human potential than the one that produces hunger. Without food, human bodies simply cannot exist. Human bodies that do not consume a sufficient quantity of food, or food that does not contain sufficient nutritional quality simply cannot function. Without enough food, the everyday processes necessary for social production and reproduction become daunting and/or impossible. Human bodies are produced through socio-metabolic processes that link their existence to external processes that produce food. The same socio-metabolic processes necessitate that human bodies produce different types of nature as the result of socio-physical processes that are themselves constituted through relations of social and political power and through a wide assortment of cultural meanings (see Haraway 1991; 1997; Chapter 1 of this book).

While in no way attempting to diminish the importance of other spatial configurations of hunger (i.e. rural hunger), *urban hunger* has received little attention for too long and requires more substantial theoretical investigation and political action. By articulating the connections, interrelations, and interdependences between the metabolism of food through the human body and the metabolism of cities, this chapter will discuss the socionatural production of urban hunger. I will draw specifically on the history of political struggles around childhood hunger in Milwaukee, Wisconsin, and more particularly a recent campaign to have the Milwaukee public school system ratify a universal free breakfast programme that would significantly reduce childhood hunger in the city. I intend to use my own engagement in activist-scholarship, or action-research (see Greenwood and Levin 1998), as a member of a group called *Voices Against Hunger* to show the link between urban political ecology and radical geographic praxis within the context of Swyngedouw's suggestion (see Chapter 2) that: "What differentiates human actants from others is their organic capacity to imagine different possible futures, to act differentially in ways driven and shaped by human drives, desires, and imaginations."

Within this chapter, I suggest that urban hunger is both a natural condition created through complex biochemical processes, as well as a social process produced through power relations dictating who eats what and how much, and who goes hungry. The urban political ecology of hunger, like all other socionatural processes, is produced through an amalgamation of biochemical processes, material and cultural practices, social relations, language, discursive constructions and ideological practices (see Swyngedouw 1999; 2004). These processes are developed and co-evolve somewhere within the tension between *consumption based on physiological requirements* and *consumption based on cultural conditions*; between *need* and *desire*.

Lefebvre (2002: 7–8) discusses the relations inherent to the tension between *need* and *desire* connected to eating/hunger [and sex] by suggesting, "it is not only the drama of the relations between one individual and another (or the other) individual which is sketched out and foreshadowed through this *need*, it is not only the drama of the link between individual and species, it is the universality which is being offered or withdrawn." This universality will be the *anchor* for my theorization of the processes that produce urban hunger, as these are the fundamental material processes that connect humans to nature, or nature to itself (as Marx would put it).

In 2003, 11.2 percent of all US households were "food insecure" because of lack of economic resources. Since 1999, food insecurity has increased by 2.1 million households nationally, including 1.1 million households with children. Furthermore, the number of families with children requesting emergency food assistance increased by 88 percent (USCM 2003). While it is difficult to determine what percentage of these men, women and children live within US cities, a survey released by the US Conference of Mayors (USCM) in 2003 provides chilling information about the proliferation of urban hunger within US urban areas. The USCM suggest that requests for emergency food assistance increased by 88 percent within the twenty-five cities surveyed between 1997 and 2002. Their report suggests that in 2003 alone, requests increased by an average of 17 percent. So while not having an absolute sense of the severity of urban hunger, the relational proliferation within the US is ominous. Given the relational hierarchy between bodily metabolization and more wide ranging urban metabolization across space, but also between/within generations, the blatant inequality associated with urban hunger, especially as it most negatively impacts children, presents substantial impediments to the production of healthy and just US urban spaces and environments.

In his poem, "The great tablecloth", Pablo Neruda (1958) helps us to empathize with the ineffable socionatural contradictions inherent in hunger. He suggests, hunger "Is hollow and green, has thorns like a chain of fish hooks" and "feels like pincers, like the bite of crabs; it burns, burns, and has no fur". Neruda produces an explicit imaginary of razor-sharp metal piercing supple intestinal lining. However, Neruda also leaves us with a hopeful message, reminding us that there is nothing inevitable about the tearstained sheets that shroud hunger, just as there is nothing inevitable about the capitalist system that produces it. Neruda's utopian call, "Let us sit down soon to eat, with all those who haven't eaten", is not just the stuff of poetry or religion, it is also a concern voiced within radical geography. As Harvey suggests (1973: 313–314):

> Many hopeful and utopian things have been written about the city throughout history. We now have the opportunity to live many of these things provided we

can seize upon the present possibilities. We have the opportunity to create space, to harness creatively the forces making for urban differentiation. But in order to seize these opportunities we have to confront the forces that create cities as alien environments, that push urbanization in directions alien to our individual or collective purpose. To confront these forces we have first to understand them.

This opportunity still exists thirty-three years after the publication of *Social Justice and the City*. Within the last three decades, we urban geographers have sharpened our theoretical tools and broadened our understanding about the dialectical processes that produce socionatural inequality and spatial forms within the city. As such, the goals of urban political ecology, especially Marxist urban political ecology, should first be to understand the socionatural contradictions that produce urban hunger, and then to act to end these contradictions.

MARXIST POLITICAL ECOLOGY OF URBAN HUNGER

Though infrequently discussed and difficult to statistically explicate, urban hunger requires an analysis separate from that of world hunger, or regional/rural hunger (see Sen 1981; Dréze and Sen 1989; Grigg 1993). As Swyngedouw and I suggest elsewhere (2003: 907): "[I]t is on the terrain of the urban that [the] accelerating metabolic transformation of nature becomes most visible, both in its physical form and its socioecological consequences." However, the politics of urban hunger are too often couched within the discursive construction of an "urban social problem". This perspective too easily obfuscates the environmental ramifications of urban hunger. There is no doubt that hunger is, as Sen (1981) suggests, a socio-economic problem. For instance, the twenty-two most food-deficient sub-Saharan African countries could meet their food needs with just 11 percent of the food surplus held by neighbouring countries (Lappé *et al*. 1998). However, hunger is also a socionatural problem that requires analysis within the context of urban metabolic processes in order to excavate the interactions and interdependencies among cities, human beings and nature.

As most explicitly discussed in Chapter 2 of this collection (and elsewhere in Peet and Watts 1993; Swyngedouw 1996; Keil and Graham 1998; Laituri and Kirby 1994), socio-spatial processes are always related to the circulation and metabolism of physical, chemical, or biological components and as such are tied to the production of material reality. These metabolisms generate both enabling and disabling socioecological conditions that often embody contradictory relations. It is within the logic of these contradictions that we must situate the political ecology of urban hunger.

At the heart of Marx and Engels' (1845: 37) conception of reality was the notion that humans must meet their material needs through the appropriation of nature to be able to "make history". Related to this, they suggest (1845) "[The] conception of history depends on our ability to expound the real process of production, starting out from the material production of life itself . . . as the basis of all history . . ."

Obviously, the fact that humans need food to survive does not mean this is all they need. However, the point is they *must* have food and many other material "things" to live full lives. The importance of a Marxist approach to urban political ecology is that it maintains the importance of the natural foundations of human life and attempts to build upon it.

As Marx's primary interest was focused on investigating capitalism, he only touched sporadically on the fundamental processes of metabolization and social reproduction, and moved more directly to the processes of commodity production and other processes inherent in capitalism. We need to return to the point of the dialectical interdependencies between nature and social reproduction (see Katz 2004; Marston 2004; Mitchell *et al.* 2003). As the demoralizing consequences of capitalism have proliferated, there is a need to take several steps back for the sake of investigating the ramifications of how material needs, metabolization and social reproduction all play out within cities.

Swyngedouw (1999) suggests the material and social conditioning of human bodies and of the metabolic transformation of nature is constituted in and through temporal/spatial social relations that operate over certain scalar extents. As such, the socioecological processes of need/desire that produce urban hunger must also be situated within an explicitly scalar context. In order to engage urban metabolization in such a way as to inform discussions both about the production of urban space and of urban nature, we must, just as Marx did from his *Economic and Philosophic Manuscripts of 1844* onward, ground our ontological and epistemological arguments in sensual bodily interaction with the world. By starting at the scale of the body, or the stomach in this case, it then becomes possible to "jump scales" (see Smith 1993) to the urban. This can then facilitate excavating the socionatural relations that fuse bodies to urban nature and provide a lens through which to understand urban environments, as both produced by, and consisting of, human bodies.

While not explicitly engaging the scalar relations of hunger, Marx (1964: 181) suggested "Hunger is a natural need; it therefore needs a nature outside itself, an object outside itself, in order to satisfy itself, to be stilled. Hunger is an acknowledged need of my body for an object existing outside it, indispensable to its integration and to the expression of its essential being." Marx's notion that "it [hunger] therefore needs a nature outside itself, an object outside itself, in order to satisfy itself . . ." has myriad scalar ramifications directly connecting bodily scales with extra-bodily socionatural scalar processes.

From this perspective, we see that socionatural nested scales of hunger explicitly link the essence of human life (food), through individual human lives, to the extra-bodily scales at which the needs of hunger can be met. Harvey (1998: 402–403) also suggests that: "the metabolic processes that sustain a body entail exchanges with its environment. If the processes change, then the body either transforms and adapts or ceases to exist. Similarly, the mix of performative activities available to the body in a given place and time are not independent of the technological, physical, social, and economic environment in which that body has its being." Thus, the power relations inherent to urban food systems cannot be divorced from the political ecological systems they operate within, nor can they be considered as only bodily processes, rather, they must be thought of as interrelated urban, regional, national, global, etc. processes. The scalar dialectics of hunger help us elucidate the socionatural relations that both produce bodies, and in turn, enable bodies to produce environments.

FROM THE BODY TO THE CITY: THE URBANISATION OF HUNGER UNDER CAPITALISM

Humans, like all mammals, live intensely at high metabolic rates and thus consume and metabolize food more quickly than cold-blooded animals. These metabolic processes arc primarily related to the physiological temperature regulation system our bodies maintain through sweating, shivering, etc. While the need for food at any given point presents the most central physiological challenge to avoiding hunger, there is also a fundamental ecological problem that results from how humans deal with future uncertainties of their food supply. According to Levins and Lewontin (1985), this presents an ongoing struggle to secure the necessary daily amounts of food, and makes humans more vulnerable to environmental viabilities. Like other mammals, humans can store extra energy through bodily fat, but because of our cultural adaptation do not tend to function this way. Instead, according to Levins and Lewontin, humans have adapted to store food physically, unlike other mammals, outside of our bodies. While arguments about the proliferation of obesity could be mounted to complicate this issue, I will leave this to Marvin and Medd's next chapter dealing with 'obe-city'.

External storage of food led to accumulation by those with the means to produce surplus foodstuffs. As such, the commodification of food, under capitalism, extends and complicates human vulnerability in two ways. First, as world markets expanded and merged into a global market, access to food became contemporarily based on what happens with food production elsewhere. As a result, a complex infrastructure of networks to get food from there to here now defines the commodification process. Furthermore, the commodification of food under capitalism, coupled with the extreme inequality it has produced, has meant that many people have increasingly less direct access to food. Marx (1867) commented on the early history of this process, by looking specifically at how capital investment into agricultural production led to sweeping peasants off the land they worked and cut off their direct access to grow and harvest foodstuffs. Here, Marx also discussed that once swept from their land, and direct access to food via agriculture, they moved to urban(izing) areas. This set of relations, that have tended to be discussed as primarily a political economic process, has profound political ecological ramifications for the urbanization of hunger. We can partially trace the contemporary lack of access to food in cities to the processes through which the commodification of food occurred, which also contributed to human alienation from nature. Ultimately, the processes of capitalism further exacerbated the environmental vulnerabilities humans face in meeting our daily food intake. To this end, Engels suggested (1880: 45):

> The new mode of production was, as yet, only at the beginning of its period of ascent; as yet it was the normal, regular method of production – the only one

possible under existing conditions. Nevertheless, even then it was producing crying social abuses – the herding together of a homeless population in the worst quarters of the large towns; the loosening of all traditional moral bonds, of patriarchal subordination, of family relations; overwork, especially of women and children, to a frightful extent; complete demoralization of the working-class, suddenly flung into altogether new conditions, from the country into the town, from agriculture into modern industry, from stable conditions of existence into insecure ones that change from day to day.

Levins and Lewontin (1985: 260) capture the political ecological complexity of hunger created through the commodification of food by suggesting, "[e]ating is obviously related to nutrition, but in humans this physiological necessity is imbedded in a complex matrix: within which what is eaten, whom you eat with, how often you eat, who prepares the food, which foods are necessary for a sense of well-being, who goes hungry and who overeats have all been torn loose from the requirements of nutrition and the availability of food." Just as Marx pointed out that humans must meet their material needs to make history, Levins and Lewontin (1985: 262) go on to suggest that throughout human history, the quantity, quality, and variety of food people have eaten has been determined by their place in their economy and the institutional structures in place within those economies to produce and distribute food.

Related to the power relations as embedded within urban food systems, Engels (1881) suggested: "[t]he Capitalist, if he cannot agree with the Labourer, can afford to wait, and live upon his capital. The workman cannot. He has but wages to live upon, and must therefore take work when, where, and at what terms he can get it. The workman has no fair start. He is fearfully handicapped by hunger. Yet, according to the political economy of the Capitalist class, that is the very pink of fairness." The imposition of these capitalist relations according to Engels is why (1958: 32) "[p]ower lies in the hands of those who own, directly or indirectly, foodstuffs and the means of production".

Since the Reagan administrations during the 1980s, the unevenness inherent in US urban food systems has increased. Substantial barriers to consuming food were one of many inequities resulting from the administration's social and economic policies (see Dirks 2003; Kahn and McAlister 1997; Levenstein 1988; 1993; Root and De Rochemont 1995). Specifically, certain urban communities lacking access to viable food sources became "food deserts" (Wrigley 2002). This inherently spatial metaphor demonstrates the problematic dilemma facing many inner-city communities seeking to meet their basic physiological needs.

The spatial restructuring of urban foodscapes has resulted in urban built environments once comprised of chain supermarkets, independent supermarkets, grocery stores, convenience stores and sit-down/specialty restaurants, now increasingly dominated by fast food restaurants, emergency food outlets, food pantries and soup kitchens (Ashman *et al.* 1993; Eisenhauer 2001). The departure of supermarkets from within inner cities, coupled with the proliferation of fast food restaurants, has produced an unaffordable, unhealthy and untenable urban foodscape (see Schlosser 2002). The inability to buy food due to lack of financial resources and spatial isolation is often times a direct cause of hunger. As such, the ways in which capitalism has led to a fundamental physical reconfiguration of urban communities is a significant factor that produces urban hunger.

While seemingly contradictory, the political ecologies of urban hunger and obesity are intimately tied to the production of a "fast-food nation" and an overall decreased quality of life within the US. This point will also be discussed in more detail by Marvin and Medd in the next chapter.

THE JUSTICE OF EATING IN THE CITY

In his trilogy, *The Principle of Hope*, Ernest Bloch (1986: 11) suggests that "Hunger, the main drive, must be worked out, and the way it proceeds to the rejection of deprivation, that is, to the most important expectant emotion: hope. A central task in this part is the discovery and unmistakable notation of the 'Not-Yet-Conscious'." So while hunger is the most fundamental impediment to the production of human history, new forms of knowledge, understanding, and political will must be realized in order to collectively reject the social production of hunger. The emphasis on excavating the power relations that fuse the "interwoven knots of social process, material metabolism and spatial form that go into the formation of contemporary urban socionatural landscapes" (Swyngedouw and Heynen 2003: 906) are explicitly engaged within Marxist urban political ecology in order to articulate who benefits and who suffers from local urban environmental metabolization. Through this critical lens, we seek to recognize the "Not-Yet-Conscious" Bloch referred to.

Through recognizing these relations inherent to the political ecology of urban hunger, we can begin to, as Harvey (1973) suggests, "confront the forces that create cities as alien environments that push urbanization in directions alien to our individual or collective purpose". The emergence of urban "food justice" as a notion within grassroots organizations, the popular press and academia is beginning to link food insecurity with other race/income/justice issues and bringing increased attention to what has been a largely unpopular and largely ignored issue within the US. The interconnectedness of necessity, desire and political will can still culminate in a utopian political ecology capable of thinking beyond the current unjust metabolic fusion of nature and society within the context of contemporary capitalism. However, this again must be explicitly connected to sensual bodily interaction with the world. Related to this, Harvey (1998: 405) suggests:

And While Marx's theorizing in Capital is often read (incorrectly as I shall hope to show) as a pessimistic account of how bodies, constructed as passive entities occupying particular performative economic roles, are shaped by external forces of capital circulation and accumulation, it is precisely this analysis that informs his other accounts of how transformative processes of human resistance, desire for reform, rebellion and revolution can and do occur.

UTOPIAN POLITICAL ECOLOGY AND URBAN HUNGER IN MILWAUKEE

Utopian moments within recent urban history can shine as a beacons of hope and serve as roadmaps to help us mobilize politics more capable, and willing, to eradicate urban hunger (and for that matter all hunger). One such utopian moment is the little-known political struggles around childhood hunger in Milwaukee, Wisconsin and the Milwaukeeans' imaginative history for carving out spaces for emancipatory socionatural change. Although so fundamental an issue, childhood hunger in Milwaukee has historically been addressed by those individuals and groups in society that are considered to be the most radical. This relational contradiction was best articulated by Brazilian Archbishop Dom Helder Camara when he suggested, "When I give food to the poor, they call me a saint. When I ask why the poor have no food, they call me a communist."

As far back as 1835 under the leadership of one of Milwaukee's founders, Solomon Juneau, who served as the "superintendent of the poor", provisions were made to those in need. Public officials recognized that it was their ethical duty to combat material inequality, but also recognized the interconnectedness of the welfare of all their residents and the larger successful functioning of society. In 1932, in the midst of the Great Depression, Milwaukee's Socialist Mayor Daniel Hoan proclaimed: "We are in the midst of a world-wide economic and social revolution that will not cease until the present industrial system, called capitalism, is entirely replaced by the next stage of human development, which is called Socialism." Loyal to his commitment to Milwaukee's poor, Hoan created one of the US's first, and most recognized public hunger relief programmes. Many praised Milwaukee's utopian vision for dealing with hunger after the *New York Times* wrote in 1930, "Dozens of jobless men today received food from 'soup kitchens' as the city opened temporary commissaries to care for hungry families. Mayor Hoan, a Socialist, ordered the old policy armory kitchen thrown open tomorrow as a municipal kitchen. Temporary headquarters gave bread, milk, cheese and coffee to the hungry today."

Another significant episode, perhaps the most important for the contemporary politics of hunger in Milwaukee, began in 1971 when three young African-American men went to Milwaukee's Cross Lutheran Church within the city's African-American core to talk to Rev. Joseph Ellwanger about implementing the Black Panther Party's *Free Breakfast for Children Program* in the basement of his church. While the church council ultimately voted against the measure, the Milwaukee Panthers and Rev. Ellwanger set into a motion a chain of events that led to the creation of what is today known as Milwaukee's *Hunger Task Force* (*HTF*; see http: //www.hungertaskforce.org/). *HTF* is Wisconsin's most important food bank and institutional advocate for hunger relief. Their success is largely due to their diverse partnerships that span through the city and state of Wisconsin. They have a small army of volunteers that organize food and fund drives in their workplaces, schools and churches. Funds donated by corporations, foundations and individuals support the *HTF's* operating costs, and ensure that their food relief programmes can be provided free of charge. The *HTF* provides free food to over 80 emergency feeding organizations and advocates that assist these groups work to end hunger.

Unlike other food banks, the *HTF* was initially established to advocate for an end to hunger within Milwaukee and across Wisconsin. Their advocacy initiatives try to foster positive socionatural changes by lobbying to sustain the vitality of programmes and public policies that exist to help the hungry. Out of the *HTF*'s advocacy tradition evolved an activist group, called *Voices Against Hunger (VAH)*, comprised of over 240 Milwaukeeans that since 2003 have been committed to ending hunger in Milwaukee. One of the group's current initiatives is useful for looking more specifically at the political ecology of urban hunger. Since 2004 *VAH* has been lobbying the Milwaukee public school system to ratify a universal free breakfast programme that could significantly reduce childhood hunger in the city. I want to mobilize Lefebvre's notion of *everyday life* as a theoretical lens through which to excavate the political ecology of childhood hunger as understood through my action-research with(in) *Voices Against Hunger*. Lefebvre (1991: 18) suggested that:

> Everyday life is made of recurrences: gestures of labor and leisure, mechanical movements . . . hours, days, weeks, months, years, linear and cyclical repetitions, natural and rational time; the study of creative activity leads to the study of re-production or the conditions in which actions producing objects and labour are re-produced, re-commenced, and re-assume their component proportions or, on the contrary, undergo gradual or sudden modifications.

Through this context, Lefebvre *began* to demonstrate that a study of the foundations of repressive society must focus on everyday life and social reproduction. He (1991: 145) suggested "the field of repression covers biological and physiological experience, nature, childhood, education, pedagogy and birth" (145). Through the *everyday*, the universal needs of eating and bodily/urban metabolization can be articulated in a way that many political economic/political ecological perspectives fail to articulate through their explicitly top-down, and narrow structural perspectives. The *everyday* socionatural relations of childhood hunger are on the one hand a result of extreme material inequality, however, stopping there without excavating what this means is only partially useful.

Part of the structural considerations inherent in the political ecology of urban hunger in Milwaukee results from Wisconsin's ranking as the 51st (worst) state in the US for low-income student participation in federally funded *School Breakfast Programs*. Only 23 percent of low-income Wisconsin children eat school breakfast, compared to 43 percent nationwide (FRAC 2004). Despite a 77 percent eligibility rate for free and reduced school breakfasts, only 16 percent of Milwaukee public school students eat breakfast.

This seemingly mundane topic receives little attention within the US. This is perhaps best illustrated by the latest US budget discussions that look to cut federal monies dedicated to nutritional programming orchestrated through the United States Department of Agriculture. On 17 March 2005, the US House of Representatives voted 218–214 to pass a budget resolution (House Congressional Resolution 95) that would lead to substantial cuts to nutrition programmes that benefit low-income children and families. This budget resolution mandates $216 billion in cuts to "domestic discretionary programmes" over the next five years. These programmes receive budgeted funding each year in order to operate and include WIC and WIC Farmers' Market, Summer Food, and the Senior Farmers' Market Nutrition Program. The House resolution would also lead to $68 billion in cuts to "entitlement programmes". These are programmes that get funding as needed to operate and are not controlled by annual budget allocations. Programmes that would likely be targeted include Food Stamps, School Lunch, and School Breakfast. Such cuts might be tolerated because their socioecological ramifications are little understood by the politicians voting on them.

Characterizing the implications of inequality regarding childhood health disparities in Milwaukee, Willis (2000) suggests that low income children in Milwaukee, which primarily means inner-city African-American children are: "almost three times more likely to experience stunted growth, and three to four more times likely to experience an iron deficiency." More specifically, the underlying biochemical physical processes at the core of childhood hunger have potentially devastating physiological consequences for childhood growth. Malnutrition impairs the body's ability to heal and decreases immune functions, leading to an infection-malnutrition cycle. Thin infants are more likely to grow up to be pre-diabetic adults. Hungry children, even when not acutely ill, become apathetic or irritable and miss critical opportunities for learning. At the very core of notions of bodily metabolization, it has been shown that eating breakfast makes a significant contribution to a child's average daily nutrient intake. This is illustrated by the fact that the average total energy intake was significantly lower for children who either skipped breakfast or who consumed breakfast at home, than for children who ate at school (Pollitt 1995). These socio-biochemical processes are the most understood component of the political ecology of childhood hunger because of the collective interest of doctors and other medical professionals.

Beyond the bodily deterioration that is a result of malnourishment and hunger, there are a host of documented behavioural relations that illustrate the externalization of childhood hunger. These socionatural relations get beyond Marx's notion that "it [hunger] therefore needs a nature outside itself", but also that hunger therefore also produces other natures outside itself. Children who eat breakfast before school begins show general increases in their math and reading scores. These children tend to have fewer discipline problems. Because they feel, and are, healthier as a result of having eaten, these children visit school nurses' offices less often (Minnesota Department of Children, Families, and Learning 1998). Children who do eat breakfast at school closer to test-taking time, perform better on tests than those who skip breakfast. Students who eat school breakfast show decreases in psychosocial problems (Murphy *et al.* 1998).

Focusing on urban childhood hunger helps to articulate the interconnected socionatural relations ranging from the scalar connection between the suffering caused by gastrointestinal acid buildup in children's empty stomachs, and how this in turn impedes their everyday lives through the development of physical, psychological and behavioural problems. These relations can be especially useful as they provide material insight into the connections between bodily metabolization and at the fundamentally socio-physiological, psychological and behavioral level of everyday life. These connections can inform the processes of urban metabolization more generally, especially since these hungry children are often spatially clustered in low-income communities within Milwaukee.

VAH has lobbied Milwaukee public school system administrators and school board members in an attempt to help them recognize the inefficiency of having hungry children at school and the learning difficulties associated with this condition. In an effort to alleviate the ramifications of having some of the most extreme racial and income inequality rates in the US, implementing the universal free breakfast programme would give Milwaukee's political leadership the chance to improve the socioecological unevenness of the city. Just like the other utopian efforts in Milwaukee, the efforts to fight urban hunger have contributed to the production of emancipatory social change. Urban activism and academic research have helped increase our collective understanding of the political ecology of urban hunger and contributed to the production of an emancipatory urban environment.

CONCLUSION

Taking impetus from Kropotkin's revolutionary notion that the most essential thing in life is food, through this chapter, I encourage increased academic attention and activism to the necessarily utopian political ecology of urban hunger. The urban struggles to feed hungry children in Milwaukee, and the attempts to understand the structural relations that have historically produced material inequality in the city, show that through struggle there is always hope. Geographers can play a leading role in promoting critical thinking through a better theorization of the roots and ramifications of urban hunger. However, where geographers often fall short is in linking our theories to praxis.

I am reluctant to conclude by summarizing, but rather want to look toward the future potential of Marxist urban political ecology scholarship and action-research to motivate more possible cases that seek to realize the "Not-Yet-Consciousness" that Bloch referred to. We must work hard to find an alternative to the contemporary mix of economic free-marketeerism and political irresponsibility as humankind continues to experience increased inequality through the proliferation of neoliberal capitalism, coupled with the rampant urbanization the world over that has contributed to the production of urban hunger.

The processes that bridge physiology and markets, or *necessity for food* and *desire for particular foods*, is neither straightforward, nor is the socionatural metabolization that underpins the politicization of urban hunger a simple matter. Instead, the production of urban hunger results from the contradictions inherent to capitalism and unimaginative political processes that tolerate urban inequality. Understanding the socio-physiology of hunger is essential for understanding the social production of urban inequality. As such, linking the scalar metabolic processes that connect individual bodies to urban social processes and spatial forms helps situate the political ecology of hunger and can facilitate a better understanding of these relations in a way that might motivate a more deliberate response to the problems of hunger.

To better articulate the ineffable contradictions produced through the synthesis of nature and society, in a way to motivate a more emancipatory politics around hunger, communicating ecology in more creative and resourceful ways than just discussing stomach contractions, gastric fluids and gastrointestinal muscle tone are imperative. It is also necessary to get beyond discussing urban hunger as just another, in a long list of "social problems". Neruda's symbolic interpretation of the body's response to hunger, "having a chain of fish hooks, trailing from the heart, clawing at your insides; hunger feels like pincers, like the bite of crabs; that it burns, burns", helps to inform our socionatural imagination in a way necessary for understanding that hunger is at the same time dialectically natural and social. Through this more crisp articulation, perhaps we can get closer to believing and acting upon Harvey's (2000) notion that "[t]here is a time and place in ceaseless human endeavor to change the world, when alternative visions, no matter how fantastic, provide the grist for shaping powerful political forces of change". As such, the problem before us is perhaps best personified within Herbert Marcuse's classic question: "What is involved in the liberation of nature as a vehicle of the liberation of man [*sic*]?"

BIBLIOGRAPHY

Ashman, L., J. de la Vega, M. Dohan, A. Fisher, R. Hippler, and B. Romain (1993) *Seeds of Change: Strategies for Food Security for the Inner City.* University of California Graduate School of Architecture and Urban Planning

Bloch, E. (1986) *The Principle of Hope*, Vol. 1. Cambridge: MIT Press

Dirks, R. (2003) "Diet and nutrition in poor and minority communities in the United States 100 years ago", *Annual Review of Nutrition*, 23(1): 81–101

Dréze, J. and A. Sen (1989) *Hunger and Public Action.* Oxford: Clarendon Press

Eisenhauer, E. (2001) "In poor health: supermarket redlining and urban nutrition", *GeoJournal*, 53(2): 125–133

Engels, F. (1872) (1958) *The Condition of the Working Class in England.* Stanford, CA: Stanford University Press

Engels, F. (1880) (1970) *Socialism: Utopian and Scientific.* Moscow: Progress Publishers

Engels, F (1881) "A fair day's wages for a fair day's work", *Labour Standard*, No. 1, May 7. Online. Available HTTP:<//www.marxists.org/archive/marx/works/1881/05/07. htm> (accessed 4 June 2004)

Food Research and Action Center (FRAC) (2004) *State of the States: 2004: A Profile of Food and Nutrition Programs Across the Nation.* Washington, DC: Food Research and Action Center

Gottlieb, R. (2001) *Environmentalism Unbound: Exploring New Pathways for Change.* Cambridge, MA: MIT Press

Greenwood, D.J. and M. Levin (1998) *Introduction to Action Research: Social research for Social Change.* Thousand Oaks, CA: Sage Publications

Grigg, D. (1993) *The World Food Problem.* Oxford and Cambridge, MA: Blackwell

Guyton, A.C. (1991) *Textbook of Medical Physiology.* Philadelphia, PA: Saunders

Haraway, D. (1991) *Simians, Cyborgs and Women. The Reinvention of Nature*. London: Free Association Books

Haraway, D. (1997) *Modest_Witness@Second_Millenium.FemaleMan_Meets_ OncoMouse™*. London: Routledge

Harvey, D. (1973) *Social Justice and the City*. Cambridge, MA: Blackwell Publishers

Harvey, D. (1998) "The body as an accumulation strategy", *Environment and Planning D: Society and Space*, 16: 401–421

Harvey, D. (2000) *Spaces of Hope*. Berkeley, CA: University of California Press

Heynen, N. and H.A. Perkins (2005) "Scalar dialectics in green: urban private property and the contradictions of the neoliberalization of nature", *Capitalism Nature Socialism*, 16(1): 99–113

Kahn, B.E. and L. McAlister (1997) *Grocery Revolution: The New Focus on the Consumer*. Reading, MA: Addison-Wesley

Katz, C. (2004) *Growing Up Global: Economic Restructuring and Children's Everyday Lives*. Minneapolis, MA: University of Minnesota Press

Keil, R. and J. Graham (1998) "Reasserting nature: Constructing urban environments after Fordism", in B. Braun and N. Castree (eds) *Remaking Reality – Nature at the Millenium*. London: Routledge

King, R.F. (2000) *Budgeting Entitlements: The Politics of Food Stamps*. Washington, DC: Georgetown University Press

Laituri, M. and A. Kirby (1994) "Finding fairness in America's cities? The search for environmental equity in everyday life", *Journal of Social Issues*, 50(3): 121–139

Lappé, F., J. Collins and P. Rosset (1998) *World Hunger: Twelve Myths*. New York: Grove Press

Lefebvre, H. (1991) *Everyday Life in the Modern World*. New York: Harper & Row

Lefebvre, H (2002) *Critique of Everyday Life*, Vol. 2: *Foundations for a Sociology of the Everyday*. London and New York: Verso

Levenstein, H. (1988) *Revolution at the Table: The Transformation of the American Diet*. New York and Oxford: Oxford University Press

Levenstein, H. (1993) *Paradox of Plenty: A Social History of Eating in Modern America*. New York and Oxford: Oxford University Press

Levins, R. and R. Lewontin (1985) *The Dialectical Biologist*. Cambridge, MA: Harvard University Press

Marcuse, H. (1972) *Counterrevolution and Revolt*. Boston, MA: Beacon Press

Marston, S.A. (2004) "A long way from home: Domesticating the social production of scale", in E. Sheppard and R.B. McMaster (eds) *Scale and Geographic Inquiry: Nature, Society and Method*. Cambridge, MA: Blackwell Publishers

Marx, K. (1844) (1964) *Economic and Philosophic Manuscripts of 1844*. USA: International Publishers

Marx, K. (1867) (1990) *Capital Vol. 1*. London: Penguin Press

Marx, K. and F. Engels (1845) (1998) *The German Ideology*. New York: Prometheus Books

Minnesota Department of Children, Families, and Learning (1998) *School Breakfast Programs Energizing the Classroom*, Minnesota Department of Children, Families, and Learning, Roseville, MN

Mitchell, K., C. Katz and S.A. Marston (2003) *Life's Work: Geographies of Social Reproduction*. Cambridge, MA: Blackwell Publishers

Murphy, J.M., M. Pagano, J. Nachmani, P. Sperling, S. Krane, and R. Kleinman (1998) "The relationship of school breakfast to psychological and academic functioning", *Archives of Pediatric and Adolescent Medicine*, 152: 899–907

Neruda, P. (1958) (2004) "The great tablecloth", in M. Eisner (ed.) *The Essential Neruda: Selected Poems*. San Francisco, CA: City Lights Books

New York Times (1930) "Milwaukee opens 'Soup Kitchens'", *New York Times*, 6 March

Peet, R. and M. Watts (1996) *Liberation Ecologies*. London: Routledge

Pollitt, E. (1995) "Does breakfast make a difference in school?" *Journal of American Dietetic Association*, 95(10): 1,134–1,139

Pothukuchi, K and J.L. Kaufman (2000) "The food system: a stranger to the planning Field", *Journal of the American Planning Association*, 66(2) 113–124

Root, W. and R. de Rochemont (1995) *Eating in America: A History*. Hopewell, NJ: The Ecco Press

Schlosser, E. (2002) *Fast Food Nation: The Dark Side of the All-American Meal*. New York: Perennial

Sen, A. (1981) *Poverty and Famines: An Essay on Entitlement and Deprivation*. Oxford: Clarendon Press

Smith, N. (1993) "Homeless/global: scaling places", in T. Bird, L. Tickner, B. Curtis, T. Putnam and G. Robertson (eds) *Mapping the Futures. Local Cultures, Global Change*. London: Routledge

Swyngedouw, E. (1996) "The city as a hybrid – on nature, society and cyborg urbanization", *Capitalism, Nature, Socialism*, 7(1): 65–80

Swyengedouw, E. (1999) "Modernity and hybridity: the production of nature, water, modernization in Spain", *Annals of the Association of American Geographers*, 89(3): 443–465

Swyngedouw, E. (2004) *Social Power and the Urbanization of Water Flows of Power*. Oxford: Oxford University Press

Swyngedouw, E. and Heynen, N.C. (2003) "Urban political ecology, justice and the politics of scale", *Antipode: A Journal of Radical Geography*, 35(5): 898–918

U.S. Conference of Mayors (2003) "Hunger and Homelessness 2003"

Willis, E. (2000) "Health disparities and children in Milwaukee", in K. Little, S.F. Battle and R. Hornung (eds) *The State of Black Milwaukee (57–80)*. Milwaukee: The Milwaukee Urban League

Wrigley, N. (2002) "'Food deserts' in British cities: policy context and research priorities", *Urban Studies*, 39(11): 2,029–2,040

9 Metabolisms of obe-city

Flows of fat through bodies, cities and sewers

Simon Marvin and Will Medd

Fats, oils, and grease aren't just bad for your arteries and waistline; they're bad for sewers, too. These substances can clog sewer pipes, leading to overflows and backups that can create health hazards, damage home interiors, and threaten the environment.

(Water Environment Federation 1999)

INTRODUCTION

While strategies to tackle obesity have led to renewed debate about the specific relationship between the body and urban form (Sui 2003) the (im)mobilities of fat through bodies, cities and infrastructure reveal a more complex web of urban metabolisms. We argue that to understand the mobilities of fat in a city context metaphors of urban metabolism become important. Urban metabolism need not refer to stable sets of relationships in which explanations of social order subsequently refer, consider for example the early work of the Chicago school (Park *et al.* 1967). However, to reject the concept of metabolism by reducing it to functionalist and teleological metaphors would be to lose the insights a reformulated concept can reveal. "Cities", Harvey (2003: 34) argues, "are constituted out of the flows of energy, water, food, commodities, money, people and all the other necessities that sustain life". Metaphors of metabolism are therefore useful for understanding such flows. The contingencies and mobilities of fat in bodies (as individuals), cities (as a collective site of action) and sewers (as infrastructure), we argue, highlights a multiplicity of urban metabolisms, each with different interconnectivities and forms of instability.

Fat poses key challenges for understanding the future of mobile society. For while fat can appear very much as a mobile fluid, travelling through global networks of food production and consumption, it also encapsulates the other aspect of contemporary networked cities – immobility. Fat literally, but also metaphorically, can also appear as static, difficult to move, and solidified. While fat provides bodies with important functions, for example cushioning, insulation and storage it is the excess of fat that causes so much concern (Ruppel Shell 2002). More specifically excess fat, particularly when bodies become obese, is associated with an impressive list of associated health problems. The World Health Organization (WHO) (1997) reports a "greatly increased risk" of diabetes, gall bladder disease, hypertension, dyslipidaemia, insulin resistance, breathlessness, sleep apnoea, "moderately increased risk" of coronary heart diseases, osteoarthritis, hyperuricaemia and gout, and "slightly increased risk" of cancers, reproductive hormone abnormalities, polycystic ovary syndrome, impaired fertility, low back pain, increased anaesthetic risk, and foetal defects arsing from maternal obesity. At a population level there is much concern about the growing levels of obesity internationally. The International Obesity Task Force (IOTF) reports data illustrating obesity concerns in a wide range of countries including Australia, Japan, Brazil, England, the United States of America, Mauritius, Kuwait and Western Samoa. The World Health Organization reports that, while in 1995 there were an estimated 200 million obese adults worldwide, by 2000 it had increased to over 300 million (WHO 1997). The very formation of the International Obesity Task Force in 1996, with the aim to "alert the world about the urgency of the problem of the growing health crisis threatened by soaring levels of obesity", signifies the growing international anxiety about the "global epidemic" of obesity (see IOTF website).

Within the context of this growing anxiety about obesity there is typically a turn to the United States of America to see where the rest of the world is going. In this chapter we also turn to the USA to examine how, in response to the rising numbers of "obese bodies", there has been the mobilization of the concept of "fat cities" involving renewed debate about the relationship between bodies and the city, provoked largely by the innovative representations of a men's fitness magazine. We shift focus in this debate to look to the problems of fat in infrastructure, focusing specifically on the experience in US cities of sewer blockages that reveal quite different sets of processes within which fat is embedded. We show how in each of these sites of intervention – the body, the city collective and the sewer – strategies of prevention, removal or acceptance each reveal a multiplicity of metabolisms as well as partial interconnections between them. Finally, we conclude with reflections on the implications of "fat" for developing our understanding of mobility.

THE EMERGENCE OF OBE-CITY

The proliferation of anxiety about the obesity epidemic is in part mobilized by the experience in the United States. In the US, it is reported that 60 percent of the population is overweight, and 21 percent obese (Revill 2003). This has more than doubled in the last two decades, with obesity now rising by 5 percent per annum. This is not evenly spread through the population. There are serious health inequalities, with more than half of black women in low socio-economic groups being obese. One in five children is overweight. Obesity is estimated to account for 12 percent of health care costs, $100 billion and rising. The US has provided the emblematic marker for the rest of the western world to signify where other countries are heading and as a context to examine the causes of obesity, the problems it generates and possible solutions. In this chapter, however, we take a different turn and examine how the crisis of fat deposition has started to raise the visibility of the interconnections between the multiple metabolisms of the body, sewer and city.

In the US the concern with the rising levels of obesity of individual bodies has sparked an interesting mobilization of the problem of obesity at the urban level. At the core of this movement has been the development of city "fatness and fitness" league tables produced by *Men's Fitness*, a leading men's health magazine. The magazine set out "to measure, city by city the relative environmental factors that either support an active, fit lifestyle, or nudge people towards a pudgier sedentary existence" (*Men's Fitness* magazine website, *Survey Methodology*). The magazine's analysis enrolled a range of existing surveys and data and involved aggregating scores based on a multiplicity of indicators, including gyms and sporting-goods stores, health club memberships, surveys of exercise, levels of fruit-and-vegetable consumption, alcohol consumption, smoking, television watching, number of junk food outlets per 100,000 population, recreation facilities, etc. Obesity levels were scored for cities by drawing on data from the Centers for Disease Control and Prevention and the Center for Chronic Disease Prevention and Health Promotion.

Consequently cities have been jostling not to be top of the league. While New Orleans was ranked as the number one fattest city in the first year of the league in 1999, in 2000 it was Philadelphia, and then Houston took the lead for the subsequent two years. These rankings were not ignored. When Philadelphia was top of the league a headline in *USA Today* ran "Blame the Cheese Steak: Unfit Philly wins flab crown" (Hellmich 1999). The article reports an initial response by a spokesperson for the city saying that: "this just proves what we've been saying all along, because Philadelphia has the best restaurants of any city in America, and apparently we've got the evidence to prove it." A more considered response followed from the new Mayor, John Street, elected in 2000, who appeared on *The Oprah Winfrey Show* to discuss Philadelphia's new status as fat capital. The Mayor enrolled the city into a strategy of collective weight loss (Calandra 2001). His initial response was to appoint a Health and Fitness Czar (or Fat Czar as the media reported it) and to initiate a programme called "76 tons in 76 days" with the support of the professional basketball team the Philadelphia 76ers. The challenge was set for the city to lose a collective 76 tons of weight in 11 weeks. The official weight was due on 3 July 2001 and 26,000 people were reported to have lost an average of 5.3 pounds each. Strategies for collective weight loss included groups for dieting and weighing in, line dancing programmes for city employees, enrolling restaurants to provide healthier food, and free fitness programmes (Twyman 2003). The Mayor's Office of Health and Fitness now reports that since the initial strategy was adopted in January 2000 a range of actors have become enrolled into a coalition for improving the health of Philadelphia: "citizens, the Department of Health, private corporate and community organizations, institutions of higher learning, public and private schools, communities of faith, hospitals, health systems, health clubs and gyms, pharmaceuticals, media outlets, local, state and federal government agencies, non-profit health agencies and sports teams" (City of Philadelphia, "Fit and Fun" website). And, it turns out, Philadelphia has moved down to seventh position in the league tables in 2004. All is not without controversy, however, and with the cities' growing fiscal problems, opposition politicians had urged that slimming the city's budget should include firing the health Czar (Twyman 2003).

Houston, by contrast, has been faced with the opposite problem to Philadelphia, captured by the ABC news headline, "Houston you have a problem: a big, fat problem" (Reynolds 2002a). Houston took the lead from Philadelphia and for 2002 and 2003 was at the top of the league table. This identity is provocative, not something the Mayor, Lee Brown, wanted "as a distinction of our city" (Reynolds 2002a). Houston's response was to get advice from Philadelphia. A fitness Czar was appointed, Lee Labrada, a former Mr Universe, who began a "Get Lean Houston" campaign. This included a Fat Drive in which participants pledged to lose weight, resulting in 17,000 pounds being lost by 2,000 participants by 2003. Again a range of different agencies was enrolled. Most notable was McDonald's as Official Restaurant Sponsor rolling out a menu of "Salad and More" across all 253 restaurants in Houston. By 2004, Houston was able to boast "Houston: We did it! We are officially no longer the Fattest City in America. Dropping to #2 behind Detroit, MI" ("Get lean Houston" website). This, according to *Men's Fitness* was the result of better scores for increased sports participation, decreased alcohol consumption and improved nutrition (*Men's Fitness* website, Houston). So pleased are "Get lean Houston" with the success that they now claim to have rolled out the campaign nationally with the launch of "Get lean America" ("Get lean Houston" website).

The mobilization by *Men's Fitness* of the metaphor Fat City has provoked a renewed interest in the relationship between bodies and urban form (Sui 2003). Papers published in *The American Journal of Health Promotion* (Killingsworth *et al.* 2003) and *American Journal of Public Health* (Jackson 2003) reported a series of studies that "are among the first to link shopping centres, lack of sidewalks and bike-trails and other features of urban sprawl to deadly health problems" (Fackelmann 2003). One report shows how "people who live in sprawling neighbourhoods walked less and had less chance to stay fit . . . people living in sprawling neighbourhoods weighed 6 pounds more on average than the folks living in compact neighbourhoods where sidewalks are plentiful and stores and shops are close to residential areas" (Fackelmann 2003). And, in such areas, obesity is reported as more prevalent.

Sui argues this literal uptake of the fat metaphor can be illuminated through debates about the body and the city:

> Fat City depends on the fatness of the body since this is the body of a coordinator: a data processor, business person, real estate agents, or urban planner . . . And the fatness of the body depends on the fatness of the city, since it develops as a result of the automobile dependent, privatized spaces of the fat city. The excess circulation of the city (roads) allows isolation of the person, which transforms again into excess circulation (blood vessels to the fat tissue). If the fat body becomes an obsession for health reasons and narcissism alike, it cannot be returned to a more sustainable size and shape as long as the city remains outside of theories of health.
>
> (Sui 2003: 82)

Raising analysis of the relationship between the body and the city in this way is important. However, the emphasis on the direct linkages between fat bodies and urban sprawl neglects the subtleties and multiplicities in the work done by *Men's Fitness* magazine. The *Men's Fitness* league tables have significantly raised the visibility of fat issues in US cities. It has mobilized in the USA the classic debate between nature versus nurture by arguing that the rapid rise in obesity in the last century could not simply be explained by genetics and must therefore be a consequence of changing environmental circumstances. In explaining their rationale for this they use the imaginary example of two identical twins living in different cities to argue that the twin living in a poor city environment – in which finding healthy food is difficult, there are few places to exercise, the weather is not conducive to exercise, where people smoke freely, commuting is a hassle, there is poor access to gyms, and health care is poor and where junk food is easy to hand – is more likely to be fat. With this simple analogy to demonstrate the issue, the magazine develops a sophisticated methodology upon which the league tables are based. The magazine displaces methodology based on simple measures of obesity prevalence based on Body Mass Index with a range of indicators that move beyond individual bodies to the wider city ecology. This is clear when one considers that, based on BMI alone, the league tables would look different. For example, in 2003, Memphis, Tennessee had the highest percentage of overweight and obese adults and yet was placed 21st position in the league tables. The effect is to move from measuring the state of obesity in a city to highlighting the processes that lead to obesity.

This identification of the "fat city", as we have seen with the examples of Philadelphia and Houston, can lead to the mobilization of representations of the city to collectively re-mobilise fat with the enrolment of a range of diverse actors into strategies to reshape the metabolism of bodies and the city. The emergence of the representation of fat cities has involved a significant move in making visible the dynamics between bodies and the city as an environment and in doing so has opened up new discourses and forms of activity around the city as a collective metabolism. While *Men's Fitness* magazine presents a sophisticated account of the relationship between fat bodies and the fat city, understanding the circulation and deposition of fat requires looking beyond the dichotomy of body and city. To look at the multiple mediations of fat embedded within other less visible urban metabolisms we need to recognize the complexities and multiplicities of the relationship between bodies and cities that includes the role of social-technical infrastructure in the distribution of fat.

FAT THROUGH INFRASTRUCTURE: THE SEWER-FAT CRISIS

The movement of discourses about fat from the individual to the city level has, as we have seen, involved the mobilization of a range of actors into programmes of slimming down American individuals and slimming down American cities. It would be a mistake, however, if the relationship between individuals and cities, as both environment and site for collective action, ignored the role of infrastructures.

Infrastructure is, by its very nature, often defined by its very hidden presence with apparent reliability and stability, as if it represents an almost neutral intermediary between the individual and the city. Yet, understanding the city as a socio-technical process points to the "constant effort" required to keep infrastructure working (Graham and Marvin 2001). The social, technical and spatial-temporal characteristics of infrastructures become more acutely revealed during times of crises when the material embodiment of different sets of social, political, economic or organizational relations are ruptured (Graham and Marvin 2001; Summerton 1994). This is true of fat and its relationship to sewerage infrastructure: the sewer-fat crisis is a crisis that points to the, otherwise largely hidden, complex interconnectivities between, for example, the food industry, local waste disposal systems, and global food oil markets. Sewers, as Gandy (1999) argues "are one of the most intricate and multi-layered symbols and structures underlying the modern metropolis, and form a poignant point of reference for the complex labyrinth of connections that bring urban space into a coherent whole" (p. 24). A crisis in the sewers is more than a technical malfunction. It is a crisis that reveals the instabilities between the geographies of city infrastructure and the unbounded interdependencies of city metabolisms.

In October 2001, Randy Southerland wrote about the growing numbers of sewer blockages and overflows across cities in the United States as restaurants and fast food chains pour cooking residue into drains while local governments lack the resources to monitor grease disposal and enforce the relevant regulations (Southerland 2002). The solidification of fat, oils and grease, he writes "choke pipelines, eventually clogging them and causing them to cough up rivers of raw sewerage". He cites how, in January 2001, the US Environmental Protection Agency sued Los Angeles for 2,000 sewer spills over five years, 40 percent of which were caused by fat. In the *Wall Street Journal* Barry Newman (2001) writes how in New York there are about 5,000 "fat-based backups a year with several big gum ups". For reasons that range from the decline in global markets for waste fat and the increased costs of fat disposal, following Mayor Giuliani's crackdown on the garbage Mafia (Methvin 2000), more grease is illegally disposed of into the sewers:

> Fat won't pollute: it won't corrode or explode. It accretes. Sewer rats love sewer fat; high protein builds their sex drive. Solid sticks in fat. Slowly, pipes occlude. Sewage backs up into basements – or worse, the fat hardens, a chunk breaks off and rides down the pipe until it jams in the machinery of an underground floodgate. That to use a more digestible matter causes a municipal heart attack.
>
> (Newman 2001)

While there are hotspots around restaurant areas there are also, however, sewer fat problems around residential areas, particularly where large numbers of multi-family units are located and where residents discharge their grease into the drain. And in some cases attention is turning to schools and prisons (Southerland 2002). Just as with obesity then, there are different distributions of sewer-fat problems across different publics.

In contrast to *Men's Fitness* who made visible the problem at the level of the city collective, making visible the problem in sewers continues to be a challenge for local authorities and utilities charged with sewer maintenance. A variety of techniques have been developed for identifying blockages, including CCTV, smoke infrared thermography, and even radar and sonics (National Association of Sewer Service Companies website). Having made fat visible, cities are developing strategies to combat the fat in the sewers. High-pressure hoses can remove blockages, but dislodged blocks of fat may then cause new problems downstream. Large vacuum trucks are also used to either suck or blow fat out of congested sewers. In New York an enzyme product that reduces grease build-up is routinely used, and a liquid emulsifier can now tackle larger build-ups (Pagano 1999). Similarly in New Bedford, bacteria developed by a bio-technological company are used to metabolize the grease, breaking it apart into water, carbon dioxide and free fatty acids, that can be washed away from metal, concrete and brick (Allen 2000).

When the crisis hits, the problems becomes all too visible in sewer overflows but prevention is more problematic. Ultimately, the concern of cities is to avoid the grease being put into the sewers and making visible the practices of restaurants and households distributed throughout cities raises different challenges. New "lean sewer" ordinances have been developed, but the cost of monitoring and enforcement is high (Southerland 2002). Not surprisingly then we see also the emergence of appeals to the city collective once more to reshape the deposition of fat. In New York, for example, the Environmental Protection Department guidance makes the following plea: "Sewer back up damages property and damages public health – the city needs businesses and individuals to do their part to maintain the system" (City of New York 2002). Indeed, across US cities brochures promoting "fat free sewers" show homeowners what they can do to keep grease out of the sewers (Water Environment Federation 1999).

Fat is not just a problem for bodies and the public health of cities, it is also a problem that involves the processes of city infrastructure. The sewers crisis has given us a pertinent example of this, but the arguments can also be understood in relation to other infrastructures. Indeed, resolving the problem of fat being disposed in the sewers can lead to problems in other infrastructures. A traffic jam in San Francisco, one of the fittest cities in the US according to the *Men's Fitness* league, was caused by a spillage of cooking oil from an overturned truck. It took cleanup crews nine hours to re-open the lanes of the highway according to the California Highway Patrol (Sfgate 2004). It was reported the truck was carrying used cooking oil collected from restaurants. Just as fat in the sewers solidifies as it cools, so too on the highway, in the cool weather, the fat had solidified into a gel making the clean up harder: "it got into the grooves (of the road) and was difficult to extract" a spokesman from the truck company said.

CITIES AS SITES FOR PARTIALLY CONNECTED METABOLISMS

The USA is currently seen as the world leader in the obesity epidemic, as "Land of the Fat" (Engel 2002). Much attention has turned to the anxiety raised by the US epidemic and the risk that the rest of the western world is heading in the same direction. While such attention is usually focused on the problems of fat bodies and associated issues of food and diet, in this chapter we have looked to the US to explore the issues of fat within a specifically urban context. Key to this shift is an understanding of the connections between the problem of fat in individual bodies and the mobilization of fat at the level of the city. Additionally, the embodiment of fat extends beyond fat bodies and fast food into infrastructure itself through the deposition of fat in the sewage network. What then can we learn by looking across these contexts?

The city is a site of multiple metabolisms, but these are always partial and selectively connected metabolisms. This chapter has not attempted to provide a complete explanation of the political-ecology of the metabolism of fat. Consequently, what is missing from our account is, for example, the global system of oil and fat production, the international movement and distribution of fats and oils, the production and distribution of food chains that incorporate fats and oils, and the social organization of waste oil and fat systems. At the same time, we have not presented a synthesis of the metabolism of fat in terms of the interconnectedness of its social, economic, spatial, bio-chemical, and cultural dimensions. Yet by using the frame of the city we can do something else. We can begin to bring into focus elements of these multiple metabolisms and the partial interconnections between them. What is particularly interesting for us is how strategies for dealing with fat at an urban level start to problematize the metabolisms of fat in quite different ways. There are three possibilities for dealing with fat, they are: the removal of fat, prevention of fat deposition and fat acceptance (see Table 9.1). These three strategies provide competing ways of seeing the metabolism of fat. Imagine looking through a "kaleidoscope" that can highlight how each of the three strategies brings into focus the disconnections and partial interconnections between the multiple metabolisms of the city, body and infrastructure. What we see is a series of changing yet always partial metabolisms and forms of connection.

Removal of deposited fat

The first strategy for dealing with the crisis of deposited fat is a strong discourse of removal. These discourses begin disdainfully with unhealthy and corpulent bodies in terms of a representation of excess and greed, are mobilized in relation to the "Fat

Table 9.1 Strategies for dealing with fat: removal, prevention and acceptance

	Removing deposited fat	Preventing fat deposition	Accepting fat deposition
Crisis	Arrived	Impending	Redefined
Problematic	Blockage	Flow	Adaptation
Intervention	Decontextualized	Contextualized	Recontextualized
Metabolism	Incomplete	Interconnected	Reconfigured
Vision	Attack	Defence	Absorption

City" and "Fat Sewers" that are to be "slimmed down" or "to get lean". And with this discourse the strategies of intervention focus on a range of technological devices used to remove fat from bodies and sewers. These range from suction devices, the physical removal of fat or the use of drugs and biotechnologies to remove fat. Specialist medical and sewer treatment technologies are mobilized to focus on highly individualized and site-specific fat removal at the level of the individual body and sections of the sewer. This strategy mainly relies on the use of specialist technical and professional expertise with relatively passive users with little collective action at the level of the city. There is a relatively limited understanding of the metabolism of fat within this discourse. Primarily, the focus is on the crisis of blockage and deposition with the problematic being one of extraction and disposal. There is less focus on the wider issues involved in the production, distribution and consumption of fats and the wider processes that shape its deposition in bodies and sewers. In this way, the body and the sewer are decontextualized from their relationships with a wider metabolism of fat. At the same time the inter-connections between the metabolisms of body and sewers are relatively weak. The problem of metabolism is deposition and disposal of waste fat. Within all these contexts, the removal of fat is difficult and its disposal expensive. Though strategies for the removal of fat do not immediately involve shaping urban form, the interdependencies across city metabolisms – from bodies, to infrastructure to the city collective – mean that removal from one location – the body or the sewer – potentially leads to a problem of re-disposition in another context. Overall, this strategy presents a partial understanding of the metabolism of fat, fails to connect the problem with wider questions about the specific contexts and environments that bodies and sewers are located within and there are serious concerns about its longer term effectiveness and viability as strategy for dealing with fat.

Preventing the deposition of fat

If the removal of fat is complex and expensive, an alternative strategy is that of preventing the deposition of fat. It is not just those with fat bodies that must be anxious about fat, it follows, but as an epidemic, like with any other disease, the public must be aware and enrolled into fitness and diet regimes, into life-long healthy lifestyles. And this is also the case with the city where there is a potential window of opportunity to maintain health, fitness and leanness before the crisis strikes. The *Men's Fitness* campaign has stimulated the mobilization of a renewed interesting in the "re-tooling of the urban environment" to mobilize bodies. The strategy becomes one oriented around the maintenance and acceleration of flows to ensure that the fat remains mobile and that the opportunities for its deposition are limited. In relation to sewers there has been the emergence of a range of promotional strategies, guidelines and ordinances all designed to reduce the deposition of fat and with that a concern to mobilize other waste infrastructures that could provide alternative routes for fat disposal. And, just as bodies trying to lose fat require the enrolment of a wide set of networked relations from food supply chains to transport infrastructures that become mobilized through the city collective, so too we find that strategies of removal of fat in sewers involves a vast array of food supply chains and transport infrastructures that include the local configuration of waste management to the global markets for oils. When we turn to prevention strategies then, we see again the complex of interdependencies between bodies, infrastructure and cities but now we also find in addition the intentional re-shaping of urban form. In this sense the need to maintain flow means that the city needs to become more malleable both physically and in terms of social practices to mobilize fat. Not only is the urban environment being actively explored in relation to its implication for the movements and feeding of bodies, but so too, the habits of bodies in houses and restaurants are being re-shaped to protect the health of the sewers.

Intervention is much more contextualized, looking at the bodies and sewers and cities in terms of their mutually shaping relations. The body has to be looked at in the context of its movement and mobility through the city, the sewers have to be looked at in relation to the practices of waste disposal by households and restaurants, and the role of the city in creating a context for collective action to mobilize bodies and action around the sewers becomes significant. These interfaces become clearer when strategies to deal with the removal, prevention and acceptance of fat are illuminated. But, significantly, it asks wider questions about the metabolism of fats. Rather than focusing on issues of deposition, it raises wider questions about bodies in contexts – what is consumed and eaten, how bodies move through cities. This in turns starts to raise wider questions about the consumption of foods and the distribution of fats. A wider metabolism of fat comes into view. But what is also significant is that the interconnections between the different metabolisms are brought into view much more powerfully in this approach. This response places bodies, sewers and the cities in relation to each other. The context needs to be reshaped to provide a context for bodies and sewers to avoid fat deposition.

Accepting the deposition of fat

The final strategy is a strategy of fat acceptance. This discourse takes as its starting point the assumption that the crisis itself needs to be redefined through a strategy of adaptation to the new conditions of fat deposition. Strategies for defending fat bodies vary. In some cases it is about acceptance, though not celebration, by learning to live with fat rather than dealing with constant anxiety of the failure to remove fat (BBC 2004). In others though it is about a celebration of fatness, as evidence by "Fat Cities – for Big Beautiful Fat Women and Men and Friends" (Fat Cities website). In this context, the strategy for intervention is to recontextualize the way in which fat bodies, sewers and cities may be understood. The acceptance of obese bodies has implications for the "re-tooling" of other urban infrastructures. Rick Hampson (2000) in *USA Today* reports how "Americans" expanding backsides are behind a trend toward wider, more comfortable seating in public areas." Here we find a discourse of acceptance emerging: "In The End, People Just Need More Room" runs the title of the article. Hampson reports how subways, theatres, airlines, sport stadiums and opera houses have responded to demands for wider seats, therefore reducing seating capacity to increase sitting room. Seating had become standardized in the US with 18 inches recognized in building codes. This is coming under question. Indeed, sponsored by the airforce and major manufacturers, a project – the Civilian American and European Surface Anthropometric Resource – is using scanning technology to measure the contours of 4,000 volunteers and develop new standards. Bodies themselves are also being recategorized with the US standards on levels of overweightness and obesity being revised upwards to normalize larger body size. At the level of the sewage infrastructure, the acceptance of fat deposition is a strategy with the sewers being enlarged to accommodate fat deposition or simply being treated with biotechnologies that eat the fat. In this sense the city is recontextualized to support the normalization of fatter bodies and sewers. The city itself is reshaped and retooled not to mobilize bodies and sewers but to accommodate the immobilities of deposited fat. If bodies and sewers are no longer malleable then the city itself has to be made malleable to fit the new bodies. This strategy does little to problematize the wider question of the extended metabolism of fat. Instead the metabolism of the city itself has to change to cope with the problems of deposition in the metabolism of bodies and sewers.

By looking across the three potential responses to fat deposition in sewers, cities and bodies, we highlight powerful affinities across metabolisms between the discourses, social practices and technologies involved in re-mobilizing fat. Further still, in each of the metabolisms, mobilizing fat in one context requires a mobilization in another: people have to be enrolled in terms of intercepting grease and fat from their homes or from restaurants, to keep the sewers mobile; people need to be mobilized to keep fat more mobile within an urban context; and cities themselves need to become retooled or reconfigured to keep the people moving. Finally, each of the strategies embodies a quite different view of the cities' reconfigured relations with fat. A strategy of removal implies that the fat is attacked to ensure that bodies, sewers and the city can be slimmed down. The strategy of prevention of fat deposition implies a defensive approach in which bodies, sewers and the city are socio-technically re-engineered to ensure that fat is kept on the move. A strategy of fat acceptance implies a sense of defeat or possibly tolerance and perhaps even celebration as society learns to live with fat bodies, sewers and cities. Each strategy brings fat into focus in quite different, but not necessarily mutually exclusive ways, which is why strategies for dealing with fat in these different metabolisms are so contested.

CONCLUSION: FAT, METABOLISM AND MOBILITY

The metabolism of fat in bodies, cities and sewers is clearly interwoven within each other yet these linkages are poorly understood. The challenge is to understand the mutually defining relations between bodies, cities and sewers. Evidently, the city is a key factor in the production and reproduction of bodies and sewers. The city provides an emerging context and coordination for bodies and sewers, the order and organization that links bodies and sewers and the framework in which fat is consumed and deposited in bodies and sewers. But the relations between these metabolisms are extremely complex. Bodies and sewers must be considered active in the production and reproduction of the city. These mutually redefining relations can be viewed as series of interfaces between metabolisms. Rather than seeing the metabolism of cities and bodies and sewers as single entities, they should be regarded as assemblages with fat capable of crossing the boundaries between metabolisms to form particular sets of linkages. This is not to stress the unity of an idealized ecological balance but a set of interrelationships that involve a series of flows that are brought together and drawn apart in a series of temporary alignments.

The aim of this chapter has not been to present a comprehensive account of the flows of fat across the metabolisms of bodies, cities and infrastructures. Rather, our approach has sought to highlight the possibilities for following fat in ways that raise the visibility of the interdependencies between bodies, city and infrastructure within mobile society and which problematize attempts to isolate interventions to particular sets of relationships without the disruption of others. Researching mobile society, our exploratory investigation into fat suggests, requires a close examination of the interconnectivities constituted in mobile society that become more acutely revealed during times of "crisis" (Graham and Marvin 2001; Summerton 1994). Nonetheless our initial insights into the travels of fat through the metabolisms of bodies, cities and sewers demonstrates the significance of developing social scientific under-standing of mobility beyond people into the "material worlds" (Urry 2003). Understanding the flows of fat in cities requires more than modelling techniques might suggest; it requires understanding of the complex configurations of bodies, cities and infrastructure within which fat emerges, transforms and, sometimes, moves on. Fat presents a case for opening up the social sciences to understanding mobility in terms of the interconnectivity of phenomena and spaces, not in a holistic sense as if all the relationship could be understood, but in terms of a sensitivity to moments of interconnectivity between the interfaces of different metabolisms. These interfaces become clearer when strategies to deal with fat are of removal, prevention and acceptance are illuminated. Whichever strategy is developed, directed towards bodies, infrastructure or the city, there are implications that go beyond the immediate boundaries of the strategy.

The problem of fat, we have argued, is one that circulates and while we have brought attention to particular relations between circulations in bodies, infra-structure and cities in the United States, to explore the mobility of fat requires further travels into a vast range of global interdependencies from the particular local challenges of sewer pipeline management to the global networks of the food production and pharmaceutical industry. Even fat acceptance then has implications. Doing nothing is not an option.

BIBLIOGRAPHY

Allen, M. (2000), "Company demonstrates product in city", *The Standard Times*, New Bedford, 22 December 2000, Online. Available HTTP: <www.bactapur.com/english/pages/bacta710.html>

BBC 1 (2004) "Victoria's big fat documentary, 2nd part", 16 January 2004. www.bbc.co.uk

Calandra, B. (2001) "Trimming the fat in Philly City on a diet", *WebMD Feature Archive*, Online. Available HTTP: <http://content.health.msn.com/content/article/12/1676_52750?printing=true>

City of New York (2002) *Waste Water Treatment: Preventing Discharges into Sewers. Guidelines for New York City Businesses.* Department Of Environmental Protection, The City Of New York, Online. Available HTTP: <www.nyc.gov/html/dep/html/grease.html.> (accessed 15 February 2004)

Engel. M. (2002) "Land of the fat", *The Guardian*, 2 May 2002

Fackelmann, K. (2003) "Studies tie urban sprawl to health risks, road danger – driving every adds up on American's waistline, and can dangerous for their health", *USA Today*, 29 August 2003

Fat Cities, Online. Available HTTP: <http://www.fatcities.com>

Gandy, M. (1999) "The Paris sewers and the rationalization of urban space," *Transactions of the Institute of British Geographers*, 24: 23–44

Get Lean Houston, Online. Available HTTP: <http: //www.getleanhouston.com/>

Graham, S. and Marvin, S. (2001) *Splintering Urbanism: Networked Infrastructures, Technological Mobilities and the Urban Condition.* London: Routledge

Grosz, E. (1995) *Space, Time and Perversion: Essays on the Politics of Body.* London: Routledge

Hampson, R. (2000) "In the end, people just need more room: Americans' expanding backsides are behind a trend toward wider, more comfortable seating in public areas", *USA Today*, 10 April 2000, Online. Available HTTP: <http://www.fatcities.com>

Harvey, D. (2003) "The city as body politic", in J. Schneider and I. Susser (eds) *Wounded Cities: Destruction and Reconstruction in a Globalized World.* Oxford: Berg

Hellmich, N. (1999) "Blame the cheese steak: unfit Philly wins flab crown. Men's magazine weighs pizza shops against health clubs", *USA Today*, 9 December 1999, Online. Available HTTP: <http://www.eatingbythebook.com/fatcity.html>

International Obesity Task Force, Online. Available HTTP: <http://www.iotf.org>

Jackson, R.J. (2003) "Editorial: the impact of the built environment on health: an emerging field", *American Journal of Public Health*, 93: 1,382–1,384

Killingsworth R. , Earp, J., Moore, R. (2003) "Introduction: supporting health through design: challenges and opportunities", *American Journal of Health Promotion*, Sept/Oct: 1–3

Mayor's Office of Health and Fitness, City of Philadelphia, Online. Available HTTP: <www.phila.gov/fitandfun/index.html>

Men's Fitness, *Survey Methodology*, Online. Available HTTP: <http://www.mensfitness.com/rankings/3>

Men's Fitness, *Houston*, Online. Available HTTP: <www.mensfitness.com/rankings/221>

Methvin, E. H. (2000) "Undercover against the mob", *Readers Digest*, May

National Association of Sewer Service Companies, Online. Available HTTP: < www.nasco.org>

Newman, B. (2001) *The Wall Street Journal*, 4 June 2001

Pagano, V. M. (1999) *Letter to Nitin Patel, RE: Zircon Industries, Inc.*, Administrative Supervisor, N.Y.C. Department of Environmental, Protection Bureau of Water and Sewer Operations, 30 June, 1999, Online. Available HTTP: <http://www.greenchem.com/citofnewyord.html>

Park, R., Burgess, E..W., Mckenzine, R.D. (1967) *The City*. Chicago, IL: Chicago University Press

Revill, J. (2003) "A deadly slice of American Pie", *The Observer*, 21 September 2003

Reynolds, D. (2002a) "Houston, you have a problem. A big, fat problem", *ABC News*, 4 January

Reynolds, D. (2002b) "Your city's fat – now what? Being America's fattest city isn't good for one's image", *ABC News*, 4 January 2002, Online. Available HTTP: <http://abcnews.go.com/sections/wnt/WorldNewsTonight/fatcity020104.htm>

Ruppel Shell, E. (2002) *Fat Wars: The Inside Story of the Obesity Industry*, London: Atlantic

Sfgate (2004) "Southbound 101 fully reopens after grease spill", *Sfgate*, 7 January, 2004

Sharpe, M.E. and Stearns, P. N. (2002) *Fat History: Bodies and Beauty in the Modern West*. New York: New York University Press

Sims, L. S. (1998) *The Politics of Fat: Food and Nutrition Policy in America*. London: M.E. Sharpe

Southerland, R. (2002) *"Sewer fitness: cutting the fat"*, *American City and County*, Online. Available HTTP: <http://www.americancityandcounty.com/mag/government_sewer_fitness_cutting/>

Sui, D. Z. (2003) "Musing on the fat city: are obesity and urban forms linked?" *Urban Geography*, 24(1): 75–84

Summerton, J. (ed.) (1994) *Changing Large Technical Systems*. Boulder, CO: Westview Press, 1–24

Twyman, A. S. (2003) "Amid fiscal crunch, street backs health czar", *The Philadelphia Inquirer*, 25 February, 2003, Online. Available HTTP: <http://www.philly.com/mld/philly/living/health/fitness/5254940,htm>

Urry, J. (2003) *Global Complexity*. Cambridge: Polity

Water Environment Federation (1999) "Fat-free sewers: how to prevent fats, oils, and greases from damaging your home and the environment", *Water Environment Federation*, Online. Available HTTP: <http://www.wef.org/publicinfo/FactSheets/fatfree.jhtml>

World Health Organization (1997) *Obesity: Preventing and managing the Global Epidemic: Interim report of a WHO consultation on Obesity*. Geneva: World Health Organization

10 The political ecology of water scarcity

The 1989–1991 Athenian drought

Maria Kaika

INTRODUCTION

This chapter examines the political ecology of the drought that hit Athens between 1989 and 1991. Firstly, the chapter explores the process though which nature became discursively constructed as a source of crisis during the drought period. Secondly, it examines how this particular discursive production of nature became central in building social consensus around a number of "emergency measures". The analysis interprets the drought as the "ferment" that expedited a set of political-economic transformations that expanded both the capital base (in the direction of liberalization of water management) and the resource base (with the construction of a new dam project).

"ABUNDANCE": WATER AS A PUBLIC GOOD AND WATER MANAGEMENT AS A SOCIAL PROJECT

The year 1985 became a landmark in the history of water supply of modern Athens. During that year, the city's main reservoir (the Mornos Reservoir) overflowed. For the first time since it was declared the capital of the modern Greek state (1834), Athens appeared to have more water available than it actually needed to sustain its life and metabolism. The event generated great civic pride and political enthusiasm and was hailed as the turning point in Athens' century old battle against water scarcity (Gerontas and Skouzes 1963; Kalantzopoulos 1964; Kaika 2005). This optimistic period coincided with the restructuring of the legal and institutional framework for water resource management in Greece, that began in earnest in 1980. Law 1068/80 (1980) merged the Hellenic Water Company (*EEY*) and the Athens Sewerage Organization (*OAΠ*) to form a unified water supply and sewerage authority, the Water Supply and Sewerage Company of Athens (*EYΔAΠ*) (Hellenic Republic 1980). *EYΔAΠ*'s sole stockholder was the Greek state and the company operated as a public utility company (*EYΔAΠ* 1995). An even greater breakthrough in the institutional reorganization of water resources management in Greece came in 1987, when the socialist government voted in Law 1739/87 (1987). This new legislation replaced circa 300 laws and decrees that used to form the main institutional framework for water resources management, some of which dated back to the 1900s. It was a bold effort to restructure this fragmented institutional framework by centralizing decision-making and planning. The new law nationalized the management of water resources, declared water a "natural gift", and secured every citizen's access to potable water as his/her "undeniable right". It also recognized the priority of domestic water supply over any other use of water, and annulled all previously existing water rights linked to land property. Moreover, it sanctioned the right of the state to expropriate land, edifices and settlements, as well as to restrain the use of water by individuals or companies (after compensation) if such measures were deemed necessary for the utilization of water resources for the benefit of the general public. Finally, the law created a new administrative framework for the management of water resources, by dividing the country into 14 hydrological departments (Hellenic Republic 1987). This legislation that centralized decision-making and power over water resource planning and allocation, *de facto* nationalized water resources and was part of a broader shift in the philosophy of planning and resource management introduced by the socialist party after it came in power in 1981. This new philosophy aspired to move away from a mechanical and technocratic approach to planning, and to introduce a *socially sensitive* approach to planning and resource management. Planning activities became imbued with socio-political meaning and rhetoric, which often exaggerated what planning was actually capable of achieving. Karydis notes that:

> Never before [1981] had the political and the urban been so organically connected; never before had urban issues been popularized to that extent; never before had the social consensus been granted so generously . . . never before had it been claimed in such an explicit way [through a deterministic environmental scheme], that the application of a new urban regulation would have the capacity to *change* a whole way of life.
>
> (Karydis 1990: 343)

It was within this broader context, that the management of water resources became the subject of new legislation and state regulation. The general optimism over the positive social effects that a national water resources management plan could deliver was reflected in the five-year project plan (1988–1993) for Athens' water supply that was issued by the Ministry of Environment, Physical Planning and Public Works (*ΥΠΕΧΩΔΕ*) in 1988. The document confidently stated that:

> The Mornos reservoir [the city's main reservoir] will continue to cover the city's needs for water . . . while there is also scope for using the existing water supply system and available resources to cover the needs of areas which are not yet connected to the network.
>
> (*ΥΠΕΧΩΔΕ* 1988: 69)[1]

According to the 1998 report, despite the continuous rise in consumption levels, which, by 1989, reached 376 million m³ per year, Athens, with a population of 3.5 million and growing, was marching towards the 1990s with full confidence in the adequacy of its water resources.

"SCARCITY": FROM PUBLIC GOOD TO COMMODITY

Only one year after the publication of the optimistic 1988 report, a severe drought period began that was to last for almost three years (1989–1991). The buoyant optimism that characterized the period before the drought turned on its head and the public discourse on water shifted dramatically: once an abundant resource, a source of national pride and hope, a vehicle for social cohesion, water became discursively constructed as a scarce resource, a source of crisis and conflict, a threat to the life and social cohesion of the "eternal" city and its citizens. The drought became one of the hottest topics and was placed at the centre of media coverage and fervent political and public debates. Great public anxiety was created when the water company decided to disrupt its services as a demand management strategy, and to launch a public awareness campaign with daily announcements counting down the 170 days of water (and life) left for the city.

As a response to the 1989–1991 drought, and after a very intense but remarkably short period of political and scientific debate, the government brought four projects before Parliament in the form of "Emergency Acts" for immediate implementation (*YΠΕΧΩΔΕ* 1990a). The acts stipulated:

1   the implementation of demand-management strategies, including: a new tariff and pricing system, increasing the price of water by up to 300 percent; a public awareness campaign; and the prohibition of heavy water-consuming activities (e.g. car washing, watering gardens, filling swimming pools);
2   the construction of a new dam at Evinos River, which would increase the quantity of water delivered to the Mornos Reservoir;
3   the undertaking of drilling works along the Mornos Reservoir and the banks of Lake Yliki; and
4   the transportation of water (by means of tankers and new pipelines) from Lake Trihonida to the Mornos reservoir.

There was no doubt that the implementation of the proposed measures would bear serious social, economic, political and environmental impacts, and thus broad consultation and political debate was in order. Nevertheless, all four Acts were brought before the Greek Parliament in the form of "Emergency Acts", a procedure which is normally reserved for moments of national crisis, and which allows only for very brief discussion in Parliament before an immediate vote is taken. For example, the water transportation Bill and the relative Bill that modified the 1987 Water Resources Law (1739/87) were voted in by the government within the record time of 16 minutes (Proceedings, PΑΓ Assembly of the Greek Parliament, 21/5/93). The justification given by the government for the treatment of these Acts as "National Emergencies" was that such a procedure was judged to be absolutely necessary in order to "save the city of thirst". Nature was causing an indisputable crisis and there was no time for debating. Immediate urgent action was the only way to overcome the crisis.

Both the necessity and the efficiency of the proposed "emergency measures" came under scrutiny by independent academics and environmental NGOs (Vasilakis and Bourbouras 1992). In fact, the very existence of a crisis of water scarcity was questioned, since even the exact amount of water available in the city's main reservoir was a matter of dispute. Each ministry published its own numbers, defending its own interests and strategies. The estimate provided by the Ministry of Agriculture was 400×10^6 m^3; that of the Ministry of Industry, Energy and Technology was 221×10^6 m^3; while the water company itself estimated the water available at Mornos reservoir at 580×10^6–630×10^6 m^3. (*ΥΠΕΧΩΔΕ* 1990a; Karavitis 1998). Thus, the threat of an imminent water scarcity crisis, the extent of the urgency of the situation, as well as the genuineness of the perilous "170 days of water" were never fully verified, despite the dramatic sensationalization of water politics. However, these debates came after the voting in of the Emergency Acts and did not stop the immediate implementation of most of them.[2]

Still, the invocation of the crisis brought about by a prolonged drought period is not enough to account for the urgency with which the situation was vested, the confusion around water availability and the dramatic shift from discourses of water abundance to discourses of water scarcity and from the depiction of water as a public good to its depiction as a scarce commodity (see Bakker 2000). Indeed, it would not be possible to comprehend and analyze the socio-political dimensions of the Athenian water crisis and the swift implementation of the disputed "Emergency Measures", without looking at the particular socio-political configuration with which the drought period coincided. For the 1989–1991 drought overlapped with one of the most turbulent periods in contemporary Greek politics. An economic and political scandal, in which the socialist government was allegedly involved while at the height of its popular support, led the country to a deep and long political crisis. Three rounds of national elections were carried out within a period of less than two years. The first round of national elections, in June 1989, coincided with the first dry summer period. The elections failed to deliver a majority to any single political party and a "transitional" government was formed to carry the country to the next round of elections. However, the second round of elections (November 1989) equally failed to deliver the parliamentary majority to any single political party. An "ecumenical" government was then formed through an unprecedented alliance between the conservatives and the left. This government stayed in power for a short period with the aim of overseeing a project of national political "catharsis". The third round of elections, carried out in April 1990 in the midst of trials of former socialist ministers, led to a conservative party victory (Chadjipadelis and Zafiropoulos 1994; Zafiropoulos and Chadjipadelis 2001). This chapter will argue that it was the fusion of this political crisis with a natural crisis that facilitated the almost uncontested implementation of the proposed "Emergency Acts", despite their questionable soundness and transparency. In what follows, I shall focus on the implementation of the first two emergency measures (price hikes and a new dam project) in order to foreground the social, political, and economic implications of their swift implementation and the role of the discursive construction of a "crisis" situation as a facilitator for the implementation of a particular neoliberal socio-environmental agenda.

NATURE'S CRISIS AS A FACILITATOR FOR EXPANDING THE CAPITAL BASE

The demand-management strategies that were implemented in Athens as part of the first Emergency Act included a public-awareness campaign, the ban on watering gardens, washing cars and filling swimming pools and, most importantly, the decision (on 8 May 1990) to enforce a *retroactive* increase in water prices, which was to take effect from 1 March 1990. The price per cubic meter paid by domestic consumers increased by between 105 percent and 338 percent and water bills increased by between 40 percent and 140 percent. This increase was brought about through a complex system of rates related to water consumption levels (see Kaika 2003a; Kallis and Coccossis 2003). According to statistical data provided by *EYΔAΠ*, an average Athenian household of four members consumes 15.6 m³ of water per month. Under the old rates, this household would have to pay 885 dr. (€2.59) per month for their water bill, but under the new rates the same bill would amount to 2,187 dr. (€6.41) However, if the same family succeeded in saving 20 percent over its previous month's consumption, they would pay a "reduced" bill of 1,733 dr. (€5.08) per month instead (still higher than the original 885 dr.). The catch to this complex pricing system was that higher-volume consumers were offered a greater price reduction for water savings than lower-volume consumers. For example, a 30 percent saving in water consumption of, say, 15 m³ per month achieved by lower-volume consumers would save them around 132 dr. (€0.38) per m³. However, the same percentage in saving – of say, 120 m³ per month – achieved by a higher-volume consumer would save them 675 dr. (€1.98) per m³.

The alchemy of the new water-pricing system and water-saving incentives was highly contested by the public and by a number of organizations, including the public water company itself, which had suggested only an 18 percent rise in water rates. The committee of the employees of *EYΔAΠ* characterized the pricing policy implemented as "inefficient, socially unjust, profiteering and perplexing" (Newspaper *Eθνος* 10 February 1993). Nevertheless, the complex pricing system did result in an average 20 percent decrease in water demand. Notably, however, the public response to the water-saving campaign and price incentives had a clear social stratification (*KEΠE* 1996). Despite the fact that the new pricing system was designed to give greatest financial incentives to heavy users, it was, in the end, the lower-volume consumers (corresponding to lower-income households and poorer urban areas) who achieved the greatest savings and reduced their consumption by up to 30 percent. Heavier users (corresponding to higher-income areas and houses with gardens and swimming pools) saved very little or nothing at all (*Proceedings* OΘ′ Assembly of the Greek Parliament, 16 February 1993). Thus, after the increase in water prices, the 3 percent highest-volume consumers ended up using 40 percent of the total water supplied (I. Tsaklidis, MP, *Proceedings*, OΘ′ Assembly of the Greek Parliament, 16 February 1993: 4,105; Newspaper *To Bήμα* 19 August 1990). These figures indicate a clear "class" stratification, not only in water consumption, but also in sensitivity expressed as public response to a call for environmental protection and demand-management.

REIFIYNG THE SOCIAL RELATIONS OF PRODUCTION OF POTABLE WATER

The new pricing system, combined with the threat of a drought-induced imminent water scarcity and a strong political rhetoric of crisis, facilitated the construction and public acceptance of water as an economic resource and, finally, as a valuable commodity (Bakker 2000; Kaika 2003a). What is important to note, however, is that the scarce character of water and the increase in its exchange value was attributed to the "natural" character of the resource, rather than to the actual institutional, economic and social organization of a produced commodity. While the "natural = scarce" equation was invoked in order to create public consensus around increasing water prices, the very process that actually makes water a commodity – that is, its *production process* – was suppressed. The social relations of the production and consumption of water remained in the background as if they were not part of the equation of water's availability, distribution and pricing. Ironically, even the president of the Union of Employees of the Water Company failed to address the fact that it is not *nature*, but the *production process* involved in the urbanization of water that lies behind changes in its availability and price. Arguing *against* price hikes and privatization, the Union's President marshalled exactly the same argument as his opponents, namely, the "natural" character of water:

> [M]arket competition cannot be applied to the case of the water company, since the product that this company delivers [i.e. water] *is not produced* as is, for example, electricity.
>
> (T. Zaharopoulos, interview, Newspaper *To Βήμα*
> 24 May 1998; emphasis added)

Perhaps more acutely aware than anybody else of the materiality, labour-power expenditure and social relations involved in water's production, the Union's President still opted to emphasize the "natural" character of water, rather than its production process, as part of his argument against the increase in water's exchange value and against the company's pending privatization. The dominant rhetoric and the *pensée unique* of market liberalization, which depicts anything produced as necessarily and automatically commodified, made the Union's president resort to using "nature" as a lever for his argument in favour of maintaining the sacred human right to water: "Water is a natural good; moreover, it is scarce; therefore, it cannot be privatized" (T. Zaharopoulos, interview, Newspaper *To Βήμα* 24 May 1998).

It becomes clear from the above analysis that the discursive construction of water as a purely natural element and the reification of the social relations of its production strip the discourse over water resource management and allocation from any social and political meaning. Once taken out of the social and political nexus in which they are positioned, the production, consumption and conservation of water can be used to support almost any argument: for or against privatization; for or against the commodification of nature. As a case in this point, it is useful to note that the argument used by the Minister of Environment Planning and Public Works *in favour* of price hikes was identical to the one used by the President of the Trade Union *against* price hikes, namely, the dependence of water availability on nature's whims:

> The sole target of the admittedly high price rises is the reduction of water consumption, since, according to the available data, the existing water reserves are enough to last only until the 2nd of November [1990]. If we achieve a 20% drop in consumption levels . . . water could last until mid-December, by which time it is hoped that it will have rained.
>
> (Newspaper *Τα Νέα*, 9 May 1990)

NATURE'S CRISIS AS A FACILITATOR FOR EXPANDING THE RESOURCE BASE

The decision to construct a dam at Evinos River was the second of the Emergency Measures that was implemented as a response to the drought crisis in Athens. The Evinos River is situated in the Aetoloakarnania region, 250 km away from Athens. The project proposal included the construction of a dam, 100 m in height with a capacity of 130 million m^3, as well as the construction of a 30-km-long channel that would transport water from the dam to the city's main reservoir (Mornos) and feed it from there into the existing network. It was estimated that the project would supplement the Mornos Reservoir with 50–120 million m^3 of water per year and cover the city's water needs until the year 2020 (*ΥΠΕΧΩΔΕ* 1990b). The cost of the project was initially estimated at 55 billion dr. (€160.8 million), and 85 percent of the total investment cost was funded through EU Cohesion Funds.

The project embodied the continuation of modern Fordist development schemes and the associated large-scale transformation of nature's water, a process that started in earnest in Greece with the construction of the Marathon Dam in 1925 (see Kaika 2005) and became an important part of Greece's postwar urban planning and development (Emmanouil 1985; Getimis 1994; Panayiotatou 1990). However, the contemporary rise in environmental awareness means that dam constructions are no longer objects of public veneration, and social consensus in favour of their construction requires more than mere references to "progress" (Roy 1999; Topping 1995). In what follows, I shall examine the mechanism through which public consensus was built around the construction of the Evinos Dam. I shall argue that public acceptance of the project was fabricated through the combination of a threat and a promise: the threat of a pending "natural disaster" and of further increases in water prices, and the promise for development and economic growth that the new dam project would bring.

Building consensus around a crisis posed by "nature"

The main justification behind the speedy implementation of the Evinos Dam project was the need to address the urgent situation posed by a drought crisis. Indeed, the report published by *ΥΠΕΧΩΔΕ* on the implementation of the project stated that

"One of the main planning parameters for the project will be the minimization of construction time" (*ΥΠΕΧΩΔΕ* 1990b: 23). However, the Evinos Dam project proposal was not entirely new. It had been suggested originally as a possible future addition to the plans for Mornos, Athens' most recent (1969–1981) dam project. Since then, it had been awaiting the right conditions to materialize (Kingdon 1984; Nevarez 1996). Indeed, in 1991, the right configurations prevailed: economic (funding available from the EU Cohesion Fund); socio-environmental (drought presented as water scarcity); and sociopolitical (consensus due to crisis and rising water prices combined with a turbulent political period). It was this socio-environmental configuration that permitted the relatively uncontested and rapid implementation of the old project proposal (Perelman 1979; Rutte *et al.* 1987). Although preliminary studies for alternative solutions were submitted to the government by environmental NGOs, academics and private engineering companies, they never received serious attention (Proceedings, IB′ Assembly of the Greek Parliament, 11 May 1990). For example, in 1990, a public awareness campaign, combined with the project to seal the sinkholes of the Lake Yliki Reservoir was proposed in Parliament as a possible alternative solution:

> On 23/1/90, we [the socialist party] asked for a campaign to raise public awareness regarding the irrational use of water, and the implementation of the project for sealing Lake Yliki, which would provide *a definitive* solution to the problem of water supply for both the Greater Athens Metropolitan area and for the irrigation of the Kopaida area.
>
> (Proceedings, IB′ Assembly of the Greek Parliament,
> 11 May 1990; emphasis mine)

Still, all alternative projects that were proposed were rejected. The government judged that the time it would have taken to assess, evaluate and implement these projects would be prohibitively long, given the supposed urgency imposed by the drought (Koutsogiannis *et al.* 1990; Kallergis and Moraiti 1991). An interviewee from the higher ranks of the Ministry of Development established a clear link between the political character of large-scale infrastructure projects and the swift character of the implementation of such projects:

> One problem when carrying out studies on water resource management during a drought period is the high cost and the time-consuming character of the studies themselves, due mainly to lack of primary data. Therefore, decisions for large scale infrastructure projects, which are clearly political, are normally favoured as the quickest and "cheapest" (in the short term) solutions.
>
> (Interview, Ministry of Development,
> Division of Management of Water Resources, 11 September 1996)

Thus, once again, a crisis posed by "natural" causes was the main justification for the implementation of the Evinos project.

Building consensus around promises of development

A second element for building public consensus for the dam project that was as important as the threat fabricated around a "natural" crisis was the political promise given to the construction industry and to the local community for major capital investment and economic growth that the project would bring. Since the end of World War II, state support for the construction industry in Greece has been an efficient and popular way of producing short-term economic growth, mainly because of the economies of scale produced by this industry and the relatively large number of people involved (Leontidou 1990; Filippidis 1990; Giannakourou 2000; Kafkalas 1985; Mantouvalou and Martha 1982; Vaiou *et al.* 2000). During the post-war period, the development of the water supply system of Athens played a pivotal role in supporting urban expansion and land speculation. The urban sprawl that occurred between 1950 and 1970 could not have happened without the expansion of the water supply network, and of the resource base, both of which were either funded or heavily subsidized by the state. Not only the supply network, but also the city's ecological footprint grew dramatically during this period. Water supply was not only a guarantor for urban sanitation; it also became a determining factor for land speculation. By securing water supply for new urban development schemes, the state in effect subsidized private developers and assisted in their land speculation practices. The construction industry maintained this pivotal position in the Greek economy in recent years. As K. Koutlas, vice president and management director of the *PROODEYTIKI* construction company (contractors of the Thisavros dam for the Electricity Company of Greece), put it:

> The Greek construction industry has taken on a strategic role in the economic development of our country, since it is the sector which provides the link between the inflow and the diffusion of European funds in the Greek economy. . . . The implementation of big infrastructure projects is linked to high-value cash flows into the Greek productive activities, which reinvigorate a series of activities in a broad range of economic sectors . . . In order for the above efforts to continue to take place, it is necessary to sustain the regular flow of European funds from the 2nd and the pending 3rd European Support Framework . . . These projects provide jobs for thousands of Greeks and their completion will permit further development of the country's resources . . . and in general the promotion of the image of our country . . . The benefits will be reaped by the country as a whole.
>
> (*TEXNIKA* 131, "Public works in Greece", October 1997: 23)

The Evinos project belongs precisely to the category of projects to which Koutlas refers in his interview – projects that will sustain economic growth, provide jobs and keep the economy ticking. Thus, despite the anticipated negative impact of the project on agriculture, cattle-grazing and natural habitats in the areas through which the river naturally flows, the proposed damming works did not raise as much opposition as one might have anticipated. Despite a militant Greek agricultural sector and a recent history rich in farmers' protests, opposition to this particular project was left almost exclusively to ecological organizations (Modinos 1990, interview, 27 July 1997). Although the political promise to keep the river's flow uninterrupted helped to create consensus among the local population, it was a more important political promise that accounted for the lack of strong civil opposition: the promise to provide employment for thousands of local people at highly attractive wages (700,000 dr. (€2,054) per month, at a time when the minimum wages in Greece were of the order of 100,000 dr. (€293.47) per month (Newspaper *Μακεδονία*, 26 September 1993)). For a region whose people struggle to survive on low incomes (mainly from agriculture) and are often structurally dependent on subsidies provided by the government or the EU, the promise of development coupled with guaranteed employment for at least five years at high wages was a very significant factor in fostering social consent (Baker *et al.* 1994).

This mechanism of consensus-building is also exemplified in an interview given by the president of St Demetrios, one of the communities located in the Evinos River valley. The president produced a long list of problems that the Evinos project would cause for the local people and the local environment: the expropriation of land with low – and, often, delayed – financial compensation; the loss of water for the area; the continuous dynamite explosions and dust local communities would have to endure for several years; and the negative impact of the dam on the area's flora and fauna. Yet he concludes by saying: "We [the local community] would be happy to consent to the implementation of the project, provided that we, as well, got something out of it" (Newspaper *Τα Νέα*, 12 June 1993). This phrase captures the "naturalization" of the clientelist character of political and economic relations involved in the process of implementation of big infrastructure projects. It also indicates how the promise of development becomes part of a consensus-generating mechanism. The response given by A. Karamanlis, the then Minister of Environment, Planning and Public Works, when questioned in Parliament about the estimated environmental and social impact of the dam project, reveals a similar position:

> I don't think that the water taken from Aetoloakarnania [the valley area of the Evinos River] is really depriving the area of this resource. In any case, the projects that are implemented there profit not only half of the population of Greece who live in Athens, but also the inhabitants of that particular area themselves . . . We are not taking water at the expense of anyone. We make use of this water according to the letter of the law, in order to provide domestic water supply. It is our *duty* to do so, dear colleagues . . . [S]hould we have said instead to the Athenians (to all four million of them) "No, we shall not bring water from elsewhere; go away from Athens?" Is this what you would have wanted us to say?
>
> (Proceedings, IB′ Assembly of the Greek Parliament, 21 May 1993: 7,035–7,036)

According to the minister's speech, the inhabitants of Aetoloakarnania had no choice, other than to consent to a project that would generate economic growth in their area. Of course, this kind of growth would be as short-lived as the construction period of the dam itself – the jobs and the "positive effects" would disappear as soon as the dam were completed. Again, however, the promise was made that the short-lived effects of this kind of development would be complemented by further funding (national or EU) that would ensure a continuous flow of projects and continuous local and national development. Similarly, when the mayor of Elatteia, a village in the area of the Evinos River valley, protested in a letter to the Minister of the Environment that the fields of the area had dried out due to the deviation of the area's water towards Athens, the government replied that the matter would be dealt with in due course, through further funding from the Second EU Support Framework (1994–1999) (Proceedings, OΘ′ Assembly of the Greek Parliament, 16 February 1993: 4,112). Thus, the solution to social and environmental problems caused by progress was looked for in promises of further development. The only way to continue to reap positive effects from this kind of development was to ensure a continuous flow of funds, which would allow for a continuous flow of activities. As the management consultant of one of the biggest private construction companies in Greece (*ATTI-KAT*) put it:

> The Greek choice for development is based on immediate and short-term efforts to implement the European Support Framework, by utilizing the funds provided … [T]he Greek construction industry is on the alert to respond to the implementation of the works with a high sense of responsibility.
>
> (*TEXNIKA* 131, October 1997)

In an era of increasing environmental awareness, Greece subscribed to a politics and practice that paid lip service to environmental protection while remaining loyal to a practice of development through big infrastructure projects. Hence, the crisis generated by the drought very soon became constructed as a *challenge* to overcome through the implementation of new projects, generating the oxymoronic positioning of nature both as a potential source of crisis and as the prerequisite for development (Smith 1984). Within this framework, the contradiction between development and environmental protection is considered to be eased away through the right management and the application of remedial technology, or, in this particular case, through reaching even further outside the urban settlement in search of new "mineable" water. This way, the solution to an "environmental" problem moved, in effect, into the domain of expert discourse within the state apparatus (Hajer 1995), and the crisis that was constructed around the drought period worked as a hegemonic tool to justify policies towards further development (Keil and Desfor 1996; Gandy 1997; Swyngedouw 1997; 2004).

CONCLUSIONS: REVERTING TO A PRE-MODERN WATER SUPPLY SYSTEM?

Of all the Emergency Bills that were implemented in 1990, only the increase in water prices had a short-term impact, resulting in a 20 percent cut in water consumption. The immediate results of the other three emergency measures were negligible: only half of the proposed drilling works were eventually carried out, as these were highly contested by local residents; the water transportation project led to great political controversy and to a political scandal before it was finally abandoned; and the Evinos Dam project only became fully operational in 2001, long after the drought period was over. Hence, the only *immediate* positive outcome from the proposed projects was a 20 percent saving in water consumption and an extra yield of a meagre 100,000–200,000 m^3 per day from the drilling works. These quantities fell far short of meeting the originally estimated extra need of almost 1,000,000 m^3 per day, suggesting a discursive, rather than a real imminent threat of water scarcity during the drought period. This argument is also supported by the fact that once all four Emergency Acts had been voted in, and the decision for the price rise was implemented, the discourse about water scarcity both in Parliament and in the media stopped abruptly, as if the mere act of adopting Emergency Measures sufficed to chase the ghost of water scarcity away. Indeed, by 19 August 1990 – only three months after passing the Emergency Acts, and before any concrete results were achieved – the apocalyptic prophecies about an imminent disaster disappeared. A few months after making the public claim that Athens was about to "die of thirst", the Minister of Environment, Planning and Public Works reassured the Athenians that Athens would not be faced with a water problem that year (Newspaper *To Βήμα,* 19 August 1990).

However, the discursive production of nature as a source of crisis, and the hegemonic construction of water as a "naturally scarce" resource, that was performed during the drought period had enduring social, political and environmental implications. The shift in the discourse and practice of water management that was forged during that period gave a decisive blow to the public and social character of the water company and contributed towards turning water in the public consciousness from a public good and a national heritage, into a commodity (Proceedings, IB′ Assembly of the Greek Parliament, 11 May 1990: 180). The public water company was vilified and accused of mismanagement, as neoliberal demand management policies and "price incentives" were marshalled to extricate Athens from the "environmental crisis" (Kaika 2003a). The value that the 1987 law (1739/87) had originally assigned to water as a "national heritage, a common good and a human right" was replaced by the assertion of its exchange value, shortly before the conservative government embarked on its programme for the "liberalization" of public utilities. Ironically, it would be the socialist government that would perform the final act of the water company's privatization when it came back to power, partly responding to EU pressures for "rationalization" of the public sector (Hellenic Republic 1999).

The focus on a "nature-induced" crisis also helped build consensus over further expansion of the resource base, namely the decision for the construction of a new dam at Evinos river. In this respect, the social, political and technical outcome of the 1989–1991 Athenian drought led to developmental practices which are reminiscent of the nineteenth and early twentieth century, a period when the expansion of the resource base was the automatic solution (funding permitting) for recurrent water shortages. The gradual de-nationalization of the "means of consumption" that we witness today is also reminiscent of earlier urban systems, that used to comprise small privately owned local enterprises that served specific areas of the city with varying water quality and prices. The nationalization of water, electricity, gas, etc. networks that took place from the middle of the twentieth century onwards, aimed precisely at rationalizing the provision of these services, making them more efficient and distributing resources in more just ways. The recent re-fragmentation of the city's veins, driven by the need to expand further the basis of the market economy, lies almost at the antipode of this modernizing dream for a highly rational, organized, controlled, urban space and a tamed urban nature (Swyngedouw and Kaika 2000), and leads to the resurgence of what Graham and Marvin (1991; 2001) term a "utility patchwork", an entangled network of regulators and private operators.

Thus, despite paying lip service to the "environmental cause", contemporary market-led institutional restructuring, along with growth-oriented development practices (e.g. dam projects) provide less of a solution to *environmental problems*, and more of a means to sustain the particular *socioenvironmental transformations* that form contemporary cities. Within this context of analysis, the water crisis that Athens faced in the early 1990s and the political decisions that were implemented to administer the crisis, cannot be seen only as the direct outcome of a prolonged dry period alone; rather, they were the outcome of the interaction between the available resources, the transformation of nature by human beings and the economics, politics and culture of water use.

ACKNOWLEDGEMENTS

This chapter is a revised version of the paper "Constructing scarcity and sensationalising water politics: 170 days that shook Athens" that was published in *Antipode*, 35(5): 919–954 (2003).

ENDNOTES

1   Unless otherwise stated, all translations from Greek are the author's own.
2   The first two emergency measures were fully implemented; the third one (drilling works) was only partly implemented; while the fourth one (transportation of water from Lake Trihonida) evoked such public outcry that it had to be abandoned in the end (see Kaika 2003a).

BIBLIOGRAPHY

Baker, S., Milton, K., and Yearly, S. (1994) *Protecting the Periphery: Environmental Policy in Peripheral Regions of the European Union*. Ilford: Frank Cass
Bakker, K. (2000) "Privatizing water: producing scarcity: the Yorkshire drought of 1995", *Economic Geography*, 76 (1): 4–27
Board of Scientific Staff of *EYΔAΠ* (1993) "Announcement of the Board of Scientific Staff of *EYΔAΠ*" *Οικονομικός*, 29 July 1993 (in Greek)

Chadjipadelis, T. and Zafiropoulos, C. (1994) "Electoral changes in Greece during 1981–1990", *Political Geography*, 13 (6): 492–514

Emmanouil, D. (1985) "Land use and housing policy in the case of urban expansion: a framework for analysis and planning for Athens", in Kafkalas, G. (ed.) *Planning: Theory, Institutions, Methodology*. Thessaloniki: Paratiritis (in Greek)

EYΔAΠ (Water Company of Athens) (1995) *Photographic Review*. Athens: Water Company of Athens (in Greek and English)

Filippidis, D. (1990) *Regarding the Greek City: Post-war Trends and Future Perspectives*. Athens: Themelio

Gandy, M. (1997) "The making of a regulatory crisis: restructuring New York City's water supply", *Transactions; Institute of British Geographers*, 22: 338–358

Gerontas, D. and Skouzes, D. (1963) *The Chronicle of Watering Athens*, Athens

Getimis, P. (ed.) (1994) *Urban and Regional Development: Theory, Analysis, and Politics*. Athens: Themelio (in Greek)

Giannakourou, G. (2000) "The institutional framework of urban development in Greece: historical transformations and contemporary demands", in Oikonomou, D. and Petrakos, G. (eds) *The Development of Greek Cities: interdisciplinary approaches for urban analysis and politics*. Volos: University of Thessaly.

Graham, S. and Marvin, S. (1991) "Cities, regions and privatized utilities", *Progress in Planning*, 51 (2): 91–161.

Graham, S. and Marvin, S. (2001) *Splintering Urbanism*. London: Routledge

Hajer, M.A. (1995) *The Politics of Environmental Discourse: Ecological Modernization and the Policy Process*. Oxford: Oxford University Press

Hellenic Republic (1980) Law 1068/80 on the "Establishment of a unified water and sewage company for Athens", *Official Gazette of the Greek State (ΦEK)*, Issue 190 A. Athens: National Press (in Greek)

Hellenic Republic (1987) Law 1739/87 on "Water resources management", *Official Gazette of the Greek State (ΦEK)* Issue 201 A. Athens: National Press (in Greek)

Hellenic Republic (1993) Law 2118 / 1993 on "Measures against water scarcity", *Official Gazette of the Greek State (ΦEK)* Issue 23 A. Athens: National Press (in Greek)

Hellenic Republic (1999) Law 2744/1999 on "Management and operation of the water and sewerage services sector, and for the regulation of the Athens' Water Supply and Sewerage Company (*EYΔAΠ* – A.E.)" Regulation of *EYΔAΠ Official Gazette of the Greek State (ΦEK)* Issue 222 A. Athens: National Press (in Greek)

Kafkalas, G. (ed.) (1985) *Planning: Theory, Institutions, Methodology*. Thessaloniki: Paratiritis (in Greek)

Kaika, M. (2003a) "Constructing scarcity and sensationalising water politics: 170 days that shook Athens", *Antipode*, 35 (5): 919–954

Kaika, M. (2003b) "The Water Framework Directive: a new directive for a changing social, political and economic European framework", *European Planning Studies*, 11(3): 299–316

Kaika, M. (2005) *City of Flows: Nature, Modernity and the City*. New York: Routledge

Kalantzopoulos, T. (1964) *The History of Water Supply of Athens*. Athens: Palamari Kathrogianni and Co. (in Greek)

Kallergis, G. and Moraiti, E. (1991) "The water problem of the Athens Basin", Conference paper, *The Water Problem of Athens*, 26 April 1991: Athens, The National Research Foundation: The Greek Geological Society and the Geo-Technical Chamber of Greece

Kallis, G. and Coccossis, H. (2003) "Managing water for Athens: from the hydraulic to the rational growth paradigm", *European Planning Studies*, 11(3): 245–261

Karavitis, Ch. (1998) "Drought and urban water supplies: the case of metropolitan Athens", *Water Policy* 1: 505–524

Karydis, D. (1990) *Reading Urban Planning: The Social Meaning of Spatial Forms*. Athens: National Technical University of Athens (in Greek)

_____ (1996) *The Socioeconomic Identity of the Water Departments*. Athens: Ministry of Development, Centre for Planning and Economic Research (in Greek)

Keil, R. and Desfor, G. (1996) "Making local environmental policy in Los Angeles" *Cities*, 13(5): 303–313

Kingdon, J.W. (1984) *Agendas, Alternatives and Public Policies*. Boston, MA: Little Brown

Koumparelis, S. G. (1989) *The History of the Water/Sewerage Works of the Capital*. Athens: The Sewerage Company of Athens and The Water Company of Athens (in Greek)

Koutsogiannis, D., Xanthopoulos, Th. and Aftias, E. (1990) *Exploration of Available Scenarios for Watering Greater Athens*, Final Report, Issue 18. Athens: National Technical University of Athens (in Greek)

Leontidou, L. (1990) *The Mediterranean City in Transition: Social Change and Urban Development*. Cambridge: Cambridge University Press

Mantouvalou, M. and Martha, L. (1982) "The economic 'wonder' and its traps", *Anti*, 199: 25–28

Modinos, M. (1990) "The excellent water", *Νέα Οικολογία* 69 (July–August 1990): 16–19 (in Greek)

Nevarez, L. (1996) "Just wait until there's a drought: mediating environmental crises for urban growth", *Antipode*, 28 (3): 246–272.

Panayiotatou, E. (1990) *Themes in Spatial Development*. Athens: Symmetria (in Greek)

Perelman, M. (1979) "Marx, Malthus, and the concept of natural resource scarcity", *Antipode*, 11 (2): 80–92.

Roy, A. (1999) *The Cost of Living*. London: Flamingo.

Rutte, C.-G., Wilke, H.-A., and Messick, D.-M. (1987) "Scarcity or abundance caused by people or the environment as determinants of behavior in the resource dilemma", *Journal of Experimental Social Psychology*, 23 (3): 208–216.

Smith, N. (1984) *Uneven Development: Nature, Capital and the Production of Space*. Oxford: Blackwell.

Swyngedouw, E. (1997) "Power, nature, and the city. The conquest of water and the political ecology of urbanization in Guayaquil, Ecuador: 1880–1990", *Environment and Planning A*, 29 (2): 311–332.

Swyngedouw, E. (2004) *Social Power and the Urbanization of Water: Flows of Power*. Oxford: Oxford University Press.

Swyngedouw, E. and Kaika, M. (2000) "The environment of the city or . . . the urbanization of nature", in Bridge, G. and Watson, S. (eds) *A Companion to the City*. Oxford: Blackwell.

Topping, A.R. (1995) "Ecological roulette – damming the Yangtze", *Foreign Affairs*, 74 (5): 132–146.

Vaiou, D., Mantouvalou, M., and Mavridou, M. (2000) "Postwar Greek planning between theory and chance", in *Planning in Greece 1949–1974 – 2nd conference of the Greek Society for Urban History and Planning*. Volos: University of Thessaly

Vasilakis, K. and Bourbouras, D. (1992) *The Diversion of Evinos River for the Water Supply of Attica: Consequences, Alternatives*. Athens: Greek Ornithological Society (in Greek)

YBET (Ministry of Industry, Research and Technology) (1988) *Law 1839/87 on Water Resource Management: Gaps, Problems and Implementation*. Athens: Water Resource Directorate, *YBET* (in Greek)

ΥΠΕΧΩΔΕ (Ministry of Environment, Physical Planning and Public Works) (1988) *Report and Five-Year Plan for Water Resources Management*, Athens: *ΥΠΕΧΩΔΕ* (in Greek)

ΥΠΕΧΩΔΕ (Ministry of Environment, Physical Planning and Public Works) (1990a) *Study for Increasing Athens' Water Supply to Cover Needs until 2030*. Part 1. Report no 8976701, Athens: *ΥΠΕΧΩΔΕ*, Directorate Δ6 (in Greek)

ΥΠΕΧΩΔΕ (Ministry of Environment, Physical Planning and Public Works) (1990b) *Preliminary Report for the Reinforcement of the Capacity of the Mornos Aqueduct System through the Evinos River Basin*. Report no 8976701, Athens: Division of Public Works, *ΥΠΕΧΩΔΕ* (in Greek)

Zafiropoulos, C. and Chadjipadelis, T. (2001) "The geography of elections, 1985–1993: A principal of component analysis", *Topos* 16: 91–110 (in Greek with an English abstract)

GREEK NEWSPAPER SOURCES ON THE WEB

Εθνος http: //www.ethnos.gr/ (in Greek)
Ελευθεροτυπία http: //www.enet.gr/online/online_p1.jsp (in Greek)
Καθημερινή http: //www.kathimerini.gr/ (in Greek)
  and http: //www.ekathimerini.com/ (in English)
Μακεδονία http: //www.hyper.gr/makthes/archive.html (in Greek)
Οικονομικος Ταχυδρόμος http: //oikonomikos.dolnet.gr/ (in Greek)
Το Βήμα http: //tovima.dolnet.gr/ (in Greek)
Τα Νέα http: //ta-nea.dolnet.gr/ (in Greek)

11 The metabolic processes of capital accumulation in Durban's waterscape

Alex Loftus

People, Planet, Profit.

(Umgeni Water's "Triple Bottom Line"; Umgeni Water 2002)

INTRODUCTION

In this chapter, I argue that the flow of potable water in South Africa's second largest city embodies many of the tensions and contradictions typical of a crisis-prone accumulation process. Developing Swyngedouw's (2004) claim that we are currently witnessing a process of the transformation of local waters into global money, I show the muddled fashion in which this is currently taking place in Durban. My primary focus is on Umgeni Water, the bulk-supplier of water to the city. As a state-owned entity with an aggressive commercial subsidiary, Umgeni Water has found it difficult to survive in the competitive world of water provision. As a result, it has sought to prise open new markets within South Africa and further north in the African continent. The entity's most secure revenues, however, are to be found in Durban. Here, it has imposed large increases (frequently well above the rate of inflation) upon the bulk water tariffs charged to the city. In response, the city has clamped down on the non-payment of bills by local residents and, in the late 1990s, it embarked upon a process of disconnecting debtors with unprecedented aggression. By 2002, one thousand households were being disconnected on a daily basis within eThekwini Municipality[1] for non-payment of bills (Macleod, personal interview, 26 March 2003; Bailey, personal interview, 5 November 2002). I frame these symptoms within David Harvey's writings on the spatio-temporal dynamics of capital accumulation (1982), work on the production of nature (Smith 1984), and more recent writings on urban political ecology (Swyngedouw and Heynen 2003). In particular, I focus on spatial and temporal fixes to Umgeni Water's financial woes and its apparent turn to the colonization of new resources formerly assumed to lie within a communal sphere. The latter may be understood as a process of accumulation by dispossession (Harvey 2003). All these strategies have been turned to in an increasingly hasty fashion, as the bulk-water supplier seems to lurch from one problem to the next. Locally, they have been met by resistance within the municipal administration and at the community level.

Durban's contemporary waterscape

> Durban is a distinctive and remarkable place in which to test propositions about the
> significance of the city and the significance of change in the city.
>
> (Freund and Padayachee 2002: 2)

As Freund and Padayachee argue, Durban is a city that is simultaneously at the
centre and the periphery of a global economy. Its waterscape reflects this ambiguous
and often contradictory position. Post-apartheid, the municipal water provider,
eThekwini Water Services (eTWS), has connected an impressive 100,000 new
households to the water network. From 1998 onwards, however, many of these new
households have been disconnected, as the city tries to balance fiscal austerity,
escalating bulk-water tariffs and the rising anger of community groups. At the same
time, eTWS has provided a universal free basic water allowance to residents,
consisting of 200 litres of "free" water per household per day. Whilst many claim
not to be receiving this water, for some it has been a genuine lifeline. Paradoxically,
however, as the free basic water policy is financed through cross-subsidization,
often large, poor households have found the restructured tariff mechanism
considerably worse. Arrears of over R10,000 are common in townships across the
municipality and flow limiters have been installed in order to restrict the supplies
of debtors to the free basic allowance (for a more detailed discussion of Durban's
water service transformations and the development of the free water allowance, see
Loftus 2004). At the centre of many of these problems, I argue, is the bulk-water
supplier to the city, Umgeni Water.

Durban's bulk-water arrangements

Durban's water supply infrastructure was originally developed through large
municipal projects in the early–mid-twentieth century. As a result, the city's
engineers have been lauded in several publications for their heroic efforts in
harnessing water for the city (see Lynski 1982). From 1984, however, under
pressure from the central government, the city was required to sell its bulk-water
infrastructure to a water board. Originally termed the Umgeni Water Board, this has
now been abridged to Umgeni Water. Later in the chapter, I will go into some of
the early debates around the establishment of this entity. It is important to note that,
from the start, the relationship between municipality and bulk-supplier has been
tense. Durban makes up approximately 85 percent of the entity's bulk-water custom.
The municipality, in short, has provided its reason for existence. The bulk-water
tariff to the city has, however, increased steadily since 1984, with several unprecedented rises between the late 1990s and 2002.

Full of ambiguities, Umgeni Water is now a state-owned entity with an aggressive
commercial subsidiary. It is a curious, part-privatized, public service provider and
(perhaps more bizarrely) a not-for profit entity whose "bottom line" in 2001 was
"People, Planet, Profit". Perhaps because of this confused identity, the organization
has been quite successful at raising finance on private markets, whilst being

singularly unsuccessful at finding profitable outlets for investing this capital. In the mid 1990s, as it became increasingly urgent for Umgeni Water to find investments for the capital it had been able to raise, it sought to exploit new opportunities within the water sector. One such opportunity lay in expanding commercial operations into other parts of the African continent. Another potential outlet for investment lay in what it hoped would be a lucrative, emerging market in the South African rural water sector. Finally, as these options proved to be slightly more insecure than had first been hoped, Umgeni Water turned, in the late 1990s, to the service from which it was best able to guarantee returns: it imposed massive tariff increases on the two municipalities it serves with bulk water – Umsunduzi (formerly Pietermaritzburg) and eThekwini. To begin to understand Umgeni Water's actions requires an understanding of the entity's role in relation to the accumulation process in South Africa. This forms the historical geographical materialist framework to my understanding of the metabolic processes of capital accumulation in Durban's waterscape. In the following section I clarify some of these theoretical foundations.

THE SPATIO-TEMPORAL DYNAMICS OF CAPITAL ACCUMULATION

One of the many original conceptual moves within *Capital* – and one central to making sense of Umgeni Water's panic-stricken actions in the early 1990s – lies in Marx's understanding of capital as value *in motion*. There appears to be no underlying motivation behind the accumulation process except the need to keep capital circulating and profits amassing through constant reinvestment. Marx contrasts the miser with "the more acute capitalist" who constantly throws money back into circulation (1976: 254–225), leading to the never-ending augmentation of exchange value. This need to keep capital in circulation is helpful in interpreting some of the restless dynamism in capitalist society. In several rich theoretical studies, David Harvey has emphasized the important implications this has for the space economy of capitalism (1982; 2003). Thus, the regular bouts of creative destruction so typical of the cityscape in advanced capitalism, and the ceaseless restructuring of scale economies are both consequences of this need for capital to be in continual motion and the contradictory requirement that buildings and infrastructure remain rooted in space. Through his integration of financial considerations and an analysis of "fictitious capital" formation, Harvey is thereby able to show the manner in which the physically rooted city is able to become an "active moment" in the process of capital accumulation. In Neil Smith's pioneering work on the production of nature (1984), it becomes clearer that the dynamics of the accumulation process are also embodied within (and productive of) our socio-natural environment (see also Harvey 1996). By going back to historical materialist basics, Smith relates the labour process to a metabolic fusion of the socio-natural. He is thereby able to theorise the production of scale and uneven development from this socio-natural base. As Swyngedouw has gone on to show, it becomes useful to conceptualize urban political ecology in terms of such metabolic processes or circulatory mechanisms (see Chapter 2).

With these points in mind, in this chapter I seek to show how the financial fortunes of Durban's bulk-water supplier are inextricably interwoven with the shape and form of the city's socio-natural waterscape. Merrifield (2003) distinguishes Harvey's understanding of the city from Castells', in terms of the former's view of the urban environment as an "active moment" in the process of capital accumulation and not merely a container for the reproduction of labour power. Similarly, I wish to emphasize both the crucial function of potable water in the reproduction of life itself, as well as its more recent emergence as an "active moment" in the accumulation of capital. Here, my argument follows that of Swyngedouw (1999; 2004) in several important ways.

The majority of this understanding focuses on the actions of the city's bulk-water supplier. It is through this entity, I argue, that we see the embodiment of the accumulation process in Durban's waterscape. As noted earlier, Umgeni Water has been seeking several routes out of its recent economic difficulties. Eager to find new investments for money borrowed in the early 1990s, the bulk-water supplier has expanded into new territories and channelled funds into new infrastructural projects. These strategies, I shall argue, should be situated within a broader understanding of the spatio-temporal dynamics of capital accumulation. Following Harvey (1982), such strategies might be understood as spatio-temporal fixes. In developing his argument, Harvey traces Marx's understanding of what he terms the "first-cut" theory of crisis. He defines this as "first-cut" because of Marx's failure to integrate all the insights worked out over the first two volumes of *Capital*. It is the task of integrating these insights that Harvey sets himself. Thus, he extends *Capital* in new and fruitful directions by arguing that overaccumulated capital can be switched to a secondary circuit and crises thereby temporarily alleviated (but not resolved). Thus, new profitable investments can be sought over a longer time period (a temporal fix) or by exporting overaccumulated capital to new investments in a different region (a spatial fix) or through a combination of the two (a spatio-temporal fix). If situated within a broader understanding of a crisis of overaccumulation within South Africa, Umgeni Water's actions may be seen as typical spatio-temporal fixes.

On top of this, Umgeni Water has embarked upon an expansion into rural water markets and struggled to raise bulk-water tariffs to the two main municipalities it serves. These latter two strategies, whilst being viewed as simultaneous with the spatio-temporal "fixes", should also be seen as part of a process of accumulation by dispossession (Harvey 2003). Accumulation by dispossession is understood as a broadening of Marx's analysis of primitive accumulation. Whereas this is classically understood as involving the enclosure of common lands and the transformation of such land into profitable investments by an emerging bourgeoisie, recent theorists have argued that such an understanding should not be isolated to an historical past (De Angelis 2002; Perelman 2000; Bonefeld 2002). Instead, enclosures – and the struggles against them – should be seen as continuous, encompassing the privatization of resources formerly considered under common ownership, such as water. Whereas Harvey understands this as intimately linked to the need for capital to seek out new areas for profitable investment, as part of a response to continuing problems of overaccumulation (here we see a form of "internalized" spatio-temporal fix), others have emphasized the importance of the reproduction of social relations in such a process. Thus, just as Marx noticed in Wakefield's modern theory of colonization the importance of the reproduction of relations of production for the lifeblood of capitalism (and thereby made the point that "capital is a social relation') (1976: ch. 33), so others have emphasized renewed efforts on the part of capital to ensure a propertyless proletariat. For De Angelis (2002), the importance of this lies in an understanding of the fact that the possibility for liberation is embodied within the dialectical negation of such a situation. This is fundamentally important for making sense of some of the immanent potentials that lie within Durban's troubled waterscape.

CAPITAL ACCUMULATION IN DURBAN'S WATERSCAPE

The unique positioning of Durban's bulk-water supplier

> His voice had taken on a kind of religious awe; it was as if he had spoken of some untouchable tabernacle which concealed the crouching greedy god to whom they all offered up their flesh, but whom they had never seen.
>
> (Emile Zola, *Germinal*)

> Umgeni Water's operations are financed mainly through the sale of potable water and to a far lesser extent, tariffs charged for the treatment of wastewater. In addition, capital is raised through the issue of stock on the Johannesburg Stock Exchange (JSE). These stocks are consolidated into two highly successful gilt megabonds trading as UG50 and UG55 on the Johannesburg Stock Exchange.
>
> (Umgeni Water 2004)

Umgeni Water occupies a vital position in the supply-chain of water to Durban. Whereas Swyngedouw (2004) makes reference to a process of local waters being transformed into global money, through processes of privatization around the world, this is occurring in Durban as Umgeni Water (ostensibly a parastatal entity) strives to ensure returns for bondholders on the Johannesburg Stock Exchange. Through peering into the dealings of Umgeni Water, we are able to chart something of a journey from the trickling taps of township residents to untouchable tabernacles hiding the crouching, greedy gods of the Johannesburg Stock Exchange. Equipped with some of the theoretical understanding acquired from the previous section, we are able to see some of the metabolic processes comprizing the socio-natural waterscape in Durban.

Umgeni Water has occupied such a crucial position since 1 January 1984. From this time on, the city has been required to purchase its bulk-water supplies from a third party. Originally termed the Umgeni Water Board, this third party was superficially created to act as a mediator between different users of the water abstracted from the Umgeni River. With Durban using 85 percent of this water, the municipality built up an extensive supply infrastructure along the river. Before the water board could take control of bulk water provision, the vast majority of this infrastructure needed to be purchased from the City Council (Lynski 1982). For decades, the municipality had been opposed to the creation of a water board, arguing that it would give a third party unnecessary influence over the cost of water in Durban (Kinmont 1959; Macleod, personal interview, 27 March 2003). A mixture of pride and pragmatics seemed to drive much of this opposition although frequently it was couched in strict financial terms. Thus, early on, the City Engineer recognized that:

> The present standing value of the assets of the Durban Water Undertaking on conservation and purification works must approximate £7,000,000; and it is difficult to visualise any Water Board which can be set up being able to purchase outright these assets. Further unless it does so, any new Board would be unable – for many years – to supply water to its consumers at a rate comparable with that which water is being retailed to the city today.
>
> (Kinmont 1959: 15)

Only two years later, the City was increasingly concerned that it should be granted permission to construct *its own* "urgent" augmentation of the city's bulk supplies. Thus, the same City Engineer was becoming far more strident in his criticism of the direction in which national government policy on water was moving:

> It has been apparent for some time that the government has no intention of allowing Durban to proceed with its own Water Scheme, and that it intends to implement its own proposals, whereby Durban will be supplied with water from the Umgeni River at the entire discretion of the Department of Water Affairs, and at prices which will be decided unilaterally by the responsible Minister.
>
> (Kinmont 1961: 5)

Amidst the rising anger within the municipality, one of the recommendations of Kinmont's report was therefore to "protest to the highest level". However, in spite of the City's protests, the Umgeni Water Board was formed on 14 June 1974. Originally only serving the Pietermaritzburg and Midland areas, by 1984 it was supplying water to Durban City Council.

In retrospect, it seems quite clear that the reasoning behind the creation of the water board was political. Since its creation, it has served to exacerbate serious tensions in the structuring of Durban's water provisioning, as well as ensuring that tariffs are up to 30 percent higher than they might be without it (Macleod, personal interview, 26 March 2003).[2] The political motive behind its formation is somewhat bizarre but worth relating. With the national government eager to ensure that greater legitimacy was afforded to its apartheid policy of creating separate Bantustans for Africans, the international status of these borders needed to be emphasized. To the national administration in Pretoria, it seemed that KwaZulu's potential future status would have been undermined if it was to be supplied with water from what was only a municipality. Instead, a third party, with the authority to sell water to separate states, would need to be created (Macleod, personal interview, 26 March 2003). As Kinmont had recognized 25 years prior to the handover of Durban's bulk supply operations, the creation of the water board really did have little, if anything, to do with the efficient management of water supply in the region. The current head of eThekwini Water Services, Neil Macleod argues that the lack of control exercised by the National Party in Durban exacerbated many of the tensions between the national and the local government. Although civic pride is often mixed with family pride in Macleod's case (it was his father who finally had to cede control of bulk supply operations), he argues that a secondary motive for the creation of the water board was a "way of stabbing the municipality in the back because it wasn't National Party controlled" (Macleod, personal interview, 26 March 2003).

In the midst of these tensions and struggles for control over bulk supplies, Durban fought hard for the highest price to be paid for the infrastructure it would eventually have to hand over. Thus, after having paid considerable amounts for the purchase of Pietermaritzburg's infrastructure, the water board reached an agreement in 1982 with Durban City Council regarding the acquisition of the Nagle and Shongweni dams.[3] The amount paid was initially set at R203 million, later rising to R274 million (Umgeni Water 1982). It is difficult to assess what a "market" price for these dams might have been, but with their respective ages being 32 years and 55 years (the Shongweni dam was "pensioned off" soon after it was purchased), it can be assumed that the amount paid was absurdly high and well above what the infrastructure would have been worth. Macleod now states (with a wry chuckle) that the national government's original valuation had been put at R14 million for the two dams (Macleod, personal interview, 26 March 2003). Durban, it seemed, had been able to salvage a little pride, even if, as Macleod notes in retrospect, "This drove an even greater wedge between the municipality and the central state" (ibid.).

Very rapidly, as Kinmont had predicted, the water board descended into serious debt. As had also been predicted, Umgeni Water was best able to recover this debt through the sale of water to the municipalities. Inevitably, Durban and Pietermaritzburg's tariffs rose rapidly. Thus, the cost of bulk supplies rose from 8.9 cents/kl at the time of the handover to a peak of R2.29/kl, before falling more recently to R2.14/kl. Much of this cost cannot, it seems, be adequately accounted for. As Macleod states:

> At Durban Heights and Wiggins [purification works owned and managed by Umgeni Water], 75 percent of water is purified in these two works, the cost of the purification comes to about 78c/kl. The extra 25 percent of Umgeni's work pushes the cost of water up to R2.29 meaning that 25 percent of their business triples the costs of the operation. Something is going terribly wrong.
> (Macleod, personal interview, 26 March 2003)

Clearly, things had gone wrong from the moment Umgeni Water was formed, and from the start, without clear government backing, it would have to fight hard for survival. Within the last decade, however, the situation has worsened. In part, this is due to the fact that, post-apartheid, new freedoms have been granted to water boards in order to allow them to compete more freely (DWAF 1997: ch. VI). The public service role has been downplayed, as water boards are now increasingly expected to bid competitively for contracts and raise more money from the private markets. In short, the water boards have been encouraged to commercialize. Thus, Umgeni has sought to respond in a far more entrepreneurial fashion to opportunities within the water sector. In its own corporate jargon:

> More recently Umgeni Water has restructured itself to better position it to meet the challenges of becoming a globally-competitive business-based organization that is responsible and adaptive to the requirements of a globalized economy and intense developmental need. Notable achievements have been the appoint-ment of a new and dynamic executive management team, and the development of a strategic plan that charts the course for the company to achieve its vision of being the number one water utility in the developing world.
>
> (Umgeni Water 2004)

It is worth charting some of these new opportunities being sought by the "globally-competitive business-based organization".

Umgeni Water's early ventures into the rural water services market

> [The 1997 Water Services Act] has created new opportunities for water boards. Water boards can become more commercially focused with a view to increasing income whilst at the same time creating increased employment opportunities. The Government welcomes water boards taking up secondary activities on a commercial basis. Umgeni Water is clearly a leader in this.
>
> (Kasrils 2000)

Post-apartheid, the new South African government was faced with the enormous challenge of providing potable water to all of its citizens. Preoccupied with haste – and to some extent the dominant global ideology (see Peet 2002) – it chose to subcontract much of the construction of new rural water projects. Under the Community Water Supply and Sanitation programme, private companies have been able to submit proposals for rural schemes and tender competitively for the construction of these (see Bakker and Hemson 2000; Hemson 2003b). The successful bidder then goes on to construct the scheme, before taking over its running for a set period. At the end of this, the water supply scheme is then handed back to the local Water Services Authority. In most cases, this authority is either the local or the district municipality. Somewhat depressingly, many of the projects have proved unsustainable and many communities have been left with defunct water infrastructure and an enforced return to alternative (and often dangerously polluted) water sources. In other cases, charges to rural residents have been set at an incomprehensibly high level. With prepayment systems being favoured on many projects, large numbers of households have found themselves simply unable to afford to pay for the new water supplies (Hemson 2003; Bond 2002; Thompson 2003). In the meantime, the national government was able to boast the 7, 8 or 9 millionth new consumer to be receiving clean water (see Kasrils 2003). For much of the 1990s, the rural water sector became a sphere in which many new companies cut their teeth in trying to make money out of the rural poor.

Umgeni Water, though not a private company, was one of those able to cash in on the small boom in government funds, which, at the time, were being soaked up by new entrepreneurs. Between 1996 and 2002, the organization was able to claim R104 million from the government in order to implement rural projects in five district municipalities in KwaZulu Natal (Umgeni Water 2002: 15). By June 2002, it was supplying water to over 27,000 metered households in the province. This was after the "hand over" of a further 15,000 customers back to eThekwini municipality (Umgeni Water 2002).

In 1997, Umgeni Water Services, a commercial subsidiary was established under the new freedoms opened up in the 1997 Water Services Act. Interestingly, the rural projects were not included under this subsidiary. Instead, these were considered to be part of Umgeni Water's primary (and therefore non-commercial) activities. On the surface, however, any commercial activities undertaken by the organization are supposed to be "ring-fenced" within the subsidiary. Thus, if these projects were considered commercial ventures, Durban's bulk-water tariffs were not to finance them. This point is important, because, from the start, Umgeni Water's rural projects tended to run seriously over budget and were often deeply inefficient. Several of their projects were also criticized for being over-engineered with little chance of ever being able to recover the costs laid out in them. The enormously costly Vulindlela water project located outside Pietermaritzburg is a case in point (Hemson, personal communication). For some in Umgeni Water, it is a sign of the organization's engineering prowess (Lusignea, personal interview, 11 December 2002), for others it is a costly white elephant. In some ways, this criticism might be considered unfair – surely rural residents are entitled to the same level of service as urban dwellers? However, when one considers that an individual network connection can cost as much as R80,000 under Umgeni Water and as little as R5,000 under eThekwini Water Services (Bailey, personal interview, 5 November 2002), the apparent inefficiencies in such projects become all the more stark. As I discuss later in this chapter, the handover of these inefficient projects back to local municipalities has generated a whole new set of tensions.

Whilst responsibility for such rural schemes remained Umgeni Water's, further debts began to mount. Desperate to see such projects running more efficiently, the entity became eager to ensure that residents were billed for the true running costs of individual schemes. Again, to a limited extent there was not necessarily anything particularly new in this. From the mid 1980s, as a water services provider in parts of KwaZulu, Umgeni Water had made steps to ensure that water would be provided for residents of informal settlements through a "kiosk system". Later, however, in 2002, such attempts to ensure payment for all costs began to reach new, far crueller heights. Thus, when Umgeni Water learnt of the vandalism of water meters in

Ntembeni, a settlement in Inadi (on the outskirts of Pietermaritzburg), managers within the organization presumed certain households were trying to subvert the payment system. Whether anyone in the community was responsible for the destruction or not, the entire settlement had its water supply disconnected. With the community lacking a clean water supply for a month, the South African Human Rights Commission initiated court proceedings against Umgeni Water (Mkhulise 2002).

Such aggressive measures to ensure full cost-recovery are often seen as a clear manifestation of a strategy of accumulation by dispossession (Harvey 2003; see also McDonald and Ruiters 2004). In the case of Ntembeni, we see the messy way in which such a process has developed. Thus, with the state opening up new opportunities for water boards to be able to bid for rural water contracts, Umgeni Water was able to provide water services to rural communities previously lacking any clean drinking water. This is a clear step forward for such communities and it would seem to matter little if this water is provided by a local municipality, a private company or a commercialized water board as long as people do retain access. In failing to make a return on its investments, however, Umgeni Water then embarked on a more overt strategy of dispossessing a community of a vital resource on which it depends but for which some are unable to pay. Thus, we do not see a dramatic transfer of the means of existence in the manner in which the Highland clearances of the nineteenth century took place (Marx 1976: ch. 27). Instead we see an encroaching process, driven less by aggressive state-backed appropriation and more by the gradual integration of water into capitalist relations of production. The state, therefore, plays an ambiguous role in this,[4] helping to prise open the rural water sector to capital (and thereby driving through the divorce of the majority from their means of existence), whilst also seeking to ensure that more and more citizens are connected to the water network. Likewise, the state has actively publicized its free basic water policy, and yet has refused to intervene when organizations such as Umgeni Water disconnect residents for non-payment (see Kasrils cited in Bond 2002: 264). Ultimately, this shows the process of struggle involved in the accumulation process. It should not be viewed in terms of well-worked out, smooth running laws, but rather a process struggled over and contested – often through such simple acts as the vandalizing of a water meter. These struggles become crystallized in the state (albeit not in a direct manner) (Poulantzas 1978), whilst, at times, the state serves as something of a mask through which such struggles are hidden from view (Abrams 1988).

In spite of Umgeni Water's apparent strategy of dispossessing people of water for non-payment in order to ensure healthier profit rates (or perhaps partly because of it), it soon began to look increasingly unlikely that the organization would ever break even with its rural water projects. The bulk-water side of operations was already heavily subsidizing the losses being made over the rural projects. Summarizing this cross-subsidization, the 2002 annual report notes that:

> It is the costs relating to these rural water schemes that have resulted in the 19.5% tariff (reduced from 22.3% due to Umgeni Water cost-cutting and

efficiency savings achievements) increase for the 2001/2002 year. Without the cost burden of these schemes, Umgeni Water would have been able to pass an increase of 8%.

(Umgeni Water 2002: 29)

Thus, it became increasingly urgent for Umgeni Water to offload the inefficient projects along with their debts. This generated a new round of tensions with eThekwini Municipality (*News 24* 2001). In a curious reversal of the position 20 years earlier, the municipality thus found itself fighting hard to ensure it did not have costly infrastructure (and its associated debts) foisted upon it (Bailey, personal interview, 14 February 2003). Much of this battle was reflected in future debates over what constituted a permissible tariff increase.

The outlines of Umgeni Water's problems

Perhaps the core problem in all this lay in Umgeni Water's skill at raising finance but its inability to find good profitable outlets in which to sink the money it was able to borrow. This, I would argue, is related to a broader crisis of overaccumulation within the South African economy (see Bond 2001). Thus, in the early 1990s, Umgeni had raised about R2.5 billion in the bond market (the UG50 and UG 55 gilt megabonds referred to previously). It then became increasingly urgent for it to find profitable outlets for the money raised here. At first, the rural projects seemed to provide just such fertile terrain. With the board of Umgeni Water increasingly comprised of staunch ANC loyalists,[5] it was possible that there was an assumption amongst the board that the national government would be quite supportive in providing subsidies to such new projects. Certainly, in this period of the 1990s, the relationship between the board of Umgeni Water and the national government was a highly cooperative one. Rumours abound that the award of the Vulindlela water supply contract – a Presidential Lead Project – was aided by such close links. When the rural projects started to falter, however, and when it looked more like they would be loss-making, rather than profit-making ventures, Umgeni Water "had R1-billion in surplus cash, which it needed to invest profitably in order to make its long-term debt repayments and mitigate its losses" (*Mail and Guardian* 2002a).

Capital, after all, as I discussed in the first section of this chapter, is value in motion. A danger arises if profitable outlets cannot be found and a glut builds up in what should be an ever-augmenting flow of capital. All too often, it seems that Umgeni Water had simply run out of useful outlets in which to sink its capital. Above all, they required new areas to be prised open for profitable investment. Whilst prising open new aspects of the socio-natural environment through the market in rural water services had proven deeply problematic, managers within the organization began to feel that, in future, its survival would be better guaranteed in competing for management service contracts for district municipalities.

Through assisting municipalities it can partner with similar organizations and offer management contracts to municipalities. District Municipalities could then hire an Umgeni Company . . . The only way Umgeni will survive is if it takes on public service provision in a commercial way.

(Cummings, personal interview, 11 December 2002)

Whilst it seems this might be Umgeni Water's best means of survival, the market for such services is by no means guaranteed within the borders of South Africa. Another possible outlet for capital is some kind of a spatial fix to local crises (Harvey 1982: 417). Thus, the possibility remained that Umgeni Water could extend its frontiers into other regions of the world.

Local waters, hemispheric ambitions

Africa is a strategic market in which significant inroads are being made . . . In support of the African Renaissance and specifically the New Partnership for African Development (NEPAD) . . . The organization now has sufficient expertise under its belt to increase its business in the area.

(Umgeni Water 2002: 4, 12)

By 2002, Umgeni Water had signed a three-year management contract with the city of Port Harcourt in Nigeria. Although originally only establishing operations in the state capital of Port Harcourt, Umgeni was quite clear about its ambition to bid for the much larger Lagos contract (see Ikeh 2002). On top of this, it was "engaged in projects and/or responding to opportunities in Algeria, Botswana, Ethiopia, Ghana, Lesotho, Malawi and Rwanda" (Umgeni Water 2002: 12). It is in this regard that a proud South African Water Minister praised the organization for working "with other African countries in bringing sustainable water services to the continent, and in some way making the African Renaissance vision of President Thabo Mbeki a reality" (Kasrils 2000).

Bond (2004), amongst others, has situated both NEPAD and the African Renaissance vision of Mbeki within what he sees as a sub-imperial strategy on the part of the South African government. Certainly, in the case of Umgeni Water, its ventures northwards must be seen as part of a search for profitability, through the expansion of its operations territorially. For many within the company, the sense of providing services to less advantaged regions will also be a factor, but at the current juncture, profitability is more urgent. It is quite simply, in one manager's words, a question of the organization's survival.

The relations between Umgeni Water's northward expansion and South African support for the continent-wide agreement embodied in NEPAD should not be seen in unidirectional terms. The South African government did not lobby so vociferously for NEPAD merely to ensure domestic capital expand northwards. Nor did Umgeni Water simply respond to the new incentives opened up by the continent-wide agreement. Instead, I would argue, NEPAD and Umgeni's search for profitability overseas are mutually constitutive. Harvey's work on the spatio-temporal dynamics of capital accumulation is, again, helpful, in this regard. In expanding upon the ideas already discussed, Harvey argues that we can distinguish between what he terms the capitalist logic of imperialism and a territorial logic. Thus, "[t]he capitalistic (as opposed to territorial) logic of imperialism has, I argue, to be understood against this background of seeking out "spatio-temporal fixes" to the capital surplus problem" (Harvey 2003: 89). He then continues by stating how he will "try to keep the dialectical relationship between the politics of state and empire on the one hand and the molecular movements of capital accumulation in space and time on the other, firmly at the centre of the argument" (ibid.). Thus, the barriers Umgeni Water comes up against in the accumulation process are shaped, to some extent, by state policy and the historically determined *territorial* logic. In seeking to overcome these barriers, individual capitals such as Umgeni Water can seek to challenge state policy. If successful, new conditions for accumulation are opened up and a new territorial strategy serves to shape the capitalist logic once more. Continually, in the case of Umgeni Water, we see such productive tensions shaping one another through their mutual interaction.

Hemispheric ambitions, local crises

Once again, however, the perceived potential in the Port Harcourt investment failed to materialize. The overcoming of what had been thought to be a barrier to profitability failed to produce the results hoped for. As the Nigerian municipality failed to make quarterly payments (*Mail and Guardian* 2002b), Umgeni Water pulled out. This generated a further loss for the organization of R14 million (Parliamentary Monitoring Group 2003). Prior to this, corporate scandals had engulfed the utility as it was accused of illegally subcontracting its debt management to a company set up by a former senior manager. Then, in 2001, the Chief Executive Officer was forced to resign amidst accusations that he had bugged the offices of board members (Zondi 2002). As Harvey (2003) notes, corporate scandals, such as the Enron debacle in the States, can often be seen as a symptom of deeper structural crises. This, I would argue, is clearly the case with Umgeni Water. Thus, in the midst of serious negative publicity, the entity still had to find somewhere to be able to invest its excess capital and, above all, somewhere to be able to garner the necessary profits to keep up with its debt repayments.

The most reliable source of income, indeed "the core of Umgeni's business operations" had, of course, always been from bulk-water sales to Durban and Pietermaritzburg. Perhaps then, it is not surprising that it was to this branch of the business that Umgeni was to turn. First, it sank some of the mounting "glut" of capital into expanding bulk supplies. Such longer-term investments allow what might be considered to be a temporal displacement of some of Umgeni's problems (see Harvey 1982). Indeed dam projects (profitable investments lasting for several decades) are one of the most frequently cited examples of such a temporal fix. A problem, however, remained: with falling demand from its main customer, Durban, there was little reason to invest in such infrastructure. In spite of this, a large inter-basin transfer scheme is in the process of construction from the Mooi to the Umgeni River in order to ensure that the two municipalities' supplies can be guaranteed over a longer period than ever before. Interestingly, the Department for Water Affairs and

Forestry (DWAF) is the body responsible for financing this expansion. However, with Umgeni Water in a far better position to raise finance quickly (to the extent that it is a little awash with such finance), the DWAF was able to secure a loan from Umgeni Water (Lusignea, personal interview, 11 December 2002). Indirectly, Umgeni Water bondholders were lending money to the government for the construction of a project from which Umgeni was set to benefit. The state was thereby able to provide something of an outlet for overaccumulated capital, whilst also ensuring that private bondholders would remain beneficiaries. Here, once again, a symbiotic or mutually constitutive relationship between state institutions and capital can be seen to have been reinforced through each one's response to a broader crisis.

Such an expansion was authorized after Umgeni Water's 1997 projections had indicated a steady increase in Durban's demand for water. At the same time, Durban Metro Water Services[6] had produced a separate forecast, suggesting that demand would remain constant for the next five years. Durban's estimate was far closer to what actually amounted to a fall in demand, but the plans for infrastructure development went ahead according to Umgeni's forecast (Gilham, personal interview, 11 December 2002). The most recent estimates are for ten years of flat demand, suggesting that the new infrastructure will remain unnecessary for at least the near future. In spite of this and in spite of the acknowledged difficulties in forecasting demand, the DWAF concurred with Umgeni Water's estimates of rising demand, before borrowing money from them and beginning construction of a vast augmentation scheme. On top of this, a vast R200 million project known as the Western Bypass Aqueduct has been mooted which would carry water directly from the Midmar to the Inanda dam. For the moment, however, this will remain a distant possibility, pending an increase in demand for water in Durban (Bailey, personal interview, 29 November 2002; Gilham, personal interview, 11 December 2002). The cost of constructing this would fall largely on Umgeni Water. Once again, however, it is the kind of long-term investment that, if allied with an increase in either water sales or the water tariff to Durban, would help to pull the bulk supplier out of its current difficulties.

The question of increasing the tariff has, over the last few years, been one of the most contentious issues over which the bulk supplier and the municipality have confronted one another. Bulk water-charges to Umsunduzi and eThekwini municipalities have been through several large increases in the past few years. Thus, in 2000, as financial difficulties worsened, Umgeni Water announced a bulk-water tariff increase to Durban and Pietermaritzburg of 13 percent. Then, in 2001, its increase was set at an even larger 22.3 percent. (Inflation over the period averaged between 7 percent and 8 percent.) When passed on to consumers in Durban, the average tariff increase would amount to 28.3 percent (Mhlange 2001; Bisetty 2001). Perhaps unsurprisingly, both Durban and Pietermaritzburg rebelled against the increases, arguing that they were both unfair and too high for them to cope with. As was quoted previously, over the same period that Umgeni Water was requesting higher charges, it had made significant profits from bulk-water supplies. Its problems lay not in the primary activity of supplying water to municipalities but in the debt it had acquired and the losses it was making in its secondary activities.

In this instance, the pressure from the municipalities succeeded in forcing Umgeni Water to limit its tariff increases to 19.5 percent and then to keep future increases fixed to the rate of inflation. Umgeni Water's wings appear to have been clipped as it is limited to much smaller tariff increases in future. However, the long-term effect of Umgeni's actions is a transformation of the terrain on which the municipal water provider in Durban is acting. The result for many local residents has been a far more severe separation from their means of existence through eThekwini Municipality's vicious disconnection programme. As the limits to various spatio-temporal fixes have been reached, Umgeni Water has turned to a classic, if indirect, strategy of accumulating through dispossession.

Throughout this, as I hope to have shown, the state's role has been at times symbiotic, whilst at others, antithetical. In the case of the raised tariffs to Durban, yet another line of friction emerged when a conflict developed between the local and the national states. As the national state (in both its apartheid and its post-apartheid incarnations) had been the original source of the fractious relationship between bulk supplier and municipality, its role had often clashed with the needs and interests of the local state. In the past, such tensions were seen as rooted in the bitter resentment between a National Party controlled central state and a potential (although often hypocritical) liberal opposition in Durban. Now, however, the conflict revolved more around the national state's opening up of opportunities for capital that were in direct opposition to the need of local constituents. Interestingly, part of the subsequent pressure on the municipality to resist these increases came from the threat (whether real or not) that a large number of Durban's textile manufacturers would almost instantly relocate if the increase was imposed (Butler, personal interview, 14 April 2003). Links between eTWS and the Chamber of Commerce had traditionally not been confrontational and, in this instance, they seemed able to form an alliance against Umgeni Water (an alliance which appeared not for the direct benefit of the poorest of the city, it should be noted). Capital can clearly not be considered as an undifferentiated bloc as we see through the conflicting accumulation strategies of Umgeni Water and *local* capitalists in Durban.

I noted at the start of this, however, that accumulation by dispossession, involved another crucial aspect, that of separating the mass of the population from the means through which they might be able to collectively reproduce. In this lies a final ambiguity. Whereas two tiers of state appear to play a role that should be seen as mutually shaped by struggles within the accumulation process, both also seemed to intervene to support a free water policy struggled for by groups actively opposed to further capitalist involvement in the water sector. On the surface this has a profoundly social-democratic flavour. Beneath, however, we may see the underhand fashion through which local capitalists have been able to shift the burden of redistribution onto individual poor households and away from a wider tax that might impinge on profit rates. Accumulation by dispossession continues through the rational mechanization of water supplies and the fencing in of supplies to individual households.

CONCLUSION

All this would seem to imply that, over the last decade, the political ecology of Durban's waterscape has increasingly come to embody the contradictory tendencies of a capitalist system of accumulation. The local waters of the city constitute a sphere in which a commercialized state entity has attempted to ensure its profitability, through fencing in something formerly considered to reside outside of capital's orbit. Simultaneously, this entity has tried to expand its operations throughout the African continent – but failed dismally. Always something of an odd entity, Umgeni Water was originally created by the South African state, in order to give added legitimacy to its creation of separate Bantustans. As a democratic government replaced its racist predecessor in 1994, the water board had to find a place for itself in a radically changed world. Competitive pressures would mean that it had to drastically restructure and find new outlets in which to make profitable investments. In the 1990s, its ability to raise cheap finance became an acute problem as debts began to mount and the availability of good investment opportunities seemed limited. Fortunately, the government passed legislation giving water boards the freedom to engage in commercial activities. Umgeni Water rose to the challenge but, at the same time, proved that it was a somewhat uncompetitive water services provider. On the one hand it sought to exploit new opportunities within the domestic South African market and, on the other, it sought to expand deeper into the African continent. When these strategies failed, it returned to its most secure asset, its bulk-water sales to Durban. Whilst investing some capital in new bulk-water infrastructure that might provide longer-term possibilities for profitable investments, the organization was also able to raise tariffs to the two municipalities it serves. The result, in Durban, was much greater emphasis on recovering the full costs of supplying water to residents and an aggressive policy of disconnecting those who were unable or unwilling to pay. Finally, as tariffs reached unmanageable levels, eThekwini Municipality rebelled. Umgeni Water's wings were clipped in subsequent negotiations and future increases will (so its bosses promise) be kept to what the municipalities consider to be more reasonable levels.

On the one hand, the opening up of our socio-natural environment to the accumulation process poses deep problems. The struggles of those fighting against disconnections are ample testimony to the need for the democratization of socio-natural processes. On the other hand, through analyzing the crisis-prone path to which the accumulation process tends, we have seen the actions of capital to be a sign of weakness and not strength. The fractious relationships between capitals and the sometimes-fraught relationships with tiers of the state show further signs of weakness. Rich potentials therefore lie in exploring an immanent (and materialist) critique of the metabolic processes that shape this waterscape.

NOTES

1   eThekwini Municipality is the name given to the administrative area of which Durban is considered a part. I use the terms interchangeably although strictly the Durban Metropolitan Area ceased to exist with the creation of the former.

2   The figure comes from an oft-cited Halcrow report on bulk water provision to Durban.
3   These two dams – constructed by the city in the 1920s and 1950s – comprised the main bulk water infrastructure for the municipality until they were superseded by the Midmar dam (constructed by the central government's Department for Water Affairs in the 1960s).
4   To some extent, again, this might be contrasted with Marx's detailing of the various laws passed by the Crown against vagabondage (1976: chapter 28).
5   To take one well-known example, the chair of the board, Omar Latiff, is the former ANC mayor of Pietermaritzburg.
6   The name previously given to eThekwini Water Services.

BIBLIOGRAPHY

Abrams, P. (1988) "Notes on the difficulty of studying the state (1977)", *Journal of Historical Sociology*, 1(1): 58–89

Bakker, K.J. and Hemson, D. (2000) "Privatising water: BoTT and hydropolitics in the new South Africa", *South African Geographical Journal*, 82(1): 3–12

Bisetty, V. (2001) "Umgeni Water seeks government bail out", *The Mercury*, 13 June 2001

Bond, P. (2001) *Against Global Apartheid: South Africa Meets the World Bank, IMF and international finance*. Cape Town: UCT Press

Bond, P. (2002) *Unsustainable South Africa*. London: Merlin Press

Bond, P. (2004) "Bankrupt Africa: imperialism, sub-imperialism and the politics of finance', *Historical Materialism*, 12(4): 145–172

Bonefeld, W. (2002) "The permanence of primitive accumulation: commodity fetishism and the social constitution", in *The Commoner*. Online. Available HTTP: <http://www.commoner.org.uk/02bonefeld.pdf > (Accessed: 26 January 2005)

De Angelis, M. (2002) "Marx and primitive accumulation: the continuous character of capital's enclosures", in *The Commoner*. Online. Available HTTP: <http://www.commoner. org.uk/02deangelis.pdf> (Accessed 26 January 2005)

DWAF (Department of Water Affairs and Forestry) (1997) *Water Services Act*. Pretoria

Freund, B and V. Padayachee (2002) *(D)urban Vortex: South African City in Transition*. Pietermaritzburg: Natal University Press

Harvey, D. (1982) *The Limits to Capital*. Chicago, IL: University of Chicago Press

Harvey, D. (1996) *Justice, Nature and the Geography of Difference*. Oxford: Basil Blackwell

Harvey, D. (2003) *The New Imperialism*. Oxford: Oxford University Press

Hemson, D. (2003) *The Sustainability of Community Water Projects in KwaZulu Natal*, report prepared for Department of Water Affairs and Forestry

Ikeh, G. (2002) "Umgeni Water eyes Nigerian water project", *Independent Foreign News Service*, 29 January 2002

Kasrils, R. (2000) "Draft Speech for Use by Mr. Ronnie Kasrils, MP, Minister of Water Affairs and Forestry for the Launch of the Commercial Sector of Umgeni Water: International Convention Centre, Durban, 10 November 2000"

Kasrils, R. (2003) "Address by the Minister of Water Affairs and Forestry, Mr Ronnie Kasrils MP, Budget Vote No. 34: Water Affairs and Forestry, 6 June 2003"

Kinmont, A. (1959) *Report on the Water Supply of Durban*. Durban: Durban City Council

Kinmont, A. (1961) *Umgeni Water Augmentation Scheme: A Report by the City Engineer*. Durban: Durban City Council

Loftus, A.J. (2004) "'Free Water' as commodity: the paradoxes of Durban's water service transformations", in D.A. McDonald and G. Ruiters (eds), *The Age of Commodity: Water Privatization in Southern Africa*. London: Earthscan

Lynski, R. (1982) *They Built a City: Durban City Engineer's Department 1882–1982*. Natal: Concept Communications Ltd

McDonald, D.A. and G. Ruiters (2004), *The Age of Commodity: Water Privatization in Southern Africa*. London: Earthscan

Mail and Guardian (2002a) "Water report highly critical of Director General", *Mail and Guardian*, 16 August 2002

Mail and Guardian (2002b) "Umgeni boss in hot water again", *Mail and Guardian*, 23 August 2002

Marx, K. (1976) *Capital Vol. I*. London: Penguin Classics

Merrifield, A. (2003) *Metromarxism: A Marxist Tale of the City*. London: Routledge

Mhlange, E. (2001) "Durban hit by 28% rise in price of water", *The Mercury*, 29 May 2001

Mkhulise, B.A. (2002) "Denying people's rights", *The Witness*, 22 August 2002

News 24 (2001) "Water price hike under fire", 11 June 2001

Parliamentary Monitoring Group (2003) "Water Affairs and Forestry Portfolio Committee: Umgeni Water Briefing. 11 June 2003". Online. Available HTTP: <http://www.pmg.org.za/docs/2003/viewminute.php?id=2906> (Accessed: 26 January 2005)

Peet, R. (2002) "Ideology, discourse and the geography of hegemony: from socialist to neoliberal development in South Africa", *Antipode*, 34 (1): 54–84

Perelman, M. (2000) *The Invention of Capitalism: Classical Political Economy and the Secret History of Primitive Accumulation*. London: Duke University Press

Poulantzas, N. (1978) *State, Power, Socialism*. London: New Left Books

Smith, N. (1984) *Uneven Development: Nature, Capital and the Production of Space*. Oxford: Basil Blackwell

Swyngedouw, E. (1999) "Modernity and Hybridity: Nature, *Regeneracionismo*, and the Production of the Spanish Waterscape, 1890–1930", *Annals of the Association of American Geographers*, 89(3) 443–465

Swyngedouw, E. (2004) *Flows of Power – The Political Ecology of Water and Urbanization in Ecuador*. Oxford: Oxford University Press

Swyngedouw, E. and N.C. Heynen (2003) "Urban political ecology and the politics of scale", *Antipode*, 35 (5): 898–918

Thompson, G. (2003) "Water tap often shut to South Africa's poor", *New York Times*, 19 May 2003

Umgeni Water (1982) *Umgeni Water Annual Report 1981–1982*. Pietermaritzburg

Umgeni Water (2002) *Umgeni Water Annual Report 2001–2002*. Pietermaritzburg

Umgeni Water (2004) Online. Available HTTP: <www.umgeni.co.za> (Accessed 1 September 2004)

Zola, E. (1995) *Germinal*. Oxford: Oxford University Press

Zondi, D. (2002) "Dismissed CEO's hearing begins", *The Witness*, 7 August 2002

12 The public/private conundrum of urban water

A view from South Africa

Laila Smith and Greg Ruiters

> Some forms of privatization may indeed change underlying political values, understandings and capacities for action in society as well as forms of claim-making.
>
> (Starr 1988: 15)

INTRODUCTION

The growth of public–private partnerships (PPPs) in the delivery of essential services to urban residents has been articulated as a form of market-based decentralized service delivery that makes services more efficient and brings governance structures closer to the people (Pirie 1988; Stoker 1989). Two intended key outcomes of PPPs aside from technical and financial objectives, are de-politicizing services and discursively constituting citizens as customers. Water is being revalued and re-presented as a scarce economic good. With this shift, the triangular relationships between the external provider, the state and the citizen – the three critical agents in the delivery of water – take on new forms with the ascent of the neo-liberal paradigm. When the external provider takes the lead in fostering a relationship with the citizen, the model of interaction is one of "customer management" and "sustainable development" in order to resolve the cost recovery constraints facing the state as well as educating users to appreciate water as a "scarce ecological resource". The relationship between the town and nature – a key focus of political ecology – gets significantly recast with the naturalization of scarcity and the commodification of water (Harvey 1996: 147–148). The outcome of this mode of governance, when examined at the local level, deepens rather than contains, the struggles for access to water.

Urban political ecology can provide useful critical tools for rethinking processes surrounding the politics of distribution and the production of water (Peet and Watts 1995; Escobar 1995; 1996; Swyngedouw 1997). In addition key questions about the socio-physical production of water as socio-nature are often ignored in distributional debates but become more evident in the critical political ecology tradition (Harvey 1996). Using this lens, this chapter examines the manner in which the triangular relationship amongst the service user, provider and state is mediated, strategized and routinized.

We ask several questions in assessing the impacts of decentralized forms of delivery. First, how has the state/citizen interaction, through the commercial delivery of public services, been transformed? Second, what are the power dynamics of distribution and how are these altered when third parties enter into the negotiation process? Third, how can an urban political ecology approach be used as a theoretical tool for analyzing how decision-makers arbitrate the distribution of urban resources and in doing so become key agents in the governance and control of the populations they serve?

The nature of conflict over access to municipal resources in cities across the Global South is particularly important in light of the constraint posed by weak and underdeveloped administrative mechanisms for distribution. This lacuna in the literature has been addressed through the recent plethora of case studies on cities in the Global South (Beal *et al.* 2002; McDonald and Pape 2002; Harrison *et al.* 2003; McDonald and Ruiters 2005). One of the challenges remaining for this evolving sub-discipline of geography is how to ground it in the reality of strategic local governance, and the politics of reproduction and the production of need. Political ecology examines not only patterns of capital accumulation, but also the controls over different populations via the administration of various modes of access to resources. Urban political ecology, as a field of study, sees power as a central theme, relevant for the study of managing urban populations (Escobar 1995; Payer 1982) within the new discourse of ecological scarcities, active citizenship and the commodification of nature. When the managerial approach to public resources becomes entrepreneurial, these three discourses are the inevitable outcomes.

In this context of the "roll back of the state", Foucault's conception of governmentality is particularly useful for analyzing the relational dynamics of power within a decentralized context of service delivery. Modes of governmentality refer to "forms of calculated practice (both in and outside government) to direct categories of social agency (Dean 2000). The practice of the everyday is normalized to conform to a particular political ecology framework. Foucault refers to this diffusion of government control via a range of practices as the "governmentalization of the state". Political power in this instance is located beyond the state, as governmentality does not confine political activity to central executive activities or formal law-making bodies (Marinetto 2003: 623). In the instance of PPPs, power assumes some degree of reciprocity, sublimating oppositional forces through "persuasion", "incentives" and "education" with the aim of earning consent.

A logical extension of Foucault's notion of governmentality to the individual is the introduction of the notion of bio-power, which can include environmental public services such as water, electricity and garbage collection. Foucault contrasts the sovereign power of death with modern bio-power, i.e. the power to foster life and care of populations. The enactment of "bio-power" is clearly seen in the everyday functions of municipal managers who face the twin challenge of producing not only bio-power (public health) but also obedient and responsive citizens. Foucault's idea of routinized regulation, surveillance and internalized forms of self-policing communities is particularly relevant in the South African municipal services context where local authorities are turning to self-managed consumers and technological solutions, such as prepaid water and electricity meters.

Within these partnerships, the social obligations for providing essential services to those who are too poor to pay is shifted away from the state to the "sovereign customer". In following this logic, the state or external provider avoids the political ramifications of disconnecting households that are too poor to pay, by simply giving them the "freedom" to self-disconnect through the introduction of pre-paid meters. The statistical visibility of neglecting the poor, i.e. municipal records that trace the total number of households that were disconnected because of non-payment, suddenly become invisible. The disappearance of this critical information handicaps the ability of the urban manager to understand, let alone monitor, the impacts of these tactics of coercion to pay on the poor. At a wider theoretical level, South Africa provides a test case for understanding governmentality and citizen compliance, or lack thereof within the field of political ecology.

In this chapter, we first examine the civil society/state relations by looking at the tensions involved in bringing populations under formal and routinized control. The control of the state is deepened by extending infrastructure like water to "unruly" populations. This mode of delivery can elicit an unruly civic response such as struggling township populations using strategies of "exit" and disengagement from the state – by seizing illegal access, sabotage or disregarding and "misunderstanding rules" (De Certeau 1984). This is a key element that shapes the modalities of access to municipal resources in South African townships. This tension in access to essential services is particularly strong in South Africa because people who have fought so long for freedom are being anointed with the regal title of "customer", along with its payment responsibilities, before having yet been granted the fruits of social citizenship.

Second, we have chosen South Africa as a country for a case study because its transition to democracy offers an important opportunity to see how the cooperation and conflict between a ruling regime and new private and quasi-market institutions of provision promoted by transnational capital are worked through within a framework of "market politics" (Leys 2001) and attempts to present nature in the mode of capital (Escobar 1995: 54–55; Bellamy-Foster 2002). Very few recent accounts of South Africa's transition at the local level of politics have addressed this issue since it is taken for granted that the state has "progressive and pro-poor" policies and that in a context of limited financial resources and weak capacity can only incrementally improve the conditions of the "poor" (Beal *et al.* 2002; Tomlinson in Harrison *et al.* 2003).

Third, we have chosen the water sector to illustrate how the state's efforts to promote private sector participation can distance it from the populations it services. The rhetoric of self-managing customers, ecological scarcity and responsible consumption of water services by the poor has played a key role in public campaigns in schools and township communities for treating water as a scarce commodity. The commodification of urban and welfare services, as Dean argues, "undermines people's sense of obligation as citizens of a state . . . has encouraged illegality and economic insecurity. Struggling to make ends meet has produced a sense of betrayal and 'depleted concepts of citizenship'" (Dean 1999: 58–59). It also promotes a supra-historical, reified conception of "nature", denying the social production of nature and new natures on the one hand and the social power relationships that infuse such transformations on the other (Harvey 1996; Swyngedouw 1999: 98).

This chapter highlights these tensions through two examples of water partnerships in South African secondary cities, i.e. small towns that have rapidly sprawled through rural–urban migration, where the local authority in each case was constitutionally mandated to address the issue of equity by improving the quality of water services to previously disenfranchized communities.

We examine the triangular power dynamics amongst the state, service provider and service user in a long-term water concession and in the second case, a management contract. The first example highlights the country's first water concession in Nelspruit and the reasons for its failure. In Nelspruit, the national government could not be seen to be losing its most significant pilot for foreign investment in the water sector and hence it provided financial support to the company while councillors sought maximum distance between the company and itself (blame shifting). The second example of a Suez-led management contract in Fort Beaufort illustrates how the state's lack of understanding of the rigours of cost recovery and its political costs when it signed the contract, led to the demise of service delivery for low-income service users. Their growing discontent and nonpayment for water services, combined with the council's inability to manage its own financial debts to the company, led to a premature collapse of the contract.

CASE STUDIES

After coming to power the ANC affirmed in wonderful prose that water is a fusion of political and physical processes:

> The history of water in South Africa cannot be separated from the history of the country as a whole and all of the many factors that went to create both one of the darkest and one of the most triumphal chapters of human experience. The history of water is a mirror of the history of housing, migration, land, social engineering and development.
>
> (DWAF 1995)

By contrast, consider how geographic and economic discourse may be used:

> South Africa is not a well-watered country . . . Water from costly schemes has been supplied only to economically viable enterprises in mining, industry and agriculture and to urban communities, which support them . . . South Africa's water supply situation is therefore one of contrasts: a first world standard in the greater urban complexes and a countryside as third world in character as many of the remote regions of the continent of which it is part.

Water mega-projects built in the heyday of Afrikaner statism and nationalism helped produce a sense of white national pride, racial superiority and conquest of nature:

[N]ew mega-dams demonstrated conquest and control of nature's most unpredictable element. They were a metaphor for social change where Afrikaner corporate society had displaced the unruly frontier . . . They celebrated a specifically Afrikaner contribution to industrial society and the modernization of agriculture. They were a major factor in shaping the country's social geography.

(Beinart 1994: 170)

With an average national rainfall of less than half of the world average, only 500 mm per annum, South Africa is regarded as a "water stressed" country. Rainfall is uneven across the country with a largely arid western region. Over 519 dams with a total capacity of 50 billion cubic meters control the seasonal flow of rivers, 53 percent of which is used for agriculture. Most of the major rivers have been dammed. Aside from mega dams built in the 1970s, the most pricey and controversial water project in recent times is the Lesotho Highlands Water Project which cost over $8 billion. The physical production of water (capture and storage) is complemented by water boards. Fifteen water boards provide bulk water to local and metro councils.

South Africa's water development has had a Promethean flavour. The country's apartheid system illustrates precisely how social and political aims can construct a second nature, a built environment and new socio-nature to realise certain social relations.

NELSPRUIT

In the dawn of the post-apartheid period, the historically small white town of Nelspruit became the capital of the new province of Mpumulanga in South Africa, in 1994. With re-demarcation, since 1994, Nelspruit inherited the former homeland of KaNgwane and the massive service responsibilities associated with this area, where most communities had hitherto relied on communal water standpipes. The Nelspruit re-scaling incorporated areas more than 20 kilometres from the town centre. The core town clearly had an unequal relationship to the surrounding poorly serviced dormitory townships (Maralack 1999).

The 1994 demarcation of Nelspruit increased the population from 24,000 to 230,000 overnight and significantly changed the profile of the communities to be serviced by local government (Kotze *et al.* 1999). Many of the newly incorporated areas had never received water and sanitation services. Although the population grew by 10 times, the total income of the new municipality had only grown by 38 percent (Kotze *et al.* 1999).

The local authority then turned to provincial and national authorities for financial assistance in meeting its service delivery obligations to township residents but was turned down. Instead, national government chose to pilot a water concession. The Nelspruit experiment could prove to become the opportunity for the ANC to promote foreign investment in South Africa's water sector. After four years of negotiations and with massive community and labour protest, a 30-year concession was signed in 1999. The primary source of financing for this concession came from the South African state through the Development Bank of South Africa (DBSA) through a R150 million loan.

The service provider: GNUC

The concession, a joint venture between Nuon, a Dutch utility company, and BiWater, a British multinational water company, has 48 percent of the shares of the Greater Nelspruit Water Company (GNUC). The early achievements of the concession included the provision of a 24-hour water supply to township communities, previously on intermittent supply. Service levels were upgraded to yard taps with outdoor flush toilets and water-borne sanitation. Despite these achievements, the lack of community support has haunted the concession since its inception. The most significant service delivery challenge facing BiWater is low levels of payment, as low as 8 percent and 35 percent respectively in Matsulu and KaNyamanzane, with 98 percent in the white town of Nelspruit (BiWater credit control officer, personal communication, 25 February 2003). BiWater has enforced strict credit control measures to improve payment levels in poor areas, not recognizing the failure to deal with the complexity of politics and poverty as one of the main reasons for non-payment. The recent credit control measures have included water cut-offs, removing meters and portions of pipes to prevent illegal reconnections, and reducing the township's 24-hour supply to intermittent hours throughout the day and night. Township residents responded to these credit control measures by reconnecting illegally and intimidating BiWater's workers when they entered the townships to maintain the infrastructure.

The low payment rates for water nearly brought the concession to a premature close in 2003, only five years into a 30-year contract. Community resistance, both non-payment and the intimidation of workers, severely limited the GNUC's cost recovery efforts, let alone its ability to operate water services properly since workers were impeded from reading water meters or opening and closing water valves. In 2003, the primary shareholder of GNUC, CASCAL, said that it would not provide more capital investments to BiWater to resume new infrastructure spending until payment levels in the townships reached 50 percent.

Service users

Issues of non-payment are rooted in economic, political and socio-ecological reasons. Regarding the economic situation of these two townships, both Matsulu and KaNyamanzane have a poverty rate of 62 percent with unemployment rates of 36 percent and 30 percent respectively (Census 1996). In many instances people are simply too poor to pay. The regressive tariff system developed by GNUC allows high volume users, predominantly white garden-owners in the town of Nelspruit, to consume between 30 and 100 kl of water for the same cost. Second, the greatest area of tariff increase has been in the second step, which affects low-income households, which cannot limit household water use to a mere six kilolitres.[1] Even though service users can access the first six kl of water free, many household water bills are still very high, at times reaching R300 to R500 a month. Given that a monthly household income below R1,100 is considered below the poverty line, a bill of R300 for just water as a proportion of overall household expenditure patterns is very high.

The company believes residents do not sufficiently understand how service delivery and billing systems work, such as what their rights and responsibilities as service users are. It certainly is possible, however, that people deliberately misunderstand or where they do understand how services work, choose not to translate this knowledge into regular payment.

Politically, many residents oppose the presence of BiWater because of the draconian credit control measures being used against service users for non-payment. Numerous household respondents interviewed for this research claimed they did not want to pay as a form of civil protest against a foreign company that was controlling the distribution of such vital resources. Others that fall within this category simply believed that water fell from the sky and was therefore a god-given right. Certain politicians opposed the concession on ideological grounds and discouraged payment for water services, thus lending legitimacy to the act of non-payment. For instance, a minority opposition party, The Pan African Congress (PAC) said that services should be free and payment is required only because of a profit-driven private company providing it (Councillor Siwela, PAC 2003). Confusion as to whether basic services should be provided by the government free of charge contributes to non-payment in the townships.

With regard to labour's views, the South African Municipal Workers Union (SAMWU) opposed the concession since its inception. The views of workers, as well as the presence of the Anti-Privatization Forum (APF) in these areas, raised public awareness about the negative aspects of water privatization. There is no evidence that either SAMWU or the APF have deliberately organized payment boycotts but they may well have influenced households' negative attitudes towards the concessionaire.

From the low-income consumer perspective, household interviews for this research show that there was a desire to pay for services if their bills were "reasonable". The inability to pay, the lack of understanding of why household bills were so high and the ability to avoid payment and still get some water illegally highlight the importance of township residents' taking a "bargaining approach" to paying. Perhaps the lesson learnt for the company is that when communities have been excluded for decades from access to quality services through methods of home-grown imperialism, and still find themselves in poverty, it may take more than a few years to ingrain a change in attitudes towards payment. Developing this level of awareness may be a much more difficult and slow task for the service provider than putting in place the infrastructure to open and close a water tap.

The state

Despite central state enthusiasm for the concession, the local councillors have been less involved than initially anticipated. The ANC-dominated city council passed on the responsibility of water governance to a private sector entity. The politicians admit that they were not been able to increase the township service users' support of partnerships nor their willingness to pay if they can afford it. Councillors have been minimally involved in working with communities to better understand the complexity behind the non-payment problem. In fact, councillors have chosen to be selective in how they announced important policy changes in service delivery in order to suit their own campaigns for upcoming elections. Township residents were not adequately informed that only the first six kilolitres of water would be provided free; amounts in excess would be charged at a higher cost through a stepped tariff. The effect of poor communication, or service users' poor understanding, left communities thinking that water was free and that they did not have to pay for their service bills. Dispelling the myth of unlimited free water proved to be a formidable task for BiWater, which has conceded that its strengths are in operations and maintenance rather than in customer relations.

At the official level, the state had little capacity to provide regulatory oversight, a problem that can be attributed to a structural flaw in the design of the contract. As is the case in many water partnerships in cities across the Global South, the local authority only takes designing sound structures of accountability through a robust regulatory regime seriously once the contract has already been signed and problems begin to arise. The city council in this case failed to take their regulatory role seriously by delegating to the town engineer the mammoth task of oversight, over and above his existing functions to manage all other services to an area of half a million people. By virtue of the city council neglecting the detail of the problematic relationship between the service provider and low-income service users, it was no more capable than BiWater of addressing the non-payment problem.

What have we learnt?

The introduction of an external service provider in a fledgling democracy may undermine an already weak local authority. The payment for services is a political issue when it comes to poor people's ability to pay. Severe credit control mechanisms to force people to pay simply backfired. Service users in this case forced the concession to its knees. Both the local authority and BiWater have learnt that the only way to salvage the concession was to change their tactics of coercion and find more democratic channels of communication, such as through the creation of a water forum. The Water Forum has become an opportunity for the service provider to cultivate greater cooperation and consent from service users. BiWater has, after five years of coercion, learnt that it must involve communities more widely in the service delivery process so that they can better understand how service delivery works, what it means to be a responsible service user and how to hold their provider accountable.

FORT BEAUFORT

Fort Beaufort in the Eastern Cape has a population of 50,000. In 1994, it became the second South African town to sign a privatization contract with a foreign firm. The ten-year contract effectively delegated the operation, maintenance and upgrading of the water and sanitation system to Suez, a private French company, which locally calls itself Water and Sanitation Services South Africa (WSSA). As in Nelspruit in 1994 when the white town and the black township were unified under a single administration, massive disparities in water access existed alongside very different socio-ecological experiences for white and black residents. Bhofolo, a black residential area, although barely a kilometer outside the white town had been part of the Ciskei Bantustan and had mostly communal water taps (upgraded to yard taps in 1993–4) and a bucket sanitation system with per capita consumption close to 20 litres per person per day. This community under apartheid paid a nominal fee for their low quality of water services.

The service provider

WSSA used the French format of "delegated management", meaning the company was a monopoly provider of water services within the municipality, but also undertook defined capital improvements. The town would get 11 percent of its water pipes replaced and 1,000 new meters installed for accurate billing of water usage. The local authority had to pay fixed fees to the WSSA, irrespective of the total water revenue collected by the municipality; a problem compounded by the fact that black areas would remain without meters and continue to pay a flat rate Such an arrangement exposed the municipality to non-payment and demand risks. By the middle of the contract term, WSSA had failed to reduce water losses, leaving the unaccounted for water rate at 34 percent.

An important motivation for the contract was to help "restore community acceptance and regular payment for services provided" in black areas (WSSA 1995a). Explicitly included in this aim was the depoliticization of water, running the service as a stand-alone business and educating consumers in commercial principles. WSSA insisted that "humane" customer management was possible even with a vigorous level of water disconnections. WSSA, however, was unable to effectively implement this principle because of a lack of political will by council officials to proceed with a rigid "credit control" system. The ANC councillors in the area were unwilling to cut off services to their constituents.

Another key failure in the Fort Beaufort contract was the underestimation of water-use projections. Fort Beaufort's actual water consumed was 28 percent above the WSSA five-year projection. But, because the contract also included volume-based charges, the charges payable to WSSA increased to levels not proportionate with municipal income from water payments. Another crucial limiting aspect of the contract was its short duration: the company had to get its money back in ten years as opposed to the usual 25 years found in a concession model (as in Nelspruit). The relatively limited capital invested by the company in pipe replacement and upgrading ultimately failed to improve the efficiency of the system. This situation was compounded by a lack of political buy-in by the newly elected ANC councillors who feared unpopularity among their constituencies when there was a 300 percent increase in rates for black areas, which were now charged at similar tariffs as former white areas. This situation contributed to a sudden build-up of arrears, exacerbating the non-payment problem.

The service provider's response to the PPP crisis was to cajole the municipality to hand over the entire "customer management" section of the council including billing, cut-offs and legal action to WSSA. On 16 November 1999, WSSA proposed "global customer management" as opposed to merely reading meters, which was the existing arrangement, as a way of resolving the crisis. The new scheme (WSSA 1999c) would require "a new governance" architecture based on updated customer data, surveys, and dividing the town into service zones with service committees in each zone. Each zone would send a representative to a central Fort Beaufort Service Forum, which the mayor, WSSA and councillors would attend. Customer centres would also be established.

The rescue plan for achieving greater efficiency was to first instal meters on all yard taps in black areas and to cut water use by 20 percent. Second, the WSSA wanted to increase the top end of tariffs to discourage excess consumption by 5 percent as it was estimated that lower raw water imports could realize R25,000 savings per month for the council (WSSA 1999c). This proposal led to a special workshop on customer services management, which agreed on a "customer charter whereby a good governance framework could be achieved" (WSSA 2000a). At this stage, the "WSSA had carried the council for several months" by re-scheduling debt, it expected the town to hand over "customer management" as a sign of gratitude and good faith. The central state (DWAF) in the meanwhile had also threatened legal action against Fort Beaufort for non-payment for raw (bulk) water.

The final blows came in August 2000 when WSSA stopped all new work and services to informal areas despite a cholera threat in the Eastern Cape (WSSA 2000b). WSSA also brought legal action against the council for failing to pay its monthly fees to the WSSA. The situation was temporally relieved on 4 October 2000, when the municipality paid an instalment of R386,000 to WSSA. By then, WSSA had more or less taken over the Fort Beaufort Financial Management Committee, which was a highly strategic decision-making platform, to "monitor progress and ensure planning was done in terms of WSSA proposal". Council responded to this unwelcome invasion into its decision-making process in 2002 by getting a court ruling to have the contract annulled.

Service users

Service users in Fort Beaufort were not passive recipients of the new water arrangements. Indeed many community meetings called by councillors and SANCO (an umbrella organization of civic associations) were well attended (interviews 1998). The community also managed to get members of the South African Communist Party to exert pressure on councillors. Lack of affordability combined with the continued use of buckets as sanitation service by WSSA and massive tariff hikes drove the community into non-payment. The close-knit Bhofolo community was able to extract concessions from councillors through repeated renegotiation of flat rates, hence averting widespread illegal connections so common in most other townships (see Nelspruit example). The councillors found the contract inflexible and were intolerant of the company putting pressure on councillors as a means of shifting the blame for consumer nonpayment onto them.

Underlying the township's resentment was the enduring racial-class geography of Fort Beaufort. Sanitation in black areas remained far below standards in the white town. The central town residents used 15 times more water than township resident on a per capita basis. About 60 percent of black township residents were not paying water bills, that would have averaged around 30 percent of their income. In the black township, arrears were close to R4,000 per household (seven times the average monthly family income). The WSSA had not been able to change the racial character of water distribution and sanitation that left the black township distinctly worse off, visibly inferior and demeaned. Worse still, as already mentioned, WSSA campaigned for metering of poor people to force a 20 percent reduction in use to help council lower the bill on raw water.

By 31 August 2000 the total outstanding consumer debt balloon to R17 million (two-thirds of the annual council budget) with no prospect in sight of this amount being recovered. The majority of debtors against whom legal action was taken were pensioners or unemployed. Some of these arrears went up to R8,000; an irrecoverable amount (FB TLC 2000a). Understandably councillors were reluctant to impose more hardship and preferred negotiated solutions to community resistance. Consumers and workers also wanted the contract revoked, and councillors concurred by openly condemning the contract, which was finally cancelled in 2000.

The state

By 1998, four years after the contract was signed, divisions between council and the company ran deep. For example, in October 1998 the Fort Beaufort councillors asked the WSSA to stop attending council committee meetings. Instead, remarkably, the council felt WSSA workers (now transferred to the private company), should attend meetings of the Standing Technical Committee. Instead of outsourcing political relations with "troublesome" unions, a major original motivation for privatization (see WSSA 1995a) the councillors sought closer ties. WSSA's ability to keep up morale and standards in the labour force was in doubt and so council had to intervene, thereby negating the "arms-length" rationale behind the contract.

In further examples of mistrust, council also demanded guarantees on new meters from WSSA, suspecting it had installed low-grade meters. On the 11 October 1999 the council resolved to do its own repairs on the sewer works. WSSA apparently refused to repair leaking yard taps, a common practice when the council ran the system. WSSA and local bureaucrats had blamed councillors for not taking tough action against consumers and encouraging a non-payment "culture" by making many concessions and soft-pedalling with township residents. When councillors eventually took tough action to recover its costs in delivering water services, it is significant that it was against squatters, the most vulnerable and politically weak segment of the local population. In an ironic letter WSSA (2000c) notes, "as per council decision, the bucket service to informal areas in Hillside will be terminated". WSSA warned that health hazards and civil unrest could follow, and it would not be responsible for council losses.

Councillors also felt overwhelmed by the complexity of the contract and the company's technical reporting. For example, the new Town Clerk, Mr Makana had very little knowledge of the contract (Personal interview 1998). Councillors did not understand how contract costs had quadrupled over six years from the initial quoted amounts (from R95,000 to R400,000). Cost and savings projections were based on unrealistic assumptions such as a 95 percent collection rate of tariffs. In the second half of 2000, council faced a calamitous R475,000 in monthly charges to WSSA, excluding penalties for late payments.

Lessons

The Fort Beaufort PPP was less a partnership than a series of escalating conflicts between the state and the service provider. The council learnt the harsh lesson that the loss of political and financial flexibility need not have happened had it chosen to restructure internally rather than outsource. Councillors felt that they were debt collectors for WSSA and were resentful at having to bear the full risk of consumer non-payment, particularly when the non-payment was in part motivated by a poor quality of service provided by WSSA. Cost creep in contract payments was a clear danger signal that WSSA charges were out of line with economic and political dynamics of the town and foreclosed politically and socially progressive alternatives. In the end, for the company, the contract may merely have been a wedge to enter the South African water business (Ruiters 2002), but it provided a salutary warning that deep socio-ecological entanglements would haunt attempts at commercial service delivery.

CONCLUSION

Market environmentalism, one approach informing water privatization, on closer examination, has overcome a number of political hurdles as it tries to present water and nature in the image of a commodity. One of the key points that emerges from the case studies is that once water is commodified through a private partnership where the state is driven by a private sector logic, it lacks a political, social and administrative salience for the people it is servicing. When a partnership uses authoritarian measures to enforce consumer compliance to the rules associated with receiving marketized services, such an approach can spur the growth of informal acts of everyday resistance by the poor (Scott 1985; Peet and Watts 1996). In many instances, these service users are simply too poor to afford the price of water and use these acts of resistance as survival tactics. Given this socio-economic reality amidst the proliferation of partnerships that commodify water, it is no accident that water issues have become the trigger for powerful new social movements (Bond 2002). Marketization is a not simply a transfer of services from one basket to another; in reality, it creates an entirely different service and produces a different environmental imaginary. Socio-political and ecological relationships of residents and workers to "resources" and to the municipality, and between residents themselves, create new discourses, institutions and rules of the game – in short new forms of governmentality, and social-ecological space (Williams 1994; Lefebvre 1996).

The struggle to win the hearts and minds of the consumer is lost when physical violence is done to the body of the consumer and the community through water disconnections. In such instances, the state mistakenly assumes it can play a passive role by decentralizing its responsibility to an apparently depoliticized external provider. It later has to assume a critical role in mediating the turbulent dynamics between an external provider and service users. Service users who feel that the manner in which services are delivered to them is unjust may on the surface appear as victims, but in fact can shift the power dynamics of the partnership in their favour by virtue of their resistance through nonpayment and illegal reconnections.

To conclude, in this chapter we have asked three questions in the hope of finding a nexus between the service delivery/partnership literature and the value that an urban political ecology theory has to offer. First, we asked how has the state/citizen interaction, through the delivery of public services, been transformed via the participation of an external service provider? We have shown that state–company–citizen relations are reconfigured in the direction of commodified relations. Significant cut-offs and restrictions of water uses by the poor occurred in all three cases. In both the Nelspruit and Fort Beaufort cases, the service provider arrogantly stepped into an area ridden with historical inequality and learnt very quickly that "managing the poor" is not easy. The reconfiguration of power when an external provider moves to the forefront of the delivery machine is constantly undermined by political factors and most often by popular resistance. We would therefore agree that privatization is more than a transfer of functions or assets, but a reconfiguration and change of the service itself (Edgell *et al*. 1996; Clarke 1996; Lee 1997; Lorrain and Stoker 1997). We would, however, modify the statement by saying that political imperatives and the poor themselves also play a major part in forcing the partial decommodification of services. The impoverished idea of citizenship when domi-nated by a cash nexus has proven hard to implement in South Africa's water services experiments.

Second, we asked about the *power dynamics* of distribution and how these dynamics are altered when third parties enter into the negotiation process? We have seen that the idea that "non-political" private firms are better positioned than municipalities to change public behaviour is simply inaccurate. In none of the cases mentioned where private sector principles have prevailed have payment levels improved. A common reason motivating the decision for local authorities choosing to partner with external operators is to avoid the virtual collapse of the delivery of essential services. When the decentralization of service delivery to external providers is motivated by such circumstances, the inherent management problems of the sector in question are not necessarily resolved because of the tension emerging from different expectations. The state expects the external provider to protect the local authority's constitutional obligations to deliver water to all, the external service provider expects that its service users will be compliant by paying, and the township service users expect an affordably priced and higher quality of service than what was provided under the apartheid regime. This chapter has illustrated that water partnerships following private sector principles have great difficulty in reconciling these differing expectations.

A concession or a long-term management contract claim to be useful service delivery alternatives by offering much-needed financial services and technical expertise to the state. The case studies above illustrate that this mantra is not necessarily true if it is at the cost of eroding the governance of the city council and undermining the ability of low-income communities to access water. Here is where the institutional model of a concession or "delegated management" presents a paradox: as a service delivery model set out to meet the needs of the poor, the logic of profit and efficiency that drive the management of these private sector models do not lend themselves to the patience and flexibility required to deliver services to poor people. Furthermore, the management style of these institutional models is to concentrate power and the decision-making processes of water distribution into an autonomous entity that is above the reach of political intervention – a trend that moves counter to the democratization processes that are vital if services are to be delivered in a socially just manner. Once a contract is signed, the municipality is at the weaker end of the bargaining relationship since it cannot easily regain its lost capacity and labour force to service new urban growth. Neither can it easily sign a different contract with a different company for newly incorporated areas, hence the long "lock-in" factor after the contract has been signed. Service providers are well aware of how contracts can operate in their financial interests once the initial turbulent dynamics settle. This confidence makes service providers akin to "growth statesmen" by taking a proactive stance in regional growth coalitions because they have sunk assets in specific areas (Logan and Moloch 1987).

Third, we have suggested that urban political ecology might benefit from more consideration of the difficulties of governing the poor by incorporating the idea that the poor often decide to exit the system when staying in becomes too onerous. This disengagement option, however, presages much wider collapses in governance and much weaker state penetration into the everyday life of the public. The weaker the state becomes, the less accountable it is. This exacerbates public scepticism and distrust and takes a costly social and political toll especially in strategic urban centres. This slide into two publics, one inside the framework of the law and the other outside, is what constitutes the current juncture of urban water distribution in South Africa today.

NOTE

1   In 2001, the Department of Water Affairs and Forestry set out national guidelines for all
     local authorities across South Africa to provide the first six kilolitres of water free in an
     effort to achieve the progressive realization of the right to water enshrined in the
     constitution.

BIBLIOGRAPHY

Beal, J. Crankshaw, O. Parnell, S. (2002) *Uniting a Divided City: governance and social
     exclusion in Johannesburg*. London: Earthscan
Beinart, W. (1994) *Twentieth Century South Africa*. Cape Town: Oxford University Press
Bellamy-Foster, J. (2002) *Ecology against Capitalism*. New York: Monthly Review
     Press
Bond, P. (2002) *Unsustainable South Africa*. London: Merlin Press
Clarke, J. (1996) "Public nightmares and communitarian dreams", in S. Edgell,
     K. Hetherington and A. Warde (eds) *Consumption Matters*. Oxford: Blackwell
De Certeau, M. (1984) *The Practice of Everyday Life*. Berkeley, CA: California University
     Press
Dean, M. (1999) *Governmentality: power and rule in modern society*. London: Sage
Dean, H. (2000) "Managing risk by controlling behaviour", in P. Taylor-Gooby, *Risk, Trust
     and Welfare*. Houndmills: Macmillan
DWAF (1995) *White Paper on Water and Sanitation*. Pretoria: Department of Water Affairs
     and Forestry
Edgell, S. Hetherington K. and Warde, A. (eds) (1996) *Consumption Matters*. Oxford:
     Blackwell
Els, A. (1996) "National report South Africa", *Water Supply*, 14(3–4): 106–108
Escobar, A. (1995) *Encountering Development*. Princeton, NJ: Princeton University Press
Escobar, A. (1996) "Constructing nature", in R. Peet and M. Watts (eds) *Liberation
     Ecologies, Environment, Development, Social Movements*. London: Routledge
Fort Beaufort TLC (1997a) "Infrastructure Assessment Plan"
Fort Beaufort TLC (1997b) "Minutes of Meeting with Sanco deputation", 4 July
Fort Beaufort TLC (2000a) Fort Beaufort Treasurer, Henk Bezuidenhout, letter 12 October
Fort Beaufort TLC (2000b) MSP discussions in Finance Management Committee (FMC)
     Minutes
Fort Beaufort TLC (2000c) "WSSA correspondence to Fort Beaufort TLC, 'Re Outstanding
     debts'", dated 28 July 2000, Minutes, 28 August 2000: 84–85
Fort Beaufort TLC (2000d) "Monthly Report of Assistant Treasurer", 14 August
Harrison, P., Huchzermeyer, M. and Mayekiso, M. (2003) *Confronting Fragmentation:
     Housing and Urban Development in a Democratising Society*. Cape Town: UCT
     Press
Harvey, D. (1996) *Justice, Nature and the Geography of Difference*. Oxford: Basil Blackwell
Kotze, R., Ferguson, A. and Leigland, J. (1999) "Nelspruit & Dolphin Coast: lessons from
     the first concession contracts", *Development. Southern Africa*, 16(4)
Lee, Y. (1997) "The privatisation of solid waste infrastructure and services in Asia", *Third
     World Planning Review*, 19(2): 139–161
Lefebvre, H. (1996) *The Production of Space*. Oxford: Basil Blackwell
Leys, C. (2001) *Market Driven Politics*. London: Verso
Logan, J. and Moloch, H. (1987) *Urban Fortunes*, Berkeley, CA: UCLA Press

Lorrain, D. and Stoker, G. (1997) *The Privatisation of Urban Services in Europe*. London: Pinter

Maralack, D. (1999) "A profile of Nelspruit, its people and its economy". Online. Available HTTP: <http://www.local.gov.za/DCD/ledsummary/nelspruit/nel03.html>

McDonald, D. and Pape, J. (2002) *Cost Recovery and the Crisis of Service Delivery*. London: Zed Books

McDonald, D. and Ruiters, G. (2005) *The Age of Commodity*. London: Earthscan

Mestrallet, G. (2000) "Speech to ordinary and extraordinary general meeting of Suez Lyonnaise Des Eaux", Suez Lyonnaise Des Eaux Website – <www.Press.Suez-Lyonnaise.Com/English/Mestrallet.Htm>

Payer, C. (1982) *The World Bank*. New York: Monthly Review Press

Peet, R, and Watts, M. (eds) (1996) *Liberation Ecologies, Environment, Development, Social Movements*. London: Routledge

Petrella, R. (1996) "Globalization and internationalization: the dynamics of the emerging world order", in R. Boyer and M. Drache (eds) *States against Market*. London: Routledge

Pirie, M. (1988) *Privatization, Theory, Practice and Choice*. Aldershot: Wildwood House

Ruiters, G. (2002) "Debt disconnection and privatization: the case of Fort Beaufort, Queenstown and Stutterheim", in D. MacDonald and J. Pape, *Cost Recovery and the Crisis of Service Delivery*. London: Zed Books

Scott, J. (1985) *Weapons of the Weak. Everyday Forms of Peasant Resistance*. New Haven, CT: Yale University Press

Sharp, J. and Robinson, J. (2000) *Entanglements of Power: geographies of domination/ resistance*. London and New York: Routledge

Siwela (Councillor for the Pan African Congress), Personal Communication (PAC), 27 February 2003

Starr, P. (1988) "The meaning of privatization", *Yale Law and Policy Review*, 6: 6–41

Stiglitz, J. (2002) *Globalization and its Discontents*. London: Penguin

Stoker, G. (1989) "Local government for a post Fordist society", in, J. Stewart and G. Stoker. *The Future of Local Government*. Houndmills: Macmillan

Swyngedouw, E. (1997) "Power, nature, and the city. The conquest of water and the political ecology of urbanization in Guayaquil, Ecuador: 1980–1990", *Environment and Planning A*, 29: 311–322

Swyngedouw, E. (1999) "Marxism and historical-geographical materialism: a spectre is haunting geography", *Scottish Geographical Journal*, 115(2): 91–102

Tomlinson, R. (2003) "HIV/Aids and urban disintegration in Johannesburg", in P. Harrison, M. Huchzermeyer, M. Mayekiso (eds), *Confronting Fragmentation: Housing and Urban Development in a Democratising Society*. Cape Town: UCT Press

Water and Sanitation Services South Africa (1995–2000) Monthly WSSA Reports to Fort Beaufort, Standing Technical Committee Minutes. Online. Available HTTP: <http://www.local.gov.za/DCD/ledsummary/nelspruit/nel03.html>

Water and Sanitation Services South Africa (1995a) "Proposal for Management, operation and maintenance of Fort Beaufort Water and Sanitation", Fort Beaufort Municipal Records, 25 April

Water and Sanitation Services South Africa (1995b) "Contract Agreement for delegated management of water and sanitation services in Fort Beaufort", Fort Beaufort, October

Water and Sanitation Services South Africa (1995c) "The Delegated Management Concept", Annexure 5, Fort Beaufort Proposal, Fort Beaufort, April

Water and Sanitation Services South Africa (1999) "Proposal for Global Customer Management, by WSSA in Fort Beaufort", Fort Beaufort, 16 November 1999

Water and Sanitation Services South Africa (2000a) "Customer Management Workshop including Minutes and recommendations", Fort Beaufort, 20 March 2000

Water and Sanitation Services South Africa (2000b) "Monthly Report, Fort Beaufort", Fort Beaufort, August 2000

Water and Sanitation Services South Africa (2000c) "Letter to Fort Beaufort TLC", Fort Beaufort, 3 April 2000

Williams, F. (1994) "Social relations, welfare in post-Fordist debate", in R. Burrows and B. Loader (eds) *Towards a Post-Fordist Welfare State*. London: Routledge

13 Inherited fragmentations and narratives of environmental control in entrepreneurial Philadelphia

Alec Brownlow

INTRODUCTION

Explorations into the urbanization of neoliberalism reveal what may be referred to collectively as the "new" fragmentations of the post-Fordist, entrepreneurial city. Here, patterns and processes of political devolution, the rescaling of social relations of power, and the "retreat of the state" are reflected within, inter alia, the emergence and structures of urban regimes and public–private structures of governance, and the expanding influences of and emphases on place, community, and "the local" to urban political economies (Swyngedouw 1989; Harvey 1996; Amin 1994; Zukin 1995; Lauria 1997; Brenner and Theodore 2002). As Brenner and Theodore (2002b) argue, however, how (and if) these new fragmentations emerge and develop in any given city is contingent and context-specific, conditioned as much by inherited geographies, institutions, structural and physical conditions of "past" (i.e., Fordist, industrial) urban development regimes as they are by any sort of top-down ideological project of neoliberal change (see also Peck and Tickell 2002). The articulation between the past and the present is commonly one of creative destruction, whereby inherited structures, institutions, and geographies are physically reclaimed and refurbished and discursively reshaped and repackaged under locally specific and historically contingent circumstances to satisfy the ever-changing, ever-expanding political and economic demands of global, mobile capital (Harvey 1989; Brenner and Theodore 2002b).

Environmentally translated, recent developments in urban environmental geography and a nascent urban political ecology have been instrumental in characterizing and revealing the transition of urban environments and environmental governance to new, neoliberal forms. For instance, studies exploring the neoliberalization of urban environments have critiqued the imposition of new environmental narratives insofar as they accompany and expedite the exclusive, entrepreneurial agendas of gentrification and urban renewal (Keil and Graham 1998; Cowell and Thomas 2002; Whitehead 2003); others reveal the emergence and the significance of new structures and coalitions of urban environmental governance and regulation (Feldman and Jonas 2000; Gibbs and Jonas 2000; Jonas and Gibbs 2003; Pincetl 2003), including (the capacity for) the production or reproduction of social marginalizations and inequities (e.g., Pincetl 2003). A growing body of research on the reclamation and repackaging of once-industrial urban brownfields and waterfronts in the interests of capital accumulation and the production of place (best illustrated by Baltimore's Inner Harbor) succeed in demonstrating the environments of creative destruction (DeFilippis 1997; Eade 1997; Cowell and Thomas 2002). Meanwhile, similar (if less critical and more normative) work by Thompson (2002), Freestone and Nichols (2004), and Karasov and Waryan (1993) illustrate efforts to redefine and reconfigure the meaning of urban parks in the face of global urban restructuring (cf. McInoy 2000). Finally, Smith's (2004; Smith and Hanson 2003) work on water privatization in post-apartheid Cape Town is an excellent example of the significance of inherited social geographies and institutional structures to the regulation and provision of urban environmental resources in cities making the neoliberal turn.

However, despite the unquestionable and profound changes to their structure, form and function over the past century of industrial urbanization (Mumford 1955; Antrop 2004), and the rapid and widespread (re-)awakening among urban decision makers, boosters, and elites of their significance to urban competitiveness and renewal (Platt *et al.* 1994; Harding 1999; Harnik 2000), material urban ecologies have received little attention in the neoliberal literature (cf. McCarthy and Prudham 2004). Specifically, there has been little discussion about how or where the new fragmentations and narratives of neoliberal urbanism – be they "new" discourses of nature and eco-modernization or regimes of urban ecological governance – articulate themselves with the inherited ecologies and social geographies of the industrial city. As I discuss below, in the post-industrial, post-Fordist cities of the north, the success of new, entrepreneurial environmental discourses are by no means assured; rather, following Brenner and Theodore (2002b), their relative success is (in part) contingent and context-specific, conditioned by their relative ability to articulate themselves with inherited, *in situ* ecological and social conditions, discourses, and geographies.

This chapter explores the articulation of "old" and "new" urban ecologies in Philadelphia. Specifically, I explore the articulation of contemporary, entrepreneurial narratives of restored nature with (a) the inherited ecologies and social geographies of the industrial city; and (b) the inherited, concomitant environmental narratives of adjacent, marginalized communities. As I demonstrate, both narratives orient themselves around discourses of control, albeit in distinctly different forms. On the one hand, *ecological* control is central to a rapidly growing and widespread urban ecological restoration discourse whose agenda of the reclamation and restoration of urban ecologies to some vision or version of normative (i.e., native, aesthetic, marketable) ecological structure and form puts it squarely in the realm of eco-modernization, as that term is described by Keil and Desfor (2003). On the other hand, among long-marginalized adjacent black communities, the decay of their local ecologies over the past several decades is widely perceived as indicative of a more widespread and historical crisis of *social* control, whereby histories of racism, isolation, poverty, and political neglect are implicated in uneven patterns of ecological blight and a pervasive fear of nature. The normative visions (or imaginaries (see Watts and Peet 1996: 263–268)) of nature that these two narratives subsequently promulgate suggest the capacities for both opportunity and struggle in the creative destruction of Philadelphia's urban ecologies.

The chapter is structured as follows: first, I discuss the fragmenting tendencies of industrial urbanization and explore the social and environmental legacies that are the inheritance of the post-industrial neoliberal city. Next, I discuss the parallel legacies of social and ecological fragmentation in Philadelphia and introduce the competing narratives of controlled nature – both entrepreneurial and inherited – that appear to currently be in the process of articulation. I argue that both narratives of control stem from and are responses to the fragmented legacies of urban industrialism in Philadelphia. Next, I discuss their contradictions and the capacities they appear to offer for both opportunity and contestation. In particular, I emphasize the potential for apolitical, ahistorical narratives of entrepreneurial nature to demean and reproduce the social inequalities of earlier (and contemporary) fragmenting processes.

THE "OLD" FRAGMENTATIONS OF THE INDUSTRIAL CITY

Industrial urbanization was a fragmenting process and, by extension, the industrial metropolis was a city of fragments whose patterns, structures, social and environmental consequences are now the inherited stuff of neoliberal articulation and creative destruction (see Harvey 1992). Few areas demonstrate this geography better than the urban park in the nineteenth-century US city. While it is beyond the scope of this chapter to provide a political and social history of the urban park movement in this country (for that I would direct readers to excellent accounts by Cranz 1982; Rosenzweig and Blackmar 1992; and Taylor 1999), it is nonetheless imperative to point out that the history and geography of urban parks in America accompanied and were heavily influenced by the intense urbanization of capital and the new patterns of uneven development and class fragmentation taking place during this period in America's urban history (Harvey 1985; 1992).

By most accounts (Cranz 1982; Stormann 1991; Rosenzweig and Blackmar 1992; Taylor 1999) urban parks arose as tools of social control – landscapes where, it was hypothesized, the values and aesthetics of nature, leisure, and recreation embraced by the country-dwelling, capital-owning middle and upper classes would trickle down to educate, placate, civilize, and (consequently) increase the productivity of the city-dwelling working class (Taylor 1999). Thus, the distribution of urban parks reflected and was intended to aid in the reproduction of the fragmented social geographies and class structure of urban industrial society. And, indeed, although their social milieu and demographic surroundings have likely changed over the course of the past century – when spatial divisions by class were overshadowed by fragmentations by race (Massey and Denton 1993; Hacker 1995) – with few exceptions, today's park geographies in the northern post-industrial cities are legacies of this industrial past.

Moreover, urban nature parks are themselves, fragments. Also referred to as islands or parcels in the language of landscape ecology (Forman 1998), urban parks and greenspaces are ecological isolates embedded within a milieu of urban growth and development; their geographies, geometries, and isolation from other natural spaces overexpose them to the myriad inputs (or disturbances) of the city, including: trash; water, air, and soil pollution; and invasive species (i.e., weeds), often escapees from yards, botanical and zoological gardens, and international seaports (see With 2002; Mooney and Hobbs 2000). As a result, urban parks – like ecological islands generally – are exceptionally prone to widespread and intense environmental change (Pickett and White 1985; Saunders *et al.* 1991); without relentless environmental management and maintenance, urban park ecosystems rapidly become dominated by non-native, weedy ecologies whose evolutionary characteristics and ecological abilities to not only survive but thrive under the persistent and often severe environmental conditions of the city have made them the bane of urban ecologists and park advocates (Harnik 2000; Falck 2002).

More often than not, this is the ecological inheritance passed on by the industrial city, and the "new" urban parks movement (e.g., Garvin and Berens 1997; Harnik 2000), with its accompanying discourse of entrepreneurial nature, has focused on these weedy ecologies as the sites of creative (ecological) destruction.[1] Towards this end, it has been accompanied by the rapid emergence of restoration ecology as a key environmental management tool in US and Western European cities during the 1980s and 1990s (see Gobster and Hull 2000; Lindig-Cisneros and Zedler 2000). As its primary objective, restoration ecology emphasizes the control and the removal of these inherited ecologies from (in this case) urban parks and their replacement with some arbitrary past, though ideologically normative and "natural", native floral structure (McGinnis and Woolley 1997; Gobster and Hull 2000; cf. Sprugel 1991; Botkin 2001).

In the remainder of this chapter, I explore the dynamics of inherited fragmentations and entrepreneurial visions of nature as they have unfolded and confronted one another in a West Philadelphia community. Narratives provided by Cobbs Creek residents and park officials are used as a means of drawing attention to the discursive and political significance attached to environmental change and narratives of nature in the local Cobbs Creek Park. They were gathered in focus groups and in-depth interviews in the summer of 2000 and the fall of 2001.

INDUSTRIAL LEGACIES IN ENTREPRENEURIAL PHILADELPHIA

Philadelphia, like so many other once-industrial cities, has had limited success redefining itself competitively in a global capital market (Adams *et al.* 1990; Bissinger 1997). Narrowly dodging bankruptcy in the early 1990s, the city has since committed itself to, inter alia, a series of place-making projects and schemes designed to (hypothetically) reverse decades of economic decline and population loss, attract economic investment, lure tourist dollars, and boost service sector job growth. New professional sports stadia, state of the art cultural and convention centers, a variety of ad campaigns and package deals intended to lure prospective tourists and conventioneers, and a proposal to reclaim the Delaware River industrial waterfront and become a "New River City" are all indicative of post-industrial

Philadelphia's determination for a new, more culturally attractive, economically lucrative, and competitive identity (see Harvey 1989).

It was only a matter of time before attention was cast on the city's Fairmount Park System which, at 8,900 acres, or 10 percent of the city's land area, is among the largest urban parks in the world. Fairmount Park was established by city ordinance in 1855 at the height of the city's industrial era and on the heels of an 1854 regional consolidation act that increased the city's area tenfold. It was the cultural centerpiece of Philadelphia's 1876 Centennial Exposition/World's Fair, trumpeted by the city's industrial and governmental elites as the hallmark of the city's arrival to world class status. Today, over 90 percent of the park system's area is encompassed in seven watersheds whose distribution, size, and shape reflect and, to a degree, maintain historical social and economic divisions across Philadelphia's landscape (see Figure 13.1). Nearly 61 percent (5,441 acres) of the park's area is dedicated to natural

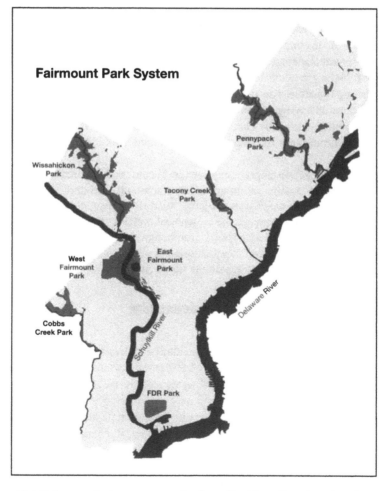

Figure 13.1 Fairmount Park system (map by Jason Davidson, Temple University)

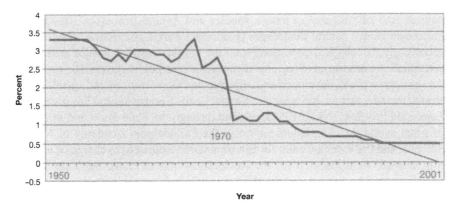

Figure 13.2 Fairmount Park budget as a percentage of Philadelphia's operating budget, 1950–2001

landscapes, primarily urban woodlands, wetlands, and meadows. Over the past three decades, the geometries and geographies of these watershed areas, combined with a generation of fiscal neglect and political ambivalence (adjusting for inflation, the park has not experienced a budget increase since 1970, and is a fraction of its pre-1970 total see Figure 13.2); full-time personnel has been cut by two-thirds; per capita park expenditures are among the lowest in the nation (Harnik 2000)), have resulted in extensive and, in places, intensive landscape level environmental changes. Soil erosion, runoff, and pollution threaten many of the streams coursing through the city's interior. An extensive, mid-successional hardwood forest is increasingly exposed to external disturbances and stressors like disease and pollution. But perhaps the most identifiable indicator of change is the expanding diversity and density of non-native, "weedy" species whose ranks now constitute one-third of the park's floral diversity.

In 1996, the William Penn Foundation (the Delaware Valley's largest and wealthiest institution of regional philanthropy and place-making investment) awarded an unprecedented $26.6 million grant to the City of Philadelphia to be used to halt or reverse decades of ecological change, restore the park's native ecology, and establish a viable environmental education programme espousing the significance and complexity of indigenous ecologies to urban life (see Goldenberg 1999). Soon thereafter, the city controller's office released a widely cited and extensively adopted document (*Philadelphia: a New Urban Direction* 1999) laying the groundwork and establishing a blueprint for urban renewal and competitiveness; in it, the restoration of Fairmount Park is posited as a fundamental and achievable goal towards the production of a more economically prosperous city future, thereby bringing to the project the extra political punch and legitimacy that it needed to gain the attention of a still-ambivalent mayor and city council. An official from the city's largest, wealthiest, and most successful business improvement district was brought in to lead the five-year restoration programme.

Restoration plans were developed for each of the seven watersheds with the overall goal of "enhancing native species and natural processes" throughout Fairmount Park's natural landscapes (Fairmount Park Commission 1999: 1–15); accordingly, "ecological condition [was] the primary justification for restoration activities". A consortium of regional institutions, both public and private, produced the restoration plans, developed and implemented their priorities and agendas. Key among this new partnership were: the Natural Lands Restoration and Environmental Education Program (NLREEP, developed through the Penn grant and in charge of grant oversight and implementation); the Philadelphia Academy of Natural Sciences and the Patrick Center for Environmental Research (in charge of restoration science, field research for the Restoration Master Plan); the Yale School of Forestry and Environmental Studies (consulting); Community Resources (a not-for-profit consulting agency); and the Fairmount Park Commission (see Goldenberg 1999).[2] Community meetings and participatory activities were used to solicit public opinion of the Restoration Master Plan and disseminate and gather popular support for the environmental imaginary that it advanced. However, any illusion the restorationists may have had about community consensus or the plan's rapid adoption dissipated when confronted with the park's inherited and fragmented geography; one that places it in a variety of communities whose capacities, abilities, and positionalities to respond to and interpret changes in their adjacent ecologies vary considerably.

Plagued by relentless patterns of white flight and the inherited racial and class fragmentations of the twentieth century, Philadelphia remains among the most segregated and racially polarized cities in the US. Indeed, its persistently high values among a variety of segregation indices identify Philadelphia as one of a handful of US cities perpetually identified as *hypersegregated* (see Massey and Denton 1989). The history of racial segregation in Philadelphia is lengthy and complex and beyond the scope of this chapter. Significantly, however, the social and economic distinctions between black and white Philadelphia are startling. Census data from 2000 indicate widening distinctions between black and white Philadelphia along a variety of social and economic axes: median family incomes differences of 35.5 percent; a three-fold difference in family poverty rate; an unemployment rate twice as high as white Philadelphia; and an annual violent crime rate (murder and rape) that is nearly twice that in white neighbourhoods. These local social conditions contextualize the significance and meaning of urban environments within or adjacent to inner-city black neighbourhoods in Philadelphia; they shape how these areas of the park system are used, accessed, and perceived and how local environmental changes and restoration efforts are interpreted.

COBBS CREEK

Cobbs Creek in West Philadelphia is among the largest black neighbourhoods in the city (see Figure 13.3). Its local social and economic conditions mirror those of Philadelphia's black community, in general. Of consequence, however, is the rapid annual growth rate of violent crime in the area, especially rape (10.7 percent) and

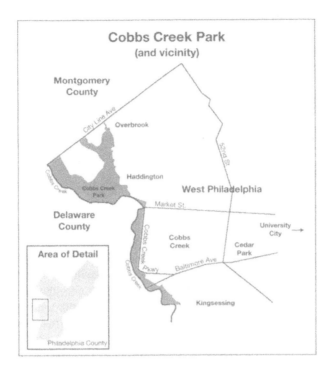

Figure 13.3 Cobbs Creek Park and vicinity (map by Jason Davidson, Temple University)

murder (4.1 percent) (1996–2000). In Cobbs Creek, these violent acts are growing at considerably faster rates than they are throughout the rest of the city (7.1 and –5.4 percent, respectively). Running along the neighbourhood's western flank is Cobbs Creek Park (est. 1904), one of the seven watersheds of the Fairmount Park System. Early in the community's black history (from the mid-1950s through the 1960s), Cobbs Creek Park – the largest public space in West Philadelphia – played an instrumental role in facilitating the production of the neighbourhood's social identity and political development, serving as a meeting place and site of local civic and civil rights activism. Today, in stark contrast, its public space attributes and opportunities have been replaced by growing violence and fear, having fallen victim to the same set of social and economic circumstances that currently afflict the larger Cobbs Creek community, what one Philadelphia columnist has called the "trashing of Cobbs Creek" (Latty 2000). Since the 1980s, violent crime has enveloped the park. In the last decade alone, more than a dozen murder victims have been pulled from the park's forested interior.[3] Most of the homicide victims have been female.

NINA (early 30s): It's getting – like they found the bodies down in the park in broad daylight. Girls getting raped. Little girls. It's sick! . . . The point is they finding girls' bodies – raped, dismantled. They found three girls' bodies in the park, right down at the bottom of the hill.

JULIA (late 60s): I know you've heard about some of the things that have happened over the last four or five years in the park. There were bodies found. There was about four or five bodies that was found, to my knowledge, in the park.

. . .

TINA (late teens): I will not go back there.

MODERATOR: Why?

TINA: Because, I think it was last year, they found a dead body back there in the creek.

AMY (late teens): There's probably dead bodies in that creek now.

TINA: It was some girl and she was back there dead in the water.

The park's social deterioration and its perception as a risky landscape have been accompanied, even expedited, by the physical ecological changes it has experienced since the arrival of blacks in the early 1960s, when budget and personnel cuts and the dissolution of the once omni-present Fairmount Park Guard (see Figure 13.4) coincided with larger demographic changes occurring throughout the city (see Figure 13.5). The park's narrow geography has resulted in severe ecological and structural changes that neither the local community nor the Fairmount Park Commission has been able to adequately respond to, either politically or economically. Today, extensive patches of non-native, weedy growth combine with the ambient risk of violence to produce a wide-ranging and increasingly gender-specific fear that is demonstrative of a perceived complete loss of social control aimed directly at the park's ecological structure (see Figure 13.6).

RACHEL (late teens): It's like there's forest back there. Once you get past the playground there's like a whole lot of trees and weeds and stuff. So once you get back there, once you get behind that part of the playground anything can happen to you!

. . .

MODERATOR: Are there certain parts of the park that you think are less safe than other parts?

TONYA (late teens): In all those invisible places.

KATE (late teens): Behind all those trees.

TYREISHA (late teens): It's like if you go back there behind the playground, the trees are not safe.

. . .

TYRONE (late 40s): The grass isn't cut. You can't see because of all the weeds. The weeds are out of control. We won't allow our kids to go but 6 or 10 feet into the [woods] because of all the weeds. We can't see them. That's so dangerous. Anything can happen. Kids can get stabbed, kidnapped, murdered.

As a result, Cobbs Creek Park has, for all intents and purposes, been abandoned by, especially, local women and children who would rather deny themselves the opportunities that come from public space access and participation than risk the possibility of violence (see Warr 1985; Valentine 1989).

Figure 13.4 "At the stables", Fairmount Park Guard, *circa* 1960

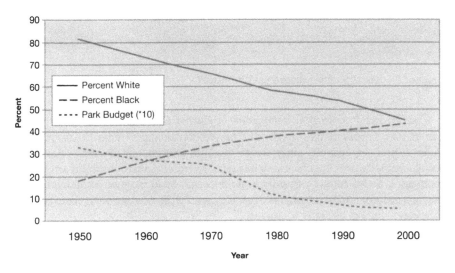

Figure 13.5 Population and park budget trends, 1950–2000 (budget numbers multiplied by 10 for purposes of illustration)

MONIQUE: I'm 34 years old and I don't even take my kids there. I don't even mess with it and I live on the top of the block. My block comes right down – at the bottom of my block I'm right at the park. I don't even walk my dog there . . . My kids used to play in the park before it got like it is now. My kids don't go in that park! My kids *do not* go down in that park [emphasis in original].

NASHEEDA (late teens): I will not go down there by myself. I will not! I see people walk down there, walking on the trail by themselves, and I'll say "Are you sure you want to walk down there?" I'm not going!

SHIRLEY (early 60s): I don't do anything over there. Absolutely nothing. And I love the park!

Figure 13.6 Ecology of fear (kudzu in Cobbs Creek)

Clearly, local ecological conditions in Cobbs Creek are interpreted by the community's residents as indicative of the greater social problems that accompany racial spatial segregation, generally; to this end, uneven, out-of-control ecologies are discursively linked to the social histories and geographies of racism, economic inequality, and a culture of political neglect that have plagued Philadelphia for the past century (Adams *et al.* 1990; cf. Heynen 2003). Uncontrolled ecologies are interpreted locally as racist ecologies.

TEWANNA (early 50s):   The interest was there back in the day, when our parents would take us up [to the park]. It was like an outing. Just to get off the street and let the kids rip and run. You didn't have to worry . . . Today, you know, *there's no control* [emphasis in original].

JAMES (early 50s):   What's going on over here at District 4 – they're undermanned; they're not responsive . . . What I hear from District 4 is their manpower is removed from this area and sent out to points in the northeast or [elsewhere]. And the reason is because there's no representation, there's no hammer, there's nobody here that will keep what we need here and make the demands and see that they're fulfilled. That's what we're dealing with . . . We need a dedicated security like they have at Pennypack or like they have at Wissahickon. They have rangers. People who know the rules as far as the park service is concerned.

CAROL (early 70s):   There has to be a reason why certain parks are so much better, or why certain parts of the city are so much better than our park . . . What we found out was that most of the grants written by Fairmount Park were designated for other sections and also that monies given to the park were not distributed equally among the various parks. And Cobbs Creek didn't get its fair portion . . . If you didn't know to ask you didn't get it. They had grants that they had written for the Pennypack, the Wissahickon – we didn't get those. So, it turned out for maybe every $6 they were spending in other sections of the park, they were spending about a dollar here.[4]

Unsurprisingly, accusations of racism and park mismanagement abound in Cobbs Creek, but the history of neglect is well understood even amongst park authorities. Indeed, the director of NLREEP reflects on the matter concisely, portraying (if not exactly displaying) the "out of sight, out of mind" philosophy that has long-governed Philadelphia's racist political geography, even in park governance.

[Cobbs Creek Park] is not as visible to most people . . . And I think the resources of [Fairmount Park], what they have, they want to go into areas that are *most visible to important people*. People that live and work downtown or people who live in [exclusively white, upper-middle class] Chestnut Hill and work downtown, or that are driving through these areas. That's where their priorities are; places that *most* people see [emphasis added].

The racist ecologies of Cobbs Creek are, it appears, the result of a deliberate campaign to steer (scarce) resources towards and emphasize the place-making process in those areas of the city widely considered to be economically lucrative and politically powerful (e.g., Center City, Chestnut Hill), disproportionately putting marginalized black communities at the (environmental) risk of violence victimization. However, results of this legacy extend beyond the production of uneven ecologies, the disproportionate distribution of risk, fear and avoidance, or the collective resentment over the politics and geographies of race-based neglect that have governed park management priorities; indeed, for many in Cobbs Creek, it seems that this collection of factors, patterns, processes, and circumstances have combined over time to condition and justify the formation of a "new" narrative of the normative, controlled landscape. Here, the antithesis to uncontrolled, racist nature – the manicured lawn, or garden – has emerged as the ideal of the well-tended, well-managed, well-groomed, equitable, aesthetic, fearless, and, most importantly, accessible urban ecology (cf. Robbins and Sharp 2003).

JULIA (late 60s): I start hollering for . . . the greenery to be cut, and for the grass to be cut and all that stuff. I found that it's not necessary to be a mover and a shaker but if you start screaming loud enough you're gonna be heard eventually. And they [the Fairmount Park Commission] knew that these things should have been done, but they're like, "Hey, if you don't care we don't care!"

LAUREN (early 50s): [comparing Cobbs Creek Park with a park surrounded by white communities] That park is so beautiful! It is so beautifully grassed all over . . . It's such a big contrast! . . . They have a spot to be proud of . . . And they have to be very restrictive about who comes in there, and they are and there result is a beautiful park . . . It's like from zero to a hundred in difference in the way it looks [compared to Cobbs Creek].

Surely, the two narratives of controlled nature (lawn vs. restored, native ecology) are poised for conflict, calling forth the age-old debate of the tamed, domestic garden vs. the untamed wilderness (Glacken 1967; Schama 1995) albeit in a distinctly different, historically contingent, urban social economic context. I close this chapter with thoughts on their possible articulation.

ANGLES OF ARTICULATION: OLD MEETS NEW IN COBBS CREEK

Judging from this historically conditioned local position, where legacies of racism, poverty, and the disproportionate distribution of risk are forefront to perceptions of the environment and interpretations of environmental change, pronouncements by restoration advocates of "helping the [Cobbs Creek] community to understand the ecological importance of what they have [through participating in restoration activities]" appear, at best, naïve and misinformed or, at worst, patronizing, unsympathetic, and uninterested.[5] Uninformed, acontextual, and socially, locally ahistorical ambitions of restoration to some "native" landscape type, and of an environmental education intended to reproduce this normative vision, are no more likely to succeed in producing locally informed and ecologically-conscious citizens in Cobbs Creek as they are to influence the racist politics or erase the racist legacies that constitute contemporary geographies of fear and exclusion (cf. Light forthcoming; Merchant 1986). Similar lessons were learned the hard way in Chicago, when ecologically minded and socially apolitical restoration interests were confronted (and eventually thwarted) by a determined local resistance embracing a very different, locally-informed and conditioned environmental imaginary (see Gobster and Hull 2000).

Despite their differences, however, overlaps between the two narratives are apparent; entrepreneurial and inherited narratives of nature are both products of and responses to earlier industrial fragmentations, both social and ecological. As I hope I have demonstrated in this chapter, however, these fragmentations are historically and discursively inseparable, paralleling one another as they did over the course of Philadelphia's developmental history; fragmentations of the one (e.g., natural landscapes) were often used to maintain, justify, and reproduce fragmentations of the other (e.g., social divisions). These inherited social and ecological divisions continue to splinter and separate urban society and ecologies (Solecki and Welch 1995).

However, the implications of decades of park neglect are no more simply ecological than they are merely social, and restoration or other eco-modernization efforts that emphasize ecological changes at the expense of their social corollaries, or that advance narratives of nature that are (socially, politically) ahistorical and thus locally meaningless or unachievable will, at best, provoke local resentment and resistance or, at worst, fail entirely, as was the case in Chicago. Alternatively, projects or programmes of, for instance, community development that fail to incorporate local ecologies and environmental histories into their relative agendas may facilitate further changes to the local environment that reproduce the kind of fractured social ecologies we see in West Philadelphia. Nonetheless, in Cobbs Creek there does appear to be an opportunity to reconcile, or articulate, entrepreneurial and inherited narratives of nature, if only partially or peripherally. Despite their differences – political, social, or otherwise – there are, nonetheless, similarities that may be the seed for further dialogue and negotiation. First, each advances a vision of controlled nature that is, in principle (if not in spirit), a response to the legacies and inequities (social, environmental, or otherwise) of industrial fragmentation. Each is prescriptive in that they advance normative ecologies whose relative success will be judged by the removal of "out of place" elements in the natural landscape – i.e., weeds. At first sight, these seem to constitute a node of discursive and political overlap and articulation around which the manipulation of local ecologies may be a means for both entrepreneurial and locally relevant ends. Unfortunately, there is little in the neoliberal literature, nor in the case study discussed here, indicating the likelihood or possibility of political and discursive agreement; the economic demands and social intolerances that underscore and advance current discourses and programs of neoliberal urbanization – including eco-modernization (Keil and Desfor 2003) – seem to hold little promise that such thoughtful and socially encompassing discussion will occur. Rather, evidence suggests that, in all likelihood, a complex, fragmented assemblage of local, historically specific environmental narratives – including those in Cobbs Creek – is beginning to emerge to contest or encourage not only the narrative of restored, entrepreneurial nature currently flowing from the Master Restoration Plan of the regional place-makers but, as well, to contest, encourage, or compete with one another for the rights and resources to advance or manage for their own colloquial visions of Fairmount Park's normative ecology. How and where the chips fall and the identity, power, and membership of new environmental coalitions remain to be seen; however, in Philadelphia, as (perhaps) in most of the post-industrial cities of the north, the inherited social and environmental fragmentations of the industrial past are the stock of the new fragmentations of neoliberalism and creative destruction.

NOTES

1 It is fair, I think, to argue that the neoliberal, entrepreneurial discourse, including some entrepreneurial narrative of nature, has been appropriated by urban parks advocates as a means to and end – that is, a renewed commitment by urban decision makers to urban parks following decades of political ambivalence and neglect (Harnik 2000). Best intentions aside, however, the move to advocate urban parks as just objects in a competitive economic landscape deserves theoretical critique and explanation.
2 NLREEP has since been absorbed into the Fairmount Park Commission. The Yale School of Forestry and Environmental Studies is involved in a number of urban ecological research projects on the east coast, including Baltimore and New Haven.
3 The two most recent victims, both women, were discovered three months prior to the writing of this article.

4   The Wissahickon and Pennypack parks are two other watershed parks. They are
    surrounded by generally white, middle to upper-middle class white communities.
5   Director, NLREEP, speaking on community participation in Cobbs Creek.

BIBLIOGRAPHY

Adams, C., Bartelt, D., Elesh, D., Goldstein, I., Kleniewski, N., and Yancey, W. (1990)
    Philadelphia: Neighborhoods, Division, and Conflict in a Postindustrial City.
    Philadelphia, PA: Temple University Press
Amin, A. (ed.) (1994) *Post-Fordism: A Reader*. Malden, MA: Blackwell
Antrop, M. (2004) "Landscape change and the urbanization process in Europe", *Landscape
    and Urban Planning*, 67: 9–26
Bissinger, B. (1997) *A Prayer for the City*. New York: Vintage
Botkin, D. (2001) "The naturalness of biological invasions", *Western North American
    Naturalist*, 61: 261–266
Brenner, N. and Theodore, N. (eds) (2002a) *Spaces of Neoliberalism: Urban Restructuring
    in North America and Western Europe*. Malden, MA: Blackwell
Brenner, N. and Theodore, N. (2002b) "Cities and the geographies of actually existing
    neoliberalism", in N. Brenner and N. Theodore (eds) *Spaces of Neoliberalism: Urban
    Restructuring in North America and Western Europe*. Malden, MA: Blackwell
City of Philadelphia (1999) *Philadelphia: A New Urban Direction*. Philadelphia: St. Joseph's
    University Press
Cowell, R. and Thomas, H. (2002) "Managing nature and narratives of dispossession:
    reclaiming territory in Cardiff Bay", *Urban Studies*, 39: 1,241–1,260
Cranz, G. (1982) *The Politics of Park Design: A History of Urban Parks in America*.
    Cambridge, MA: MIT Press
DeFillipis, J. (1997) "From a public re-creation to private recreation: the transformation of
    public space in South Street Seaport", *Journal of Urban Affairs*, 19: 405–417
Eade, J. (1997) "Reconstructing places: changing images of locality in docklands and
    Spitalfields", in J. Eade (ed.) *Living the Global City: Globalization as Local Process*.
    London: Routledge
Fairmount Park Commission (1999) *Fairmount Park Ecological Restoration Master Plan*.
    Philadelphia, PA: Fairmount Park Commission
Falck, Z.J.S. (2002) "Controlling the weed nuisance in turn-of-the-century American cities",
    Environmental History, 7: 611–631
Feldman, T.D. and Jonas, A.E.G. (2000) "Sage scrub revolution? Property rights, political
    fragmentation, and conservation planning in Southern California under the federal
    Endangered Species Act", *Annals of the Association of American Geographers*, 90:
    256–292
Forman, R. (1998) *Land Mosaics: The Ecology of Landscapes and Regions*. Cambridge:
    Cambridge University Press
Freestone, R. and Nichols, D. (2004) "Realising new leisure opportunities for old parks: the
    internal reserve in Australia", *Landscape and Urban Planning*, 68: 109–120
Gandy, M. (2002) *Concrete and Clay: Reworking Nature in New York City*. Cambridge,
    MA: MIT Press
Garvin, A. and Berens, G. (eds) (1997) *Urban Parks and Open Space*. Washington, DC:
    Urban Land Institute
Gibbs, D. and Jonas, A.E.G. (2000) "Governance and regulation in local environmental
    policy: the utility of a regime approach", *Geoforum*, 31: 299–213

Glacken, C.J. (1967) *Traces on the Rhodian Shore: Nature and Culture in Western Thought from Ancient Times to the End of the Eighteenth Century*. Berkeley, CA: University of California Press

Gobster, P.H. and Hull, R.B. (eds) (2000) *Restoring Nature: Perspectives from the Social Sciences and Humanities*. Washington, DC: Island Press

Goldenberg, N. (1999) "Philadelphia launches major restoration initiative in park system", *Ecological Restoration*, 17: 8–14

Hacker, A. (1995) *Two Nations: Black and White, Separate, Hostile, Unequal*. New York: Ballantine

Harding, S. (1999) "Towards a renaissance in urban parks", *Cultural Trends*, 35: 3–20

Harnik, P. (2000) *Inside City Parks*, Washington, DC: Urban Land Institute

Harvey, D. (1985) *The Urbanization of Capital*. Baltimore, MD: Johns Hopkins University Press

Harvey, D. (1989) "From managerialism to entrepreneurialism: the transformation in urban governance in late capitalism", *Geografiska Annaler*, 71B: 3–17

Harvey, D. (1992) "Capitalism: the factory of fragmentation", *New Perspectives Quarterly*, Spring

Harvey, D. (1996) *Justice, Nature, and the Geography of Difference*. Malden, MA: Blackwell

Heynen, N.C. (2003) "The scalar production of injustice within the urban forest", *Antipode*, 35: 980–998

Jonas, A.E.G. and Gibbs, D.C. (2003) "Changing local modes of economic and environmental governance in England: a tale of two areas", *Social Science Quarterly*, 84: 1,018–1,037

Karasov, D. and Waryan, S. (eds) (1993) *The Once and Future Park*. Princeton, NJ: Princeton Architectural Press

Keil, R. and Graham, J. (1998) "Reasserting nature: constructing urban environments after Fordism", in B. Braun and N. Castree (eds) *Remaking Reality: Nature at the Millenium*. New York: Routledge

Keil, R. and Desfor, G. (2003) "Ecological modernization in Los Angeles and Toronto", *Local Environment*, 8: 27–44

Kipfer, S., Hartmann, F., and Marino, S. (1996) "Cities, nature and socialism: towards an urban agenda for action and research", *Capitalism, Nature, Socialism*, 7: 5–19

Latty, Y. (2000) "Trashing Cobbs Creek", *Philadelphia Daily News*, 23 February

Lauria, M. (ed.) (1997) *Reconstructing Urban Regime Theory: Regulating Urban Politics in a Global Economy*. Thousand Oaks, CA: Sage

Light, A. (forthcoming) "Restorative relationships", in R. France (ed.) *Healing Nature, Repairing Relationships: Landscape Architecture and the Restoration of Ecological Spaces*. Cambridge, MA: MIT Press

Lindig-Cisneros, R. and Zedler, J.B. (2000) "Restoring urban habitats: a comparative study", *Ecological Restoration*, 18: 184–192

Massey, D. and Denton, N. (1989) "Hypersegregation in United States metropolitan areas: Black and Hispanic segregation along 5 dimensions", *Demography*, 26(3): 373–391

Massey, D. and Denton, N. (1993) *American Apartheid: Segregation and the Making of the Underclass*. Cambridge, MA: Harvard University Press

Mayer, M. (1994) "Post-Fordist city politics", A. Amin (ed.) *Post-Fordism: A Reader*. Malden, MA: Blackwell

McCarthy, J. and Prudham, S. (2004) "Neoliberal nature and the nature of neoliberalism", *Geoforum*, 35: 275–283

McGinnis, M.V. and Woolley, J.T. (1997) "The discourses of restoration", *Restoration and Management Notes*, 15: 74–77

McInroy, N. (2000) "Urban regeneration and public space: the story of an urban park", *Space & Polity*, 4: 23–40

Merchant, C. (1986) "Restoration and reunion with nature", *Restoration and Management Notes*, 4: 68–70

Mooney, H.A. and Hobbs, R.J. (eds) (2000) *Invasive Species in a Changing World*. Covelo, CA: Island Press

Mumford, L. (1955) "The natural history of urbanization", in W.L. Thomas (ed.) *Man's Role in Changing the Face of the Earth*. Chicago, IL: University of Chicago Press

Peck, J. and Tickell, A. (2002) "Neoliberalizing space", in N. Brenner and N. Theodore (eds) *Spaces of Neoliberalism: Urban Restructuring in North America and Western Europe*. Malden, MA: Blackwell

Pickett, S.T.A. and White, P. (eds) (1985) *The Ecology of Natural Disturbance and Patch Dynamics*. London: Academic Press

Pincetl, S. (2003) "Nonprofits and park provision in Los Angeles: an exploration of the rise of governance approaches to the provision of local services", *Social Science Quarterly*, 84(4): 979–1,001

Platt, R.H., Rowntree, R.A., and Muick, P.C. (eds) (1994) *The Ecological City*. Amherst, MA: University of Massachusetts Press

Robbins, P. and Sharp, J. (2003) "The lawn-chemical economy and its discontents", *Antipode*, 35: 955–980

Rosenzweig, R. and Blackmar, E. (1992) *The Park and the People: A History of Central Park*. Ithaca, NY: Cornell University Press

Saunders, D.A., Hobbs, R.J., and Margules, C.R. (1991) "Biological consequences of ecosystem fragmentation: a review", *Conservation Biology*, 5: 18–32

Schama, S. (1995) *Landscape and Memory*. New York: Knopf

Smith, L. (2004) "The murky waters of the second wave of neoliberalism: corporatization as a service delivery model in Cape Town", *Geoforum*, 35: 375–393

Smith, L. and Hanson, S. (2003) "Access to water for the urban poor in Cape Town: where equity meets cost recovery", *Urban Studies*, 40(8): 1,517–1,548

Solecki, W.D. and Welch, J.M. (1995) "Urban parks: green magnets or green walls?", *Landscape and Urban Planning*, 32: 93–106

Sprugel, D.G. (1991) "Disturbance, equilibrium, and environmental variability: what is "natural" vegetation in a changing environment?", *Biological Conservation*, 58: 1–18

Stormann, W.F. (1991) "The ideology of the American urban parks and recreation movement: past and future", *Leisure Sciences*, 13: 137–151

Swyngedouw, E. (1989) "The heart of the place: the resurrection of locality in an age of hyperspace", *Geografiska Annaler*, 71B: 31–42

Taylor, D.E. (1999) "Central Park as a model for social control: urban parks, social class and leisure behavior in nineteenth-century America", *Journal of Leisure Research*, 31: 420–477

Thompson, C.W. (2002) "Urban open space in the 21st century", *Landscape and Urban Planning*, 60: 59–72

Valentine, G. (1989) "The geography of women's fear", *Area*, 21: 385–390

Warr, M. (1985) "Fear of rape among urban women", *Social Problems*, 32: 238–250

Watts, M. and Peet, R. (1996) "Towards a theory of liberation ecology", in R. Peet and M. Watts (eds) *Liberation Ecologies*, New York: Routledge

Whitehead, M. (2003) "(Re)Analysing the sustainable city: nature, urbanization, and the regulation of socio-environmental relations in the UK", *Urban Studies*, 40: 1183–1206

With, K.A. (2002) "The landscape ecology of invasive spread", *Conservation Biology*, 16: 1192–1203

Zukin, S. (1995) *The Cultures of Cities*, Malden, MA: Blackwell

14 Transnational alliances and global politics

New geographies of urban environmental justice struggles

David N. Pellow

INTRODUCTION: MODERNITY AND URBAN POLITICAL ECOLOGY

Modernity is almost universally equated with the degree to which a nation and its constituent urban centers are integrated into the global economy. One of the key paths to this integration is via a commitment to technology and infrastructure associated with the "new economy". Modernity and information technology are now considered symbiotic (Wankade and Argawal, forthcoming). The "digital divide" between those with or without access to this technology has become a public policy concern in recent years. Global North nations are heavily "wired", with more than two-thirds of adults in the US actively using computers and the Internet. Access to this technology in most nations of the South is minimal, although growing. In the North, our transportation, commercial, educational, military, media, and governmental infrastructures all rely on computerization for data management, processing, and storage, as well as for private and mass communication. The various sectors that comprise the high technology industry form the largest manufacturing effort in the world. This industry is also one of the most rapidly changing in human history. For example, the speed of the microchip – the "brain" in most computerized devices – has doubled every 18 months for more than three decades. This also means that consumer electronics – particularly personal computers – are becoming obsolete quite rapidly. What happens to all those computers and other electronics goods – also known as "e-waste" – once they are discarded? They are often shipped to urban areas and villages across Asia, Africa, and Latin America, where residents/workers disassemble them for sale in new manufacturing processes. Because each computer contains several pounds of highly toxic materials, this practice creates a massive transfer of hazardous waste products from North to South, and is responsible for impacting public health and the integrity of watersheds in numerous nations such as India, Pakistan, China, the Philippines, and Taiwan, for example. This process has also transformed rural or semi-rural regions into emerging urban spaces and created greater population and toxic pollution densities in heavily urbanized locations.

While the problem of electronic waste reflects the uneven social position of urban dwellers in the North versus those in the South, it is also emblematic of new ways in which the material, discursive, and cultural infrastructures of urban centres are dependent upon (and yet destructive of) natural resources (Swyngedouw and Heynen 2003). The ways in which computerization and digitization have impacted the ecological and structural basis of urban social inequalities are too numerous to name here (see Pellow and Park 2002), but leave no doubt that the theses of post-materiality and "weightless" economies in the Digital Age are unsubstantiated (Keil 2003: 729). Through the electronics revolution, for example, income inequalities between workers and managers in computer firms have increased steadily as has the drain on natural resources required to fuel the production of electronics commodities for global markets. These technologies also facilitate the development and dissemination of various discourses and cultural imaginings, the most effective of which reinforce capitalist and racist hierarchies within and across societies – although, as I show in this chapter, resistance movements are producing discourses and seeking structural changes to counter this hegemony. Thus, placing this case study within the larger context of this book, the struggle over the socio-environmental impacts of computerization reflects the view from scholars of urban political ecology (UPE) that urban spaces are sites of contestation over structural/physical/natural *and* discursive/cultural/symbolic goods (Peet and Watts 2004).

In this chapter I examine the origins of this problem and how the "low-tech" side of "high tech" impacts public and environmental health in urban centres around the globe. There is a sophisticated grassroots transnational effort to document these problems and activists have had success at changing corporate environmental policies and passing local, national, and international legislation to address the worst dimensions of the "e-waste" issue. I will consider these dynamics in the context of urban political ecology, social movement theory, and current debates on the power of resistance in a global political economy. I argue that urban ecological crises are best addressed through a combination of movement strategies that target both states and large private corporations.

SOCIAL MOVEMENTS IN A GLOBAL POLITICAL ECONOMY

Social movements result when networks of actors relatively excluded from routine decision-making processes engage in collective attempts to change "some elements of the social structure and/or reward distribution of society" (McCarthy and Zald 1977: 1,217). The critical components of most successful social movements include: organizing and mobilizing resources, framing their grievances and goals, and engaging the political opportunity structures that constrain or enable social change. In this chapter, I will focus mainly on organizing/mobilizing and political opportunity structures.

Local and domestic/national social movements have been the primary scholarly emphasis since the study of movements by sociologists began in the nineteenth century. But with the continued proliferation of transnational movements, recent scholarship has paid more attention to this phenomenon than ever before (Smith *et al.* 1997). When it becomes clear that one society's actions impact another society, or an organization (e.g. a corporation) based in one nation affects people's lives and environment in another, we must expand our scope of study to include the transnational nature of movements in the modern era. Thus transnational social movement organizations (TSMOs) specialize in "minding other people's business" (Tarrow 1998: 189).

Social movements have become quite sophisticated in their understanding of globalization and in their efforts to combat its associated negative social impacts at multiple levels. Some scholars view the structure of globalization itself as facilitative of the development of global forms of resistance (Hardt and Negri 2000). What tactics and strategies are movements developing in a globalizing world? How effective are transnational social movements at producing change? What are the targets of social movements and why does this matter? How do race, class, and nation intersect to produce political barriers and openings for TSMOs? How does expanding urbanization impact and link peoples of the global North and South? These are some of the questions I explore in this chapter and that scholars of movements have only just begun to consider.

TSMOs are largely based in the Northern Hemisphere for reasons of access to global decision-makers in world cities such as London, New York, and Washington, D.C., but also because the telecommunication and transportation infrastructure in these nations are often more supportive of rapid and intense utilization by activists who may need to communicate or mobilize on very short notice. And while the leadership of most TSMOs is native to the global North, there are also a number of activists from the global South who live in the North and work for TSMOs there. There are also a growing number of TSMOs in the South, particularly in urban centers like New Delhi, Manila, Durban, Penang, and Chiang Mai. Northern and Southern environmentalists have long acknowledged that, in order to successfully engage the political and economic proponents of globalization, they must collaborate.

Social movements must mobilize resources – funds, technology, people, ideas, etc. – to achieve their goals. TSMOs are rarely successful if we narrowly define success as a major change in a specific policy within a nation state (Keck and Sikkink 1998; Smith *et al.* 1997). But they are increasingly relevant in international policy debates, as they seek to make not only policy changes in international law and multilateral conventions, but also to change the terms and nature of the discourse within these important debates. These conventions include, for example: the Montreal Protocol (on the production of ozone-damaging chemicals), the Kyoto Protocol (on global warming), the Basel Convention (on the international trade in hazardous wastes), and the Stockholm Convention (on the production and management of persistent organic pollutants). In each of these cases, TSMOs are often a critical source of information for governments seeking to learn about a problem, and their presence raises the costs of failing to act on certain issues, thus increasing the possibility of government accountability. In a global society where a nation-state's reputation on a range of matters can be tarnished in international political and media venues, TSMOs can have real impacts. Specifically, when information disseminated becomes a part of common wisdom, such "popular beliefs . . . are themselves material forces" (Gramsci 1971: 165). That is, meaning systems can support or challenge systems of structural and material control (Moore 1993). This is a critical point because, as urban political ecologists have argued, social movements are struggling over cultural meaning systems as much as they are fighting for improved material conditions and needs (Escobar 1992).

Resource mobilization is only part of the story. One must also have (or create) an opening in the political process in order to realize a movement's goals. In most of the research on social movements, the political opportunity structure consists mainly of the following dimensions:

1 The relative openness or closure of the institutionalized political system.
2 The stability of that broad set of elite alignments that typically undergird a polity.
3 The presence of elite allies who sympathize with or support social movements.
4 The state's capacity and propensity for repression.

(McCarthy 1997: 255)

Keck and Sikkink (1998: 7) define the political opportunity structure as "differential access by citizens to political institutions like legislatures, bureaucracies, and courts". I contend that this model is rooted in a "state-centric" perspective (McAdam 1996: 34). Political process research tends to present the state as the primary movement target or main vehicle of reform. Many movements indeed share this state-centric orientation. However, numerous movements today increasingly view the nation-state as weak – no longer a natural ally or enemy – and sometimes as just one of many players (if not a marginal player) in their struggles for change. Thus, the primary targets and major sites of reform are no longer centred solely within traditional political institutions. The state and the political elite are only one component of the political process and are sometimes circumvented by movements seeking to challenge more powerful institutions, such as large corporations.

During the 1990s, several international free trade agreements among nations emerged for the purpose of removing obstacles to trade and commerce and allow the unencumbered mobility of commodities and currencies around the globe. The North American Free Trade Agreement (NAFTA) and the World Trade Organization (WTO) produced governance structures that would successfully weaken environmental laws (among others) among member nations. Under such a system, state sovereignty and policy-making capacity are formally subverted as non-state institutions (i.e. supranational trade organizations heavily influenced by corporations) are authorized to rule on a vast range of laws on behalf of private investors and corporations. These global structures have direct impacts on citizens and social movements across nations, independent of elected officials' efforts to close or open the traditional political process. These developments present major challenges to social movements and nation-states. Thus, in the wake of these free trade agreements, if movements (particularly labour and environmental) continue to focus on the nation-state (and its sub governments) as their sole target, they will have misdirected much of their energy.

In light of this changing political and economic terrain, I have proposed a modification of the political process model – a *political economic process* perspective (Pellow 2001). This extension of the political process model recognizes the close links between traditional political institutions (e.g. states and legislative bodies) and economic organizations (e.g. corporations and international financial institutions) and their interactions with social movements (McAdam 1996). It emphasizes the power of private capital over nation-state governance, and acknowledges that corporations are often the primary targets of social movements, both domestic and transnational. In other words, movements must confront both state and corporate entities, because these are the primary institutions shaping the system of power in a global society. As Peet and Watts (1993: 240) point out, scholars of political ecology have paid considerable attention to this phenomenon, as they are "concerned with institutions and organizations in the context of shifting configurations of state and market roles".

For urban political ecologists, this is all part of a much broader on-going discussion and analysis of neoliberalism and its effects on urban spaces and politics. Smith (2002: 433), for example, argues that while there is no question that the state's economic power is eroding, its political and cultural power may not be. Activists have made use of this shift in creative ways. The political power of states is critical, for example, when considering efforts to achieve social change at the global scale. Hence the intense level of resources many TSMOs have put into negotiating international environmental treaties – which are *supra*national institutional configurations.

The state's precarious economic position may provide other spaces for grassroots movement innovation as well. Consider Isin's (1998), view that neoliberalism is not just a form of state retreat, but also a complex set of changing technologies of power "whereby citizens are redefined as clients and autonomous market participants who are responsible for their own success, health, and well-being" (Keil 2002: 582). If we accept this model, then it makes perfect sense for citizen-activists and social movements to engage in "venue *shopping*" (Baumgartner and Jones 1991: 1050) for the institutions whose reform will give them the most bang for their political "buck". This is precisely what transnational social movement organizations do – targeting corporations, international financial institutions, and other venues – when the state fails to provide access to paths for reform. To a great degree, these dynamics have been with us for a long time. For example, the central challenge for movements in this context is best summed up in the following question: "What role can civil society play in a world where governments and corporations have reigned supreme for centuries?" (Burbidge 1997).

Why does all of this matter? The importance of understanding the shift from state to corporate power in national and global politics is critical because states are a form of governance (at least in democracies) that are supposed to be accountable to the citizenry. Corporations are not. The only constituents to whom they are accountable are the firm's stockholders. In fact, as long as they are in compliance with local and national laws, they are free to make decisions that exploit and abuse workers, communities, and consumers, as long as their shareholders are satisfied. So if institutions of this ilk are making decisions about the public good and the general welfare of a population and its environment, we are faced with a troubling reality. This is particularly acute in urban areas where poor, working class, immigrants, and people of colour are concentrated in economically deprived and politically less powerful enclaves that are frequently burdened with the environmental cast-offs of production and urban consumption patterns. No form of pollution embodies this dilemma of urban environmental racism better than electronic waste.

€L€CTRONIC WAST€

Electronic waste (e-waste) encompasses a broad and growing range of electronic devices ranging from large household appliances such as refrigerators, washers and dryers, air conditioners, hand-held cellular phones, fluorescent lamp bulbs, and personal stereos. Where once consumers purchased a stereo console or television set with the expectation that it would last for a decade or more, the increasingly rapid evolution of technology has effectively rendered everything disposable. Consumers no longer take a malfunctioning toaster, VCR or telephone to a repair shop. Replacement is often easier and cheaper than repair. And while these ever-improving gadgets – faster, smaller, and cheaper – provide many benefits, they also carry a legacy of waste (SVTC 2001: 2). The most visible and harmful component of e-waste today is the personal computer.

E-waste is the most rapidly growing waste stream in the world. It is a crisis not only of quantity but also one born of toxic ingredients – such as the lead, beryllium, mercury, cadmium, hexavalent chromium, and brominated-flame retardants that pose extraordinary occupational and environmental health threats (BAN and SVTC 2002: 1). Computer and television displays (CRTs) contain an average of 4 to 8 pounds of lead each. The estimated 315 million computers that became obsolete between 1997 and 2004 contain a total of more than 1.2 billion pounds of lead. Monitor glass contains about 20 percent lead by weight. When these components are illegally disposed of and crushed in landfills, the lead is released into the environment, posing a hazardous legacy for current and future generations. Consumer electronics already constitute 40 percent of lead found in landfills. About 70 percent of the heavy metals (including mercury and cadmium) found in landfills come from electronic equipment discards. These heavy metals and other hazardous substances found in electronics can contaminate groundwater and pose other environmental and public health risks. Lead can cause damage to the central and peripheral nervous systems, blood system and kidneys in humans. Lead accumulates in the environment, and has highly acute and chronic toxic effects on plants, animals and microorganisms. Children suffer developmental effects and loss of mental ability, even at low levels of exposure. Computers and other electronics constitute a significant component of the physical and communicative infrastructure of cities. Given the level of natural resources required to produce these commodities, and the ecological damage that results from their production and disposal, e-waste is a symptom of the problematic of the human–nature interactions inherent in urban spaces (Keil 2003).

An estimated 80 percent of the US's computer waste collected for recycling is exported to Asia, where it is known to be dumped and recycled under very hazardous conditions (GrassRoots Recycling Network 2003a). Environmental activists have called this "toxic colonialism" and a "global environmental injustice" (Puckett, personal communication, 5 March 2002).

ENVIRONMENTAL INJUSTICE AND THE POLITICAL ECONOMY OF E-WASTE

The high correlation between poverty, race, and pollution in the United States has been variously referred to as environmental racism, environmental inequality, and environmental injustice (Bullard 2000). This is also a problem with a global reach.

It has become clear that, like many other forms of pollution, e-waste also follows the "path of least resistance" and finds its way into nations that are poor and largely populated by non-European peoples – a form of global environmental racism. Governments and industry leaders allow (if not encourage) these practices because they facilitate profit-making and the creation of low-wage work for many residents of economically desperate communities. Ravi Agarwal, an activist with Toxics Link India, states: "As developing countries become cleaner and it becomes very expensive to dispose of waste because of rising environmental standards and labor standards, such waste finds itself in places like India and South Asia, and South East Asia" (Asia Pacific nd).

Jim Puckett, one of Agarwal's transnational movement collaborators at the Seattle-based Basel Action Network (BAN), concurs. He argues that the global trade in e-waste is a shady business that "leaves the poorer peoples of the world with an untenable choice between poverty and poison" (Gough 2002). These two activists and their organizations have joined forces across thousands of miles to ensure that there is some accountability on the part of industries and consumers in urban centres of the global North whose waste ends up in the urban centres of the South. Puckett states:

> This mentality perpetuated now by the United States is an affront to the principle of environmental justice, which ironically was pioneered in the United States and championed by the EPA domestically. The principle states that no people because of their race or economic status should bear a disproportionate burden of environmental risks. While the United States talks a good talk about the principle of environmental justice at home for their own population, they work actively on the global stage in direct opposition to it.
>
> (Basel Action Network & Silicon Valley Toxics Coalition 2002: 29)

Activists from other SMOs that have worked on the e-waste problem echo these sentiments. As a communiqué from the GrassRoots Recycling Network declared in response to the USEPA's recent decision to allow e-waste exports, "Asian peoples are now asked to accept pollution that we have created simply because they are poorer" (GrassRoots Recycling Network 2003). Responding to similar reports, Von Hernandez of Greenpeace International stated, "Asia is the dustbin of the world's hazardous waste" (Vidal 2004). Each of the above-mentioned social movement organizations has collaborated to challenge global environmental racism (in the form of e-waste dumping) at the discursive/symbolic level as well as through efforts to reform the material/structural practices that embody environmental inequality. This is the essence of the transnational social movement form and the most effective way for resistance to take root in the context of neoliberal regimes.

€-waste in Asia

While there are documented cases of e-waste dumping in Mexico, West and East Africa, and elsewhere, the majority of this waste ends up in Asia. This is largely because of the deep commitment of many Asian governments to – and the success of many Asian corporations in – high technology and electronics. In urban centres in India, Pakistan, China, Vietnam, Malaysia, and the Philippines there has been a marked growth in the development of an infrastructure to support the electronics industry. Since so many of the electronics products sold on the global market are produced in Asia, e-waste dumping from markets abroad in Asia is viewed as logical, because the disassembled products can more efficiently be incorporated back into production there.

A recent report from the Basel Action Network (BAN) and Silicon Valley Toxics Coalition (SVTC) titled *Exporting Harm: The High-Tech Trashing of Asia*, documented the international trade in toxic electronic waste from the United States to China, India and other Asian nations (Basel Action Network & Silicon Valley Toxics Coalition 2002). Computer monitors, circuit boards and other electronic equipment collected in the US – sometimes under the guise of "recycling" – are regularly sold for export to Asia where the products are handled under deplorable conditions, creating tremendous environmental and human health risks. Workers – including children – use their bare hands, hammers, propane torches, and open acid baths to recover small amounts of gold, copper, lead and other valuable materials. What is unused is dumped in waterways, fields, and open trenches, or simply burned in the open air.

In a recent report by Toxics Link India titled *Scrapping the Hi-tech Myth: Computer Waste in India*, it was revealed that the "disposal and recycling of computer waste in the country has become a serious problem since the methods of disposal are very rudimentary and pose grave environmental and health hazards" (Toxics Link 2003: 5).

The import of hazardous waste into India is actually prohibited by a 1997 Indian Supreme Court directive, which reflects the Basel Ban (the international convention prohibiting the export of hazardous waste from OECD to non-OECD nations). Northern nations, however, continue to export e-waste to Southern nations like India, rather than managing it themselves. So the trade in e-waste is camouflaged and is a thriving business in India, conducted under the pretext of obtaining "reusable" equipment or "donations" from industrialized nations (Toxics Link 2003: 6).

Since 1995, the village of Guiyu in China's Guangdong province has been transformed from a poor, rural, rice-growing community to a booming e-waste processing centre. While rice is still grown in the fields, virtually all of the available building space has given way to providing many hundreds of small and often specialized e-waste recycling shelters and yards. One impact that has not gone unnoticed has been the deterioration of the local drinking water supply. The local residents claim that the water has become foul tasting and have it trucked in from 30 kilometers away. Water and soil sample tests confirmed that lead and heavy metal contamination was much higher than levels allowed by the USEPA and the World Health Organization (Basel Action Network & Silicon Valley Toxics Coalition 2002: 22).

Child labour is widespread in the e-waste workshops in China. The Chinese press has estimated the number of people employed in the sector to be as high as 100,000. A 60-year-old resident of the region told a reporter, "For money, people have made a mess of this good farming village . . . Every day villagers inhale this dirty air; their bodies have become weak. Many people have developed respiratory and skin problems. Some people wash vegetables and dishes with the polluted water, and they get stomach sickness" (BBC News 2002). The export of hazardous e-waste is rooted in a simple, albeit brutal, calculus. As the Silicon Valley Toxics Coalition's Ted Smith put it, "The reason is because it is ten times cheaper to send it to China than to recycle it here" (personal communication, 18 March 2002).

THE GLOBAL MOVEMENT FOR EXTENDED PRODUCER RESPONSIBILITY

As noted earlier, many TSMOs are based in urban centres of the global North. This is strategic in large part because so much of the political-economic power and decision-making authority rests in these spaces and the institutions located there. Thus, one would likely have a greater chance of changing a transnational corporation's practices in many global South nations by targeting that firm's headquarters in the North. And this is precisely what TSMOs have done, whether by targeting corporate practices directly, or via government legislation. This is the *political economic process* approach that activists have adopted to strategically apply pressure where the decision-making power rests. Sometimes the power is located primarily within state apparatuses, while other times it is within corporate boardrooms. Still other times it may be a combination of the two. Hardt and Negri (2000) conclude that the exercise and location of power under globalization has become quite diffuse, yet this system may create opportunities for resistance against it. Perhaps it follows then that in a global system such as this, consumers – a widely diffuse group of individuals – must be mobilized to produce social change. The grassroots pressure that TSMOs mobilize comes not only from their own membership and personnel, but also from the consumers of these technologies. Given that such a high percentage of citizens in the North are consumers of computer and electronic products, this is a considerable pool of potential activists who could be mobilized to pressure or even boycott any number of companies.

Four US-based SMOs have been critical to the success of various campaigns to reform electronics manufacturer practices. They are the Basel Action Network (BAN), the GrassRoots Recycling Network (GRRN), the Silicon Valley Toxics Coalition (SVTC), and the Texas Campaign for the Environment (TCE). Together, these SMOs created the Computer TakeBack Campaign, a national effort to make computer producers responsible for their products at the end of life. Two of these organizations are actually transnational social movement organizations – SVTC and BAN – and have worked successfully at getting international legislation passed in the European Union (EU) as well as having great success in shaping international environmental treaties such as the Basel Convention. They also work closely on e-waste recycling campaigns with SMOs in other nations, including: Clean Production Network (Canada), Greenpeace International, Greenpeace China, Toxics Link India (India), Shristi (India), and SCOPE (Pakistan). The mission statements of each of the US-based leading SMOs on e-waste issues reveal their understanding that environmental justice must be approached through the intersection of markets *and* politics rather than one or the other. They also under-score the integration of discursive practices and efforts to effect material change (Escobar 1992). Consider the following:

> Texas Campaign for the Environment is dedicated to *informing and mobilizing* Texans to protect the quality of their lives, their health and the environment. We believe that people have a *right to know and a right to act* on issues that fundamentally affect our lives and future. TCE cannot compete with *corporate polluters when it comes to writing checks for the election campaigns of politicians who make the laws.* However, we win when we organize at the grassroots level and gain strength in numbers from people like you who get involved and support our work.
>
> (www.texasenvironment.org)

> Silicon Valley Toxics Coalition (SVTC) is a diverse grassroots coalition that engages in *research*, advocacy, and *organizing* around the environmental and human health problems *caused by the rapid growth of the high-tech electronics industry.*
>
> (www.svtc.org)

The GrassRoots Recycling Network's

> mission is to eliminate the waste of natural and human resources – Zero Waste. We utilize classic activist strategies to achieve corporate accountability for and public policies to eliminate waste, and to build sustainable communities . . . *We prioritize corporate accountability because global corporations are the primary engines of environmental and social destruction.* The key tool to achieve corporate accountability for waste and ultimately Zero Waste, is

extended producer responsibility (EPR) – the principle that manufacturers and brand owners must take responsibility for the life cycle impacts of their products, including take back and end of life management.

(GrassRoots Recycling Network 2003b)

The Basel Action Network (BAN) is

an international network of activists seeking to put an end to economically motivated toxic waste export and dumping – particularly hazardous waste exports from rich industrialized countries to poorer, less-industrialized countries. The name Basel Action Network refers to an international treaty known as the Basel Convention. In 1994, a unique coalition of developing countries, environmental groups and European countries succeeded in achieving the Basel Ban – a decision to end the most abusive forms of hazardous waste trade. Unfortunately, very *powerful governments and business organizations* are still trying to overturn, circumvent or undermine the full implementation of the Basel Ban and in general seek to achieve a "free trade" in toxic wastes.

(www.ban.org)

The four SMOs above, all collaborated to create the Computer TakeBack Campaign:

The goal of the Computer TakeBack Campaign is to protect the health and well being of electronics users, workers, and the communities where electronics are produced and discarded by requiring consumer electronics manufacturers and brand owners to take full responsibility for the life cycle of their products, through effective public policy requirements or enforceable agreements.

(Computer TakeBack Campaign 2003)

Separately and in combination, these social movement organizations and their campaign(s) focus on decision-makers in governments and business organizations, revealing a broader *political economic process model* of movement targets. The centrality of the links between corporations and the state is critical here. These links are most visible and effective in urban centres like Silicon Valley, San Francisco, Austin, Boston, and increasingly points in Europe, Asia, and Latin America. Urban centres provide the critical infrastructure needed to produce high tech products and to reach dense consumer markets. These movement groups also link the discourse of environmental justice, a citizen's right to know, and corporate accountability with structural reforms aimed at state, corporate, and supranational institutional practices and policies.

One of the Southern partners in the global campaign to combat e-waste is Toxics Link India. They work with SVTC, BAN, and other Northern groups to highlight the growing e-waste crisis, and their mission underscores the political economic process approach as well:

Toxics Link's goal is to develop an *information exchange mechanism* that will strengthen campaigns against toxics pollution, help *push industries towards cleaner production* and link groups working on toxics issues. We are a group of people working together for environmental justice and freedom from toxics. We have taken it upon ourselves to collect and share both information about the sources and dangers of poisons in our environment and bodies, and information about clean and sustainable alternatives for India and the rest of the world.

(Toxics Link 2003: 2)

In addition to emphasizing the corporate role in the global e-waste crisis, Toxics Link is explicit about the need to "link groups working on toxics issues" around the globe.

The above US-based movement groups and their international partners joined forces to produce *Exporting Harm: The High-Tech Trashing of Asia* in 2002 (BAN and SVTC 2002). After months of strategizing with SMOs to gain access to sensitive sites and interview workers in China, Pakistan, and India, they released the report and it sent shock waves through the electronics industry and was picked up by nearly every major media outlet in North America, Europe and Asia.

Immediately after *Exporting Harm* was released, the government of China responded with stepped up inspections at its major ports and a declaration that it will not accept e-waste smuggling. In September of 2003 the Governor of California signed landmark legislation establishing a funding system for the collection and recycling of certain electronic wastes. The largest computer manufacturers and retailers like Hewlett-Packard, Dell, and Sony have all launched computer recycling programmes in the US and other nations in response to the TakeBack Campaign. The most visible of these was the struggle to push the Dell Corporation toward extended producer liability.

The Dell campaign

The Computer TakeBack Campaign emphasized Extended Producer Responsibility (EPR), the emerging global framework that holds producers and brand owners financially responsible for the life cycle impacts of their products, with particular emphasis on product take back and end of life management. By shifting the costs of managing discarded products away from taxpayers and local governments and onto producers and brand owners, EPR creates a powerful market incentive for brand owners to reduce those costs by re-designing products to be less toxic, more re-usable, and more easily recycled. The goal statement of the campaign captures this nicely: "Take it back, Make it clean, and Recycle responsibly." Since its inception, the CTBC has pursued a deliberate dual strategy: (1) Build sustained consumer and market pressure on Dell Computer Corporation, and (2) Build informed public support for regulatory reforms embracing producer responsibility. The CTBC's founding organizations quickly recognized the power inherent in a markets campaign (sometimes called "corporate campaigns") paired with a policy campaign focused on state legislative change. This dual approach nicely embodies the "political economic process" model (Pellow 2001).

Dell's business model is fundamentally about cultivating a relationship with their customer base that builds long-term loyalty. Fortunately, for environmental justice activists, that business model uniquely positions the company within the computer industry to develop and successfully implement a comprehensive, national computer take back system. Dell is the only computer company in the US (if not the world) that knows all of its customers by name, mailing address, e-mail address, phone number, date of purchase, product specification and more – exactly the kind of information a company would need to design a system to recover its obsolete products. If activists could force Dell to use its customer networks to design a computer take back system it would be the most efficient and far-reaching recycling effort to date. The creative use of Dell's business model as a means of organizing resistance was an example of how economic globalization processes unintentionally create opportunities for their reform, if not undoing (Hardt and Negri 2000).

Furthermore, Dell is not so much a manufacturing company as it is a marketing company. Dell assembles made to order computers from parts supplied to it and attaches its logo (and has no union). The Campaign believed that Dell, as a marketing company, was particularly susceptible to a strategy and associated tactics that attacked its brand name. The company bears the name of its founder, Michael Dell, who continues to be the CEO, a major stockholder and the most visible personality of the company. This also provides activists with opportunities to personalize the issue vis-à-vis Michael Dell who takes credit for the company's direction and success. For example, in a humorous effort to reach a younger audience of students and other computer consumers, the CTBC referred to Mr Dell as the "Toxic Dude" (Schatz 2003).

The CTBC issued Report Cards that tracked Dell's progress on recycling, which were very effective at capturing press headlines and getting consumers involved. The CTBC used the Report Cards to pit Dell against its competitors like Hewlett-Packard, who, at that time, was doing a better job of recycling its products. Dell provided the Campaign with other targeting opportunities, including the company's decision to partner with UNICOR, the federal prison industries, as its primary recycling partner. For Dell, selecting UNICOR was a matter of driving down costs. The Campaign's concern was that reliance on prison labour undercut development of the "free market" infrastructure necessary to operate a robust, national e-waste collection and recycling system. Moreover, because prisoners are not covered by the same worker health and safety protections as regular employees, incarcerated populations are more endangered by the toxic materials contained in discarded computers and electronics. Activists charged that Dell's recycling programme was "a high tech chain gang" not much better run than efforts studied in Chinese villages (Texas Campaign for the Environment 2003a). Barely a week after the CTBC announced its concerns about Dell's use of prison labour, Dell Computer announced that it would no longer rely on prisons to supply recycling workers for its programme (*San Diego Union Tribune* 2003). The fact that Dell cancelled the contract one week after the CTBC made its objection to these practices public is a testament to the real power this TSMO exercises against global corporations.

In its 2003 annual report card, the Computer TakeBack Campaign assigned failing grades to Micron Technology, Gateway, and Hewlett-Packard and others, but emphasized the particular shortcomings of the Dell Computer Corporation. The Report Card stated, "The Dell position on e-waste is a stain on the soul of Dell – the company and its founder . . . Michael Dell and his wife, Susan, make generous donations to children's health and environmental charities in the US but ignore the health and environmental impacts of e-waste on children and adults" (Associated Press 2003). Activists were relentless in their public criticism of the Dell Corporation and of its founder.

Activists in the US had a clear sense of Dell's operations around the globe and began to question the company's dual recycling systems – one for the US and one for Europe. The CTBC received data and details on this "double standard" from the partner SMOs across the Atlantic and made it known publicly. As Robin Schneider, Director of the Texas Campaign for the Environment stated,

> We want Dell Computer to take the same degree of responsibility for used and obsolete personal computers here in the US as the company does in European countries. In Europe, Dell takes back old equipment free of charge from all consumers. European producer responsibility laws require this of the company, ensuring that the products are kept out of landfills and incinerators and their valuable materials are reused or recycled. Our simple question to Dell Computer is "Why do American consumers and the American environment deserve second-class treatment?'
>
> (Computer TakeBack Campaign 2002)

In July 2003 recycling activists leafleted the Dell shareholders meeting in Austin, Texas and delivered a load of Dell computers collected from western cities, as part of the "hard drive across America". The Dell corporate staff accepted the waste computers and joined activists placing them into a truck to be hauled to Image Microsystems, an Austin-based recycler that replaced the UNICOR prison labour contractor that previously managed Dell's waste. "We're not exactly holding hands and singing "Kumbaya," but this is more common ground on recycling than we've had for a while", said Robin Schneider of the Texas Campaign for the Environment. Image Microsystems also agreed to honour the Recycler's Pledge of Stewardship, developed by the Computer TakeBack Campaign. Included in the pledge is the requirement that toxic waste not be sent overseas to global South nations or dumped in domestic landfills. "We consider this a big win", said Ted Smith of SVTC. "But we're also very mindful that Dell only recycled 2 percent of the 16 million computers they sold last year. Those uncollected computers will still end up in landfills or be shipped overseas" (Texas Campaign for the Environment 2003b).

I want to underscore that, in addition to high profile activists' efforts, consumers were critical to the Dell campaign, as the CTBC persuaded thousands of Dell computer owners and students to write emails, letters, and faxes to Michael

Dell urging him to institute a responsible recycling programme that would take back computers and recycle them without prison labour. Students were particularly critical in this effort, having raised funds to place open letters in full-page advertisements in newspapers in Austin, Texas, Dell's hometown headquarters. Students also successfully negotiated a meeting with Michael Dell (that was telecast nationally) where he responded to a list of concerns and demands.

Finally, much of the waste from companies like Dell and others ends up being exported and dumped in Pakistan, India, and China, creating further social and ecological havoc. E-waste recyclers face significant health hazards while disassembling the waste from global North and middle-class consumers. They too are part of the chain of production, consumption, and disposal, and activists were clear that they deserve our attention, access to basic rights, environmental protections, and a living wage. After all, without ties to social movement organizations in these Southern nations, the information about e-waste dumping might never have surfaced.

In 2004, Dell and Hewlett-Packard, the nation's largest personal computer makers, reported that they were increasing computer recycling and taking more of the financial burden for the recycling of used computers off consumers and local governments. The Computer TakeBack Campaign timed the pledges by Dell and Hewlett-Packard to the release of another annual "report card" of corporate environmental behaviour. HP received the highest rating and Dell moved up to second place on the 2004 report card. In a statement that reveals the power of "venue shopping" by movements engaging the political economic process, activist Ted Smith stated, "We believe the companies have to set up these systems, not governments" (Flynn 2004). Yet during 2003 alone, more than half the states in the US introduced some kind of electronic waste legislation (Bartholomew 2004). This is testimony to the dual-pronged approach that activists have taken to ensure that the state and corporations behave in ways that might ensure environmental justice.

DISCUSSION AND CONCLUSION

Ultimately, this is a story about how ordinary people are attempting to maintain control over spaces and bodies that are at risk. A growing number of scholars are concerned with the role of place/space in today's globalizing world. With global communications, trade, and transportation systems circling the planet like never before, with the Internet, fibre optic telecommunication, satellite technologies, real time text messaging, fax machines, FedEx, computer generated automation, and cities that are awake and operating twenty-four hours each day, our sense of place, space, and time have changed dramatically in the last two decades (Lipsitz 2001; Sassen 1998, 2001; Swyngedouw and Heynen 2003). They have all become compressed because it simply takes less time to get from one place to another and places seem less distinct, less foreign and more accessible than in the past, as McDonalds, Disney, Sony, and Nike make their mark around the planet and allow us to walk into a familiar scene in virtually any major city in Europe, Asia, and Latin America. Place seems to matter less and less.

Saskia Sassen (1998, 2001) takes on this notion that place – particularly urban places – matter less. She argues that cities are the key places and primary nodes for making the global economy possible. That is, cities provide the crucial infra-structure, the physical material and technology, immigrant economies, and sites of production and services that make the global economy function. Lipsitz (2001) also addresses the question of place but from another scale. He considers the idea that nation-states matter less in a global economy. Although it may appear that the power and influence of transnational corporations have superseded the nation-state, the state continues to be an indispensable component of the global system. The state serves as a crucial resource to multinational corporations by supplying mechanisms for capital accumulation and technological innovation through direct investment by governments, indirect support for research and development, tax abatements, and R&D spending on the infrastructure of global capital, for example high tech and the Internet, which were originally developed by the US military. The state also supplies transnational corporations with political regulation through direct repression of insurrections and strikes as well as through agreeing to international treaties like the WTO that deprive citizens of the power to use politics to challenge corporate power, environmental pollution, labour exploitation, and monopolies. The state helps discipline the labour force and imprisons surplus labour. The elite classes and racial/ethnic groups in nation-states benefit from this dynamic, from Venezuela to Malaysia. The very existence of nation-states encourages a cessation of internal hostilities among domestic populations and the projection of anger and resentment against outside enemies rather than internal elites (Chua 2003; Lipsitz 2001): consider the current situation where social inequality between the classes and races in the US is staggering and growing steadily, the labour movement is nearly crushed, environmental protection is all but dismantled, and the rights of citizens are being stripped away while nearly all of the political focus is on immigrants as potential terrorists and on alleged terrorist threats and rogue regimes abroad. Other scholars have considered the adaptive role the state plays in the face of powerful forces of capital, including "variegated" forms of flexible regulation and zoning (Ong 2004) and the creative reorganization of state intervention vis-à-vis neoliberalism (Brenner and Theodore 2002: 345). That being said, the power of transnational corporations is considerable and must be seriously considered and theorized vis-à-vis social movements.

The scholarly literature on transnational social movements, advocacy networks, and global civil society is marked by the absence of a serious integration of political economy into models of opportunity structures. It would seem that this is an important observation if we are considering the role of transnational corporate activities involving the export and dumping of hazardous wastes from Northern nations to Southern nations. This point is also noteworthy because it shapes the political economic environment in which local and global movements for environmental justice operate.

As powerful as nation-states and transnational corporations may be, it is always critical to remember that political economic opportunities are "not only perceived and taken advantage of by social movements, but they are also created" (Khagram *et al.* 2002: 17). For example, one of the most effective ways to create access to domestic political systems is through international pressure. When local activists and advocates abroad make an issue visible to the rest of the world it can create a "boomerang effect", which "curves around" local nation-state indifference and oppression to place pressure on states for policy changes (Keck and Sikkink 1998: 200). This is what the international political opportunity structure frequently looks like: the combination of closed structures domestically with more accessible structures elsewhere in places where TSMOs may be based. Through the Computer TakeBack Campaign, social movement groups in the South and the North were able to change state and corporate practices in the US and Europe that are expected to improve the situation on the ground in nations where e-waste dumping actually takes place.

Consistent with the political economic process model, environmental justice activists working on e-waste have not limited themselves to a state-centric approach, with regard to movement targets. In addition to pushing states like California and Maine to pass laws that ban e-waste from landfills, they have also successfully pressured the electronics industry and specific companies to change their practices. A political economic process approach that targets states and private corporations responsible for pollution is critical for achieving local and global environmental justice.

Related to considerations of environmental justice is the literature on urban political ecology, in which scholars concerned with urban spaces emphasize the interactive relationship between cities and nature. Theorizing urban spaces as socionatural spaces forces us to acknowledge that not only are the state and capital interdependent with one another, they are both entirely dependent upon and regularly shape the planet's natural resource base (Swyngedouw and Heynen 2003). Thus the pollution of urban areas is not fundamentally distinct from the despoliation of rural spaces because they are part of the same process and reflect the urbanization of nature on a global scale. What this means for social movements for environmental justice is that distinctions between urban or rural spaces or local and global geographies should matter less because the principal issue at hand is the phenomenon of socioenvironmental harm, which cuts across all of these constructed divides. Social movements that successfully bridge those divides may be much more effective. As Peck and Tickell (2002: 401) argue, movement "campaigns of disruption" must be accompanied by a "reform of macroinstitutional priorities and the remaking of extralocal rule systems". With its focus on changing the discourse and material nature of transnational corporations' policies and international law/multilateral conventions, the transnational environmental justice movement has done just that.

BIBLIOGRAPHY

Asia Pacific (nd) *India: Fears that IT hub becoming electronic waste dump*. Online. Available HTTP: <http: //www.abc.net.au/ra/asiapac/programs/s886414.htm> (accessed on 29 July 2003)

Associated Press (2003) "Computer makers hammered for their poor handling of 'e-waste'",
Houston Chronicle, 10 January 2003

Bartholomew, D. (2004) "E-commentary: computer makers tackle e-waste",
IndustryWeek.com. 1 January

Basel Action Network and Silicon Valley Toxics Coalition (2002) *Exporting Harm: The
High-Tech Trashing of Asia*. Seattle: BAN [www.ban.org]

Baumgartner, F. and Jones, B. (1991) "Agenda dynamics and political subsystems", *Journal
of Politics*, 53: 1,044–1,074

BBC News (2002) "China: Hi Tech Toxics", Online. Available HTTP: <http://news.
bbc.co.uk/hi/english/static/in_depth/world/2002/disposable_planet/waste/chinese_
workshop/> (accessed October 2004)

Brenner, N. and Theodore, N. (2002) "Preface: from the 'New Localism' to the spaces of
neoliberalism", *Antipode*, 341–347

Bullard, R. (2000) *Dumping in Dixie: Race, Class, and Environmental Quality*. Boulder,
CO: Westview Press

Burbidge, J. (ed.) (1997) *Beyond Prince and Merchant: Citizen Participation and the Rise
of Civil Society*. New York: PACT Publications

Chua, A. (2003) *World on Fire: How Exporting Free Market Democracy Breeds Ethnic
Hatred and Global Instability*. New York: Doubleday

Computer TakeBack Campaign (2002) "Dell shareholders urged to examine all aspects of
company's performance", Press Release, 1 May, Austin, Texas

Computer TakeBack Campaign (2003) "About the campaign", Online. Available HTTP:
<http://www.computertakeback.com/about/index.cfm>

Escobar, A. (1992) "Culture, economics, and politics in Latin American social movements
theory and research", in Escobar, A. and Alvarez, S.E. (eds) *The Making of Social
Movements in Latin America*. Boulder, CO: Westview Press

Flynn, L. (2004) "2 PC makers favor bigger recycling roles", *New York Times*, 19 May

Gough, N. (2002) "Garbage in, garbage out: castoffs from the computer age are a financial
windfall for Chinese villagers. But at what cost?", *Time Asia*, 11 March 2002

Gramsci, A. (1971) *Selections from the Prison Notebooks*. Hoare, Q. and Nowell-Smith,
G. (eds), London: Lawrence and Wishart

GrassRoots Recycling Network (2003a) "EPA: Keep toxic PC's out of Asia", Madison, WI:
GRRN

GrassRoots Recycling Network (2003b) "GRRN Mission", Online. Available HTTP: <http:
//www.grrn.org.> (accessed 6 July 2003)

Hardt, M. and Negri, A. (2000) *Empire*. Cambridge, MA: Harvard University Press

Isin, E.F. (1998) "Governing Toronto without government: liberalism and neoliberalism",
Studies in Political Economy, 56: 169–191

Keck, M.E. and Sikkink, K. (1998) *Activists Beyond Borders: Advocacy Networks in
International Politics*. Ithaca, NY: Cornell University Press

Keil, R. (2002) "'Common-Sense' neoliberalism: progressive conservative urbanism in
Toronto, Canada", *Antipode*, 578–601

Keil, R. (2003) "Progress report: urban political ecology", *Urban Geography*, 24: 723–738

Khagram, S., Riker, J.V. and Sikkink, K. (2002) "From Santiago to Seattle: transnational
advocacy groups restructuring world politics", in Khagram, S., Riker, J.V., and Sikkink,
K. (eds) (2002) *Restructuring World Politics: Transnational Social Movements,
Networks, and Norms*. Minneapolis, MN: University of Minnesota Press

Lipsitz, G. (2001) *American Studies in a Moment of Danger*. Minneapolis, MN: University
of Minnesota Press

McAdam, D. (1996) "The framing function of movement tactics: strategic dramaturgy in the American civil rights movement", in McAdam, D., McCarthy, J. and Zald, Z. (eds) *Comparative Perspectives on Social Movements*. Cambridge: Cambridge University Press

McCarthy, J.D. (1997) "The globalization of social movement theory", in Smith, J., Chatfield, C. and Pagnucco, R. (eds) (1997) *Transnational Social Movements and Global Politics: Solidarity Beyond the State*. Syracuse, NY: Syracuse University Press

McCarthy, J.D. and Zald, M. (1977) "Resource mobilization and social movements: a partial theory", *American Journal of Sociology*, 82: 1,212–1,241

Moore, D.S. (1993) "Contesting terrain in Zimbabwe's eastern highlands: political ecology, ethnography, and peasant resource struggles", *Economic Geography*, 69: 380–401

Ong, A. (2004) "The Chinese axis: zoning technologies and variegated sovereignty", *Journal of East Asian Studies*, 4: 69–96

Peck, J. and Tickell, A. (2002) "Neoliberalizing space", *Antipode*, 380–404

Peet, R. and Watts, M. (1993) "Introduction: development theory and environment in an age of market triumphalism", *Economic Geography*: 227–253

Peet, R. and Watts, M. (2004) *Liberation Ecologies: Environment, Development, and Social Movements*. New York: Routledge

Pellow, D.N. (2001) "Environmental justice and the political process: movements, corporations, and the state", *The Sociological Quarterly*, 42: 47–67

Pellow, D.N. and Park, L.S. (2002) *The Silicon Valley of Dreams: Environmental Injustice, Immigrant Workers, and the High-Tech Global Economy*. New York: New York University Press

San Diego Union Tribune. (2003) "Dell cancels contract for inmate labor", 5 July

Sassen, S. (1998) *Globalization and its Discontents*. New York: The New Press

Sassen, S. (2001) *The Global City: New York, London, Tokyo*. Princeton, NJ: Princeton University Press

Schatz, A. (2003) "Dell changes recycle vendors", *Austin American-Statesman*, 4 July 2003

SVTC (Silicon Valley Toxics Coalition) (2001) *Poison PCs and Toxic TVs*. San Jose, California: SVTC [www.svtc.org]

Smith, J., Chatfield, C. and Pagnucco, R. (eds) (1997) *Transnational Social Movements and Global Politics: Solidarity Beyond the State*. Syracuse, NY: Syracuse University Press

Smith, N. (2002) "New globalism, new urbanism: gentrification as global urban strategy", *Antipode*, 427–450

Swyngedouw, E. and Heynen, N. (2003) "Urban political ecology, justice and the politics of scale", *Antipode*, 35: 898–918

Tarrow, S. (1998) *Power in Movement: Social Movements and Contentious Politics*. New York: Cambridge University Press

Texas Campaign for the Environment (2003a) "An open letter to Michael Dell", 14 July

Texas Campaign for the Environment (2003b) "Dell accepts e-waste collected by activists", Computer TakeBack Campaign Press Release, 19 July, Austin, Texas

Toxics Link India (2003) *Scrapping the Hi-tech Myth: Computer Waste in India*. New Delhi, India

Vidal, J. (2004) "They call this recycling, but it's really dumping by another name", *The Guardian*, 21 September

Wankade, K. and Argawal, R. (forthcoming) "Hi tech heaps and forsaken lives: e-waste in New Delhi", in Smith, T., Sonnenfeld, D., and Pellow, D. (eds) *Challenging the Chip: Labor Rights and Environmental Justice in the Global Electronics Industry*. Philadelphia, PA: Temple University Press

15 Urban metabolism as target

Contemporary war as forced demodernization

Stephen Graham

INTRODUCTION

> If you want to destroy someone nowadays, you go after their infrastructure.
>
> (Agre 2001: 1)

> There is nothing in the world today that cannot become a weapon.
>
> (Liang and Xiangsui 1999: 5)

> Real security cannot be cordoned off. It is woven into our most basic social fabric. From the post office to the emergency room, from the subway to the water reservoir.
>
> (Klein 2001: 21)

Increasingly, both formal and informal political violence centre on the deliberate destruction, or manipulation, of the everyday urban infrastructures that are necessary to sustain the circulations and metabolism of modern urban life. As urban life becomes ever more mediated by fixed, sunken infrastructures, so the forced denial of flow, and circulation, becomes a powerful political and military weapon.

Since the devastating attacks on New York and Washington on 11 September, 2001, most attention has focused on the ways in which the banal systems of urban mobility – airlines, postal systems, water networks, commuter trains – can be instantly harnessed by non-state terrorists to produce sites of mediatized, mass death (Graham 2001; Luke 2003). Similarly, a powerful discourse has emerged in the past fifteen years suggesting that advanced, computerized societies are inevitably going to be attacked by coordinated "cyber-terror" attacks (see, for example, Verton 2003; Pineiro 2004; Rattray 2001; Debrix 2001). Here, the implication is that the everyday technics of western, urban life, based on computerized code, will be manipulated en masse from afar, producing catastrophe and death as airline, medical, financial and utility systems collapse through the pushing of a few key strokes.

Beyond such burgeoning debates in the west about "asymmetric" and "infrastructural" terrorism, the forced demodernization of societies through *state* infrastructural warfare is emerging as a central component of contemporary military strategy (Graham 2004). Largely unreported in the popular press and mainstream media, intensive military research and development efforts are fuelling a widening range of "hard" and "soft" anti-infrastructure weapons. These are being carefully designed to destroy, or disrupt, the multiple, networked infrastructures that together facilitate the continuous circulations and metabolisms necessary to sustain modern urban life (Graham and Marvin 2001). The Israeli Defence Forces' strategy of systematic infrastructural demodernization in the Occupied Territories since the mid-1990s presents the most visible example of such military strategy (Graham 2003).

The chapter's theoretical starting point, in keeping with the theme of this book, is that urban "technological networks (water, gas, electricity, information etc.) are constitutive parts of the urban. They are mediators through which the perpetual process of transformation of Nature into City takes place" (Kaika and Swyngedouw 2000: 1). With the massive technical infrastructures that sustain urban metabolism the target of increasingly sophisticated strategies of political violence, this chapter seeks to probe into the political ecologies, and political economies, of forced demodernization. That is, it explores the deliberate disruption of this "transformation of Nature into City" through infrastructure as a strategy of political violence. It attempts to lay bare the doctrines and theories through which the urban metabolism of targeted cities is analyzed and conceptualized by military architects of forced demodernization. Finally, the chapter analyzes how the deliberate targeting of urban technics in political violence impacts on the political ecologies and urban metabolisms of targeted cities.

In particular, this chapter explores the emerging strategies, doctrines, techniques and discourses that surround state-backed infrastructural warfare. In doing this, what follows is an attempt to develop a preliminary geopolitics of forced demodernization in the contemporary world. Such a perspective is necessary for two reasons. First, the centrality of infrastructural demodernization in contemporary war is scarcely ever addressed in critical urban social science. Second, the geographical, political science and international relations writers who dominate the study of political violence and geopolitics remain preoccupied by the abstract machinations of *nation states*. Such researchers rarely address the concrete materialities, and processes, that emerge when weapons, doctrines and strategies intersect with the often hidden materialities of urban life in episodes of (attempted) demodernization, designed to achieve political, or geopolitical, ends.

The chapter has three parts. In the first, I attempt to place forced demodernization via infrastructural warfare and disconnection within a theoretical perspective. This stresses the connections between infrastructure, urban metabolism, and geopolitical power. Second, I analyze a case study of state infrastructural warfare. This focused on the efforts of a leading exponent of forced demodernization as national strategy – the United States – to systematically demodernize Iraqi cities through war, sanctions, and more war since 1991. Finally, in the chapter's conclusion, I reflect upon the geopolitics of forced disconnection, and demodernization, within contemporary war and strategy.

URBAN METABOLISM AND THE CHANGING NATURE OF WAR

Collective facilities and networked infrastructures that are used by civilians and military alike have been a central target in war for as long as they have existed (Graham 2004). From the water poisonings of medieval urban sieges to the attempts at total urban annihilation in World War II, means of movement, communication, obtaining water, disposing of waste, sustaining biospheres, and obtaining fuel and energy have been at the heart of struggles for geopolitical and military power between enemies (Thomas 1995). Such strategies have grown more sophisticated and more carefully orchestrated over time (see Pape 1996). As well as the 24-hour carpet-bombing of whole cities, for example, the mobilization of operations science by Allied bombing strategists in World War II allowed the Allies, within the constraints of bombing accuracy at the time, to try and systematically target Germany's "industrial web". This entailed an attempt at the systematic degradation of whole systems of lines of communication and transport, electrical power, and oil and chemical supply (Rattray 2001: 272).

Two groups of factors are currently leading to a proliferation in the range, frequency and sophistication of attacks on the networked infrastructures that sustain every aspect of the functioning and development of contemporary urban societies.

The vulnerabilities of urban metabolisms to forced demodernization

First, the mediation of contemporary urban societies by vast arrays of technological, computerized systems of flow means that small disruptions and disablement can have enormous, cascading, effects (Zimmerman 2001; Little 2002). As societies urbanize and modernize, so their populations become ever-more dependent on complex, distanciated systems for the sustenance of the political ecological arrangements necessary to sustain life (water, waste, food, medicine, goods, commodities, energy, communications, transport, and so on). With pervasive, but uneven, computerization, software systems increasingly provide the functionalities that enable these multiple, networked systems to operate. This tends to accentuate the vulnerability of such "big" socio-technical systems, because the code can be easily manipulated from afar (Thrift and French 2000).

Disruptions to this palimpsest of everyday technics fleetingly reveal the critical importance of infrastructural systems which, when they function normally, tend to be ignored (or, in sociological parlance, "blackboxed") by their users. "The normally invisible quality of working infrastructure becomes visible when it breaks: the server is down, the bridge washes out, there is a power blackout" (Star 1999: 382).

Thus, when infrastructure networks "work best, they are noticed least of all" (Perry 1995: 2). Catastrophic failures, then, serve to fleetingly reveal the utter reliance of contemporary urban life on networked infrastructures. This is especially so where the entire economic system in advanced industrial societies is being been reconstructed around highly fragile networks of computers and information technology devices working on "just in time" principles of fluid and continuous synchronization across space (see Rochlin 1997).

The pervasive importance of twenty-four-hour systems of electrically powered computer networks, in supporting all other infrastructures, makes electrical power cuts and outages particularly fearful. The explosive recent growth of electronic commerce, consumption, and distribution and production systems – infrastructures that are mediated at every level by electrically powered computer and telecommunications – means that these days we are all, in a sense, "hostages to electricity" (Leslie 1999: 175).

As Tim Luke has argued, "some small groups of human beings maybe still can live pre-machinically, like the Kung of the Kalahari, Haitian fishermen on Hispanola, Mongol herdsmen in Siberia, or even the House Amish of Pennsylvania" (2004: 109). However, Luke (2004: 109) argues that:

> Many more human beings live highly cyborganized lives, totally dependent upon the Denature of machinic ensembles with their elaborate extra-terrestrial ecologies of megatechnical economics. This is as true for the Rwandans in the refugee camps of Zaire as it is for the Manhattanites in the luxury coops of New York City. Without the agriculture machine, the housing machine, the oil machine, the electrical machine, the media machine, or the fashion machine, almost all cyborganized human beings cannot survive or thrive, because these concretions of machinic ensembles generate their basic environment.

All of which means that, more than ever, the collapse of functioning infrastructure grids now brings panic and fears of the breakdown of the functioning urban social order. "Fear of the dislocation of urban services on a massive scale", writes Martin Pawley, is now "endemic in the populations of all great cities" (1997: 162).

The changing nature of war

> Strength and weakness, threat and security have become now, essentially, *extraterritorial issues that evade territorial solutions.*
>
> (Bauman 2002: 83; original emphasis)

Second, the above two sets of transformations need to be seen in the context of radical changes in the nature of war and organized political violence since the end of the Cold War. Some of these post Cold-War changes are now very familiar:

- the proliferation of asymmetric struggles pitching US and western forces, with their monopoly of high-tech, precision targeting, against low-tech adversaries;
- a greatly reduced frequency of state-vs-state wars with standing armies, and a consequent huge growth of civilian casualties in relation to military ones;
- an increasingly hegemonic presence of US forces around the world who seek to consolidate and protect the resource and geopolitical underpinnings that

sustain the network power of the globe-spanning US-dominated "Empire" (Hardt and Negri 2000);

- a sustained effort by the US military to utilize its hegemonic position in the strategic use of information technologies, and intelligent machines and orbital power to radically reorganise the reach and power of its strike capabilities (the so-called Revolution in Military affairs or RMA). This is being justified so that the United States military can achieve "Full Spectrum Dominance" whilst fighting a range of adversaries in major "theatre wars" simultaneously (Project for the New American Century 2000; Hirst 2001); and
- an increasing emphasis in geopolitical struggle on what Tim Luke (2003) has called "culture war". This emphasizes global media representation, propaganda, signs, and the 24/7 consumption of war by quasi-voyeuristic viewers over TV and internet networks (Der Derian 2001). Dochterman (2002: 1) argues that this new type of war boils down to an "information war, in which the television and internet become something akin to live, continuous and violent advertisements for the power of the military-technological apparatus of the United States and its allies".

With the on-going urbanization of terrain in many geopolitical conflict zones, we should also stress that, increasingly, wars are being fought out, within, and through the domestic spaces, and infrastructures, of everyday urban life. "Today, wars are fought not in trenches and fields, but in living rooms, schools and supermarkets" (Barakat 1998: 11). This is crucial because, as we have noted, civilian urban populations are especially vulnerable to war because their lives are sustained by multiple, networked systems which continuously link them to distant sources of food, energy, water and other goods and services. Cities are especially vulnerable to the stresses of conflict, suggests Sultan Barakat (1998: 12):

> City-dwellers are particularly at risk when their complex and sophisticated infrastructure systems are destroyed and rendered inoperable, or when they become isolated from external contacts.

Finally, war and geopolitical struggle are increasingly being fought *through* the infrastructures of everyday urban life. The 9/11 attacks are, of course, the paradigmatic case here, as they involved the instant transformation of banal capsules of everyday interurban mobility into mechanisms of mass, mediatized, murder (Luke 2003). But, as we shall see, a proliferating range of strategies of state violence are being developed which seek to project political pressure through the systematic demodernization of networked urban life in adversary societies and cities.

Theoretically, civilians are protected from being directly targeted in war (for example, by Article 8(2)(b) (iv) of the International Criminal Courts Statute). However, in effect, the widespread targeting of dual-use infrastructures denies this legal protection. "Such discrimination turns to fiction when extended to electrical grids, water supplies, and other infrastructure that are the sinews of everyday life" (Smith 2002: 361).

Whilst the military and civilian casualties during formal times of war may be reducing, the long-term civilian deaths which result from attacks on the crucial infrastructures that sustain urban life – what Blakeley (2003a) has called "structural violence" or the "war on public health" – are increasing. As Smith (2002: 362) argues, "while the security community views sanctions and attacks on infrastructure as limited remedies, students of human rights find them drastic indeed". King and Martin (2001: 2) suggest that "the growing centrality within war of targeting everyday infrastructures are making war safer for soldiers and much riskier for civilians. The problem is not badly aimed guns [i.e. "collaterial damage"], but rather the increasingly severe public health consequences of war" (2001: 2). And Ashford believes that these military strategies of "bomb now, die later" (Blakeley 2003a: 3) conveniently hide the long-term degradation and killing from the capricious gaze of the global media:

> The insidious effects of destroying the water supply, sewage system, agri-culture, food distribution, electricity, fuel systems and the economic base for an entire country are not obvious until starvation and disease create a humanitarian crisis that cannot be ignored. In fact, far from sparing the innocent, this deliberate strategy disproportionately kills the very young, the very old, and the very weak.

"YOU WANT 1389 ? WE CAN DO THAT !" INFRASTRUCTURAL WARFARE AND US GLOBAL HEGEMONY

> It should be lights out in Belgrade: every power grid, water pipe, bridge, road and war-related factory has to be targeted . . . We will set your country back by pulverizing you. You want 1950 ? We can do 1950. You want 1389 ?[1] We can do that, too!
> (*New York Times* columnist, Thomas Friedman, 23 April, 1999, calling for a massive US strike on Serbian urban infrastructure (cited in Skoric 1999))

> We need to study how to degrade and destroy our adversaries' abilities to transmit their military, political, and economic goods, services and information . . . Infrastructures, defining both traditional and emerging lines of communication, present increasingly lucrative targets for airpower. [The vision of] airmen should focus on lines of communications that will increasingly define modern societies.
> (Felker 1998: 1–20)

Befitting its hegemonic status currently, one state – the United States – currently dominates the practice of state-backed infrastructural warfare. Whilst it would be an oversimplification, it is possible to argue that the geopolitical and military strategy currently being developed to maintain the United States' power as a global hegemon rests on a simple, two-sided idea. On the one hand: develop information-technology-based networked capabilities of control, surveillance and targeting to a level which allows attempts at globe-spanning dominance based on a near-monopoly of space and air power. This is being termed "network-centric warfare" and the "Revolution in Military Affairs" or "RMA" (see Harris 2003; Dillon 2002). On the other: develop the tools and technologies that can disconnect, demodernize, and immobilize adversary societies at any time or place deemed necessary.

Such a transformation in US military doctrine is extremely contested – especially after the disastrous occupation of Iraq (Harris 2003). It involves complex institutional politics within the vast array of military departments, political agencies, and defence and media industry groups involved. And care must be taken not to exoticize the transformation as science-fiction (as, arguably, Paul Virilio has often done – see, for example, Virilio 2002). But the combination of the near informational and infrastructural omnipotence of US forces, and the systematic demodernization of adversary forces (and, often, societies) – particularly those in the United State's key geopolitical target areas of central Eurasia – is a central axiom of the RMA.

Underpinning this double-edged strategy is the notion of the "enemy as a system." This was devised by a leading US Air Force strategist, John Warden, within what he termed his *strategic ring theory* (1995). This systematic view of adversary societies, which builds on the "industrial web" theorizations of US air power strategists in World War II, provides the central US strategic theorization that justifies, and sustains, the rapid extension of that nation's infrastructural warfare capability (Rizer 2001). The theory has explicitly provided the basis for all major US air operations since the late 1990s.

"At the strategic level," writes Warden, "we attain our objectives by causing such changes to one or more parts of the enemy's physical system" (1995). This societal "system" is an attempt to portray the socio-natural constructions, and metabolisms, of urban life which actually sustain national geopolitical power. This is done so that these constructions can be systematically demodernized in the projection of aerial, geopolitical power. In Warden's theory, each enemy societal "system" is seen to have five interconnected parts or "rings." These are the leader-ship or "brain" at the centre; organic essentials (food, energy, etc.); infrastructure (vital connections like roads, electricity, telecommunications, water etc.); the civilian population; and finally, and least important, the military fighting force (Felker 1998) (see Figure 15.1). Rejecting the direct targeting of enemy civilians, Warden, instead, argues that only "indirect" attacks on civilians are legitimate. These operate through the targeting of societal infrastructures and urban metabolisms – an alleged means of bringing intolerable pressures to bear on the nation's political leaders. This doctrine now officially shapes the projection of US aerial power and underpins the key US Air Force Doctrine Document – 2–1.2 – published in 1998 (USAF 1998).

Kenneth Rizer, another US air power strategist, recently wrote an extremely telling article in the official US Air Force Journal *Air and Space Power Chronicles*. In it, he seeks to justify the direct destruction of dual-use targets (i.e. civilian infra-structures) within US strategy. Rizer argued that, in international law, the legality of attacking dual-use targets "is very much a matter of interpretation" (2001: 1).

Rizer writes that the US military applied Warden's ideas in the 1991 air war in Iraq with, he claims, "amazing results". "Despite dropping 88,000 tons in the 43

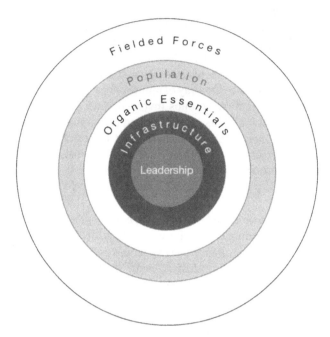

Figure 15.1 John Warden's (1995) Five-Ring Model of the strategic make-up of
contemporary societies – a central basis for US military doctrine and strategy
to coerce change through air power

Source: Felker 1998: 12

day campaign, only 3000 civilians died directly as a result of the attacks, the lowest
number of deaths from a major bombing campaign in the history of warfare" (2001:
10). However, he also openly admits that – as we shall soon discuss – the United
State's systematic destruction of Iraq's electrical system in 1991 "shut down water
purification and sewage treatment plants, resulting in epidemics of gastro-enteritis,
cholera, and typhoid, leading to perhaps as many as 100,000 civilian deaths and the
doubling of infant mortality rates" (2001: 1).

Clearly, however, such "indirect" deaths, as urbanites slowly succumb to the
deliberate annihilation of the infrastructures that sustain their fragile lives, are of
little concern to US Air Force strategists. For Rizer (2001: 10) openly admits that:

> The US Air Force perspective is that when attacking power sources, trans-
> portation networks, and telecommunications systems, distinguishing between
> the military and civilian aspects of these facilities is virtually impossible. [But]
> since these targets remain critical military nodes within the second and third
> ring of Warden's model, they are viewed as legitimate military targets . . . The
> Air Force does not consider the long-term, indirect effects of such attacks when
> it applies proportionality [ideas] to the expected military gain.

More tellingly still, Rizer goes on to reflect on how US air power is supposed to influence the "morale" of enemy civilians if they can no longer be carpet-bombed. "How does the Air Force intend to undermine civilian morale without having an intent to injure, kill, or destroy civilian lives ?" he asks:

> Perhaps the real answer is that by declaring dual-use targets legitimate military objectives, the Air Force can directly target civilian morale. In sum, so long as the Air Force includes civilian morale as a legitimate military target, it will aggressively maintain a right to attack dual-use targets.

(ibid.: 11)

In 1998 Edward Felker – a third US air power theorist who, like both Warden and Rizer, is based at the US Air War College Air University – further developed Warden's model. This was based on the experience of "Desert Storm", and drew on Felker's argument that infrastructural networks, rather than a separate "ring" of the "enemy as a system", in fact pervaded, and connected, all the others to actually "constitute the society as a whole" (Figure 15.2). "If infrastructure links the subsystems of a society," he wrote, "might it be the most important target ?" (1998: 20).

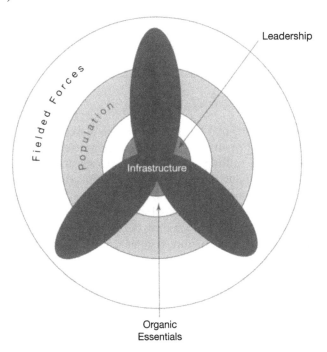

Figure 15.2 "A new model for societal structure": Edward Felker's adaptation of Warden's Five-Ring Model (see Figure 15.1), stressing the centrality of infrastructural warfare to post Cold War US airpower doctrine.

Source: Felker 1998: 12

Table 15.1 Patterson's models of the first-, second- and third-order effects of disrupting
an enemy's critical infrastructure systems: the example of power

First-order effects	Second-order effects	Third-order effects
No light after dark or in building interiors	Erosion of command and control capabilities	Greater logistics complexity
No refrigeration	Increased requirement for power-generating equipment	Decreased mobility
Some stoves/ovens non-operable	Increased requirement for night vision devices	Decreased situational awareness
Inoperable hospital electronic equipment	Increased reliance on battery-powered items for news, broadcasts, etc.	Rising disease rates
No electronic access to bank accounts/money	Shortage of clean water for drinking, cleaning and preparing food	Rising rates of malnutrition
Disruption in some transportation and communications services	Hygiene problems	Increased numbers of non-combatants requiring assistance
Disruption to water supply, treatment facilities, and sanitation	Inability to prepare and process some foods	Difficulty in communicating with non-combatants

Source: (Patterson 2000: 6)

By constructing linear, and non-linear models of the "cascading" first-, second-, and third-order impacts of destroying key parts of the networked infrastructure of an adversary society, US military planners have started to develop a complex military doctrine underpinning the extension of US infrastructural warfare. This centres on organized, systematic demodernization not just of the military forces of those deemed to be enemies, but of their civil societies as well (Patterson 2000; see Table 15.1). Indeed, US military analysis here is now concentrating on finding the "tipping points" in critical infrastructure systems that will lead to the non-linear, spiral effects that will most rapidly induce complete, societal chaos (Felker 1998).

Whilst the abstract theorization of doctrine and tactics outlined in this section are revealing, the centrality of infrastructural warfare to emerging US geopolitical strategy can only be fully understood when specific case studies are investigated. In what follows the war–sanctions–war experience in Iraq between 1991 and 2004 is used to demonstrate the devastating impacts of this US military doctrine of systematic demodernization on the poor, urban societies that are unlucky enough to fall in the cross-hairs of US military intervention or invasion.

"BOMB NOW, DIE LATER": THE "WAR ON PUBLIC HEALTH" IN IRAQ – 1991–2004

> No violence short of a nuclear explosion has been as intense as the air onslaught unleashed upon Iraq.

> Destroying the means of producing electricity is particularly attractive because it can not be stockpiled.

> The bottom line in comprehensive infrastructure bombing is that it kills civilians whilst the adversary's military power remains largely intact.
>
> (Bolkcom and Pike 1993: 2, 6)

The First Gulf War, 1991

The First Gulf War of 1991 was – then – the biggest conventional aerial assault in history. It was also one of the most unbalanced uses of military force in history. As with the later Kosovo assault, the Desert Storm bombing campaign was targeted heavily against so-called "dual-use" urban infrastructure systems, a strategy that Ruth Blakeley has famously termed "Bomb Now, Die Later" (2003a). Moreover, the reconstruction of these life-sustaining infrastructures was made impossible by the sanctions regime that was imposed by the US and UN between 1991 and 2003. This combination of events, it is now clear, created one of the largest, engineered public health catastrophes of the late twentieth century.

Because Iraq's actual military targets were so easily annihilated, it is crucial to realize that what happened in Desert Storm was that a very large percentage of strategic aerial missions were targeted against industry, power generation, roads and bridges, rather than military assets (Bolkcom and Pike 1993: 3). The military planners, and lawyers, behind Desert Storm, made the most of the unprecedented unevenness of the forces, and the resulting lack of opposition to Allied air and space forces, in their target planning.

Along with military and communication networks, urban infrastructures were amongst the key targets receiving the bulk of the bombing. One US air war planner, Lt. Col. David Deptula, passed a message to Iraqi civilians via the world's media as the "planes started going in: 'hey, your lights will come back on as soon as you get rid of Saddam!'" (cited in Rowat 2003). Another, Brigadier General Buster Glosson, explained that infrastructure was the main target because the US military wanted to "put every household in an autonomous mode and make them feel they were isolated . . . We wanted to play with their psyche" (cited in Rowat 2003). As Colin Rowat (2003: 3) suggests, even in the immediate aftermath of the war, for at least 110,000 Iraqis, this "playing" was ultimately to prove fatal. Bolkcom and Pike recall the centrality of targeting "dual-use" infrastructures in the planning of Desert Storm:

> From the beginning of the campaign, Desert Storm decision makers planned to bomb heavily the Iraqi military-related industrial sites and infrastructure,

while leaving the most basic economic infrastructures of the country intact. What was not apparent or what was ignored, was that the military and civilian infrastructures were inextricably interwoven.

The political rationale of "turning the lights off in Baghdad" generated much debate amongst Gulf War bombing planners (Blakeley 2003a: 25). The US Military's Gulf War Air Power Survey (GWAPS), completed by the US Defence Department at the formal end of the war, revealed that:

> there was considerable discussion of the results that could be expected from attacking electric power. Some argued that . . . the loss of electricity in Baghdad and other cities would have little effect on popular morale; others argued that the affluence created by petro-dollars had made the city's population psychologically dependent on the amenities associated with electric power.
>
> (Keaney and Cohen 1993: vol. ii, part ii, ch. 6: 23,
> footnote 53, cited in Blakeley 2003a: 25)

Thus, the systematic annihilation of infrastructures that were used by both military and civilians alike – to disable Iraq's war machine and influence civilian morale – led, indirectly, to mass civilian casualties, as Iraqi urban society was ruthlessly demodernized. "On the whole, civilian suffering is not caused by near misses [collateral damage], but by direct hits on the country's industrial infrastructure" (Bolkcom and Pike 1993: 2).

A prime target of the air assault was Iraq's electricity generating system. During Desert Storm, the allies flew over 200 sorties against electrical plants. The destruction was devastatingly effective:

> almost 88 percent of Iraq's installed generation capacity was sufficiently damaged or destroyed by direct attack, or else isolated from the national grid through strikes on associated transformers and switching facilities, to render it unavailable. The remaining 12% was probably unusable other than locally due to damage inflicted on transformers and switching yards.
>
> (Keaney and Cohen 1993: vol. II, part II, ch. 6: 20,
> cited in Blakeley 2003a: 20)

Bolkcom and Pike (1993: 5) add that:

> More than half of the 20 electrical generator sites were 100 percent destroyed. Only three escaped totally unscathed . . . The bombing of Iraq's infrastructure was so effective that, on either the sixth or the seventh day of the air war, the Iraqis shut down what remained of the national power grid. It was useless.

The surfeit of armed aircraft, combined with a paucity of real targets (and a very poor or non-existent enforcement of international law) led to a total overkill in the process of demodernizing Iraq by bombs. As Bolkcom and Pike admit, in this type of overwhelming, and totally uneven, aerial onslaught, an extremely wide range of targets were attacked, not because they needed to be, but because they *could be*. An ever-lengthening list of targets was sanctioned simply because of the unopposed air power and ordinance that was available, literally hanging around Iraqi airspace, looking for things to destroy. Bolkcom and Pike offer the example of al-Hartha power plant in Basra. First attacked on the first night of the bombing:

> The initial attack shut down the plant completely, damaging the water treatment system and all four steam boilers. During the course of the conflict, al-Hartha was bombed 13 times, even though there would be little opportunity to repair the power station during a major war. The final attack bounced the rubble a half hour before the cease fire on February 28, 1991 . . . Reportedly, the power plant was bombed so frequently because it was designated a backup target for pilots unable to attack their primary targets . . . The goal of multiple bombings late in the war was to create postwar influence over Iraq. It is very difficult to repair a power generator, for example, when the repair personnel have no power.
>
> (Bolkcom and Pike 1993: 5)

Another reason for the savagery of the demodernization was a failure to enforce even the extremely questionable guidelines for infrastructural bombing adopted in the planning of Desert Storm. These clearly stated that, in the case of electricity, "only transformer/switching yards and control buildings were to be targets and not generator halls, boilers and turbines" (Blakeley 2003a: 20). The reasoning behind this was that it would take much longer to repair the latter to be reconstructed whilst the former could be repaired relatively easily, cheaply and quickly.

In practice, such guidelines were largely ignored. The Gulf War Air Power Survey concluded that "the self-imposed restrictions against hitting generator halls or their contents was not widely observed in large part because the planners elected to go after the majority of Iraq's 25 major power stations and the generator halls offered the most obvious aim points" (Keaney and Cohen 1993, cited in Blakeley 2003a: 20).

It is no surprise, then, that, at war's end, Iraq had only 4 percent of pre-war electricity supplies. After 4 months only 20–25 percent of pre-war levels had been attained, a level of supply "roughly analogous to that of the 1920s before Iraq had access to refrigeration and sewage treatment" (Bolkcom and Pike 1993: 5). The devastation of the generator halls and turbines would have condemned Iraqi society to a largely non-electric future for years to come, even if Western or Russian technological and financial assistance had been possible in rebuilding.

The UN under-secretary general Martti Ahtisaari, reporting on a visit to Iraq in March 1991, was clearly shaken by what he had seen. "Nothing that we had seen or read had quite prepared us for the particular form of devastation that has now befallen the country," he wrote:

> The recent conflict has wrought near-apocalyptic results upon an economically mechanized society. Now, most means of modern life support have been

destroyed or rendered tenuous. Iraq has, for some time to come, been relegated to a pre-industrial age, but with all the disabilities of post-industrial dependency on an intensive use of energy and technology . . . Virtually all previously available sources of fuel and power, and modern means of communication are now, essentially, defunct . . . there is much less than the minimum fuel required to provide the energy needed for movement or transportation, irrigation or generators for power to pump water or sewage.

> (reported in Perez de Cueller 1991, cited in Blakeley
> 2003a: part 8)

Even immediately after the war's end, the UN reported that:

> Iraqi rivers are heavily polluted by raw sewage, and water levels are unusually low. All sewage treatment plants have been brought to a virtual standstill by the lack of power supply and the lack of spare parts. Pools of sewage lie in the streets and villages. Health hazards will build in weeks to come.
>
> (De Cueller 1991, cited in Blakeley 2003a: 25)

Post-war sanctions and bombing, 1991–2003

Such predictions proved highly prescient. The most devastating impact of mass de-electrification was indirect. Iraq's water and sewage systems, relying completely on electrical pumping stations, completely ground to a halt. Prospects of repair, as with the electrical system, were reduced virtually to zero. This was because of the US Coalition's punitive regime of sanctions that were introduced, with the help of UN resolutions, just before the war. As a result, virtually any item or supply required for infrastructural repair was classified, and prohibited, as a "dual-use" item with military potential – ironically, the slippery legal jargon that had legitimized the massive infrastructural destruction in the first place.

Now-declassified documents from the US Defense Intelligence Agency (DIA) demonstrate the degree to which the US military were aware of the terrible impacts of the combination of aerial demodernization and sanctions on public health in post-war Iraq. Thomas Nagy (2001) has demonstrated that DIA memos in early 1991 clearly predicted what they called "a full degradation of Iraq's water system". The memos argued that a failure to get hold of embargoed water treatment equipment would inevitably lead to massive food and water shortages, a collapse of preventive medicine, an inability to dispose of waste, and a spread of epidemics of disease like cholera, diarrhoea, meningitis, and typhoid.

These, in turn, it was predicted, would lead to huge casualty rates, "particularly amongst children, as no adequate solution exists for Iraq's water purification dilemma [under sanctions]" (cited in Nagy 2001). The memo titled "Disease Outbreaks in Iraq," dated 21 February 1991 (DIA, 1991), stated that "conditions are favourable for communicable disease outbreaks, particularly in major urban areas affected by coalition bombing" (cited in Nagy 2001). Despite all this, planners went ahead with the imposition of the sanctions.

By 1999, these predictions had come true. Drinkable water availability in Iraq had fallen to 50 percent of 1990 levels (Blakeley 2003b: 2). Colin Rowat, of the Oxford Research Group, has calculated that:

> the number of Iraqis who died in 1991 from the effects of the Gulf war or postwar turmoil approximates 205,500. There were relatively few deaths (approximately 56,000 military personnel and 3,500 civilian) from direct war affects. The largest component of deaths derives from the 111,000 attributable to postwar adverse health effects.
>
> (Rowat 2002)

Using a longer time frame, UNICEF (1999) reported that, between 1991 and 1998, there were, statistically, over 500,000 excess deaths amongst Iraqi children under five – a six-fold increase in death rates for this group occurred between 1990 and 1994. Such figures mean that, "in most parts of the Islamic world, the sanctions campaign is considered genocidal" (Smith 2002: 365). The majority of deaths, from preventable, waterborne diseases, were aided by the weakness brought about by widespread malnutrition. The World Health Organization reported in 1996 that:

> the extensive destruction of electricity generating plants, water purification and sewage treatment plants during the six-week 1991 war, and the subsequent delayed or incomplete repair of these facilities, leading to a lack of personal hygiene, have been responsible for an explosive rise in the incidence of enteric infections, such as cholera an typhoid.
>
> (Cited in Blakeley 2003a: 23)

The Second Gulf War, 2003/4: "Welcome to the republic of darkness and unemployment"

Not surprisingly, the second, even more savage onslaught of aerial bombing that Iraq was subjected to in 2003–4 – organized as it was after 12 years of systematic demodernization and impoverization through sanctions and continued bombing – led to an even more complete demodernization of everyday urban life in the country. This has occurred even though key centralized infrastructure nodes were targeted less extensively than in 1991. This time, the bombing strategy was ostensibly designed to "avoid power plants, public water facilities, refineries, bridges, and other civilian structures" (Human Rights Watch 2003). But new weapons, including electromagnetic pulse (EMP) cruise missiles, were used for the first time to comprehensively "fry" "dual-use" communications and control equipment (Smith 2003; Kopp 1996).

Nevertheless, dual-use systems such as electrical and power transmission grids, media networks, and telecommunications infrastructures were still substantially targeted and destroyed. Media installations and antennae were destroyed by new CBU-107 Passive Attack Weapons – non-explosive cluster bombs, which rain metal rods onto sensitive electrical systems that are nicknamed "rods from God" by the US Air Force (Human Rights Watch 2003: 3). In addition, more traditional bombs were used to destroy Al-Jazeera's office in Baghdad on 8 April, killing several journalists (Tahboub 2003). This was because the Pentagon considered the highly successful, independent channel's coverage of the dead civilians that resulted from the bombing was undermining its propaganda (or PSYOPS) campaign aimed at asserting information dominance (Miller 2004). As Miller suggests, in current US geopolitical strategy, "the collapse of distinctions between independent news media and psychological operations is striking" (Miller 2004: 24).

Finally, as in 1991, carbon "soft" bombs were once again widely used on electricity distribution systems. The resulting fires completely ruined many newly repaired transformer stations, creating, once again, a serious crisis of water distribution because of the resulting power blackouts (Human Rights Watch 2003: 3). The resulting supply crises, and the inability of the US Occupying Authorities to bring back reliable power sources, severely exacerbated Iraq's slide into violence, resistance and looting after the formal "end" to the war was declared by George W. Bush in May 2003. Citing a piece of Baghdad graffiti, Salam Pax (2003), the famous web blogger, noted that Iraq had emerged, by mid August, as a "republic of darkness and unemployment". Meanwhile, Human Rights Watch researchers found in al-Nasiriya that "in many places people had dug up water and sewage pipes outside their homes in a vain attempt to get drinking water". Not surprisingly, large numbers of waterborne intestinal infections were reported after the formal end of the invasion part of the war, a direct result of the targeting of electrical distribution systems (Enders 2003).

CONCLUSION

> Infrastructure – the boring stuff that binds us all – is not irrelevant to the business of fighting terrorism. It is the foundation of our future security.
>
> (Klein 200: 23)

Three conclusions are evident from this chapter. The first is that the everyday systems of urban infrastructure upon which all urbanites continuously depend have now become central sites of geopolitical struggle. From the 9/11, Madrid, and London terrorist attacks; through the (still largely chimerical) discourses of cyberterror; to the central place of forced demodernization within contemporary US and Israeli military doctrine: the ability to use, and pervert, everyday urban technics through political violence has become a driving force in contemporary political strategy. Forcibly destroying or manipulating the complex connectivities that urban technics sustain, the doctrines of both terrorists and state military theorists increasingly centre on the coercive powers of forced demodernization. Given the central role of urban infrastructures in mediating processes of urban metabolism – in allowing Nature to be continually metabolized into Society and Culture – the effects of such targeting, whilst usually less obvious than those of bombing – are dramatic and deadly.

The proliferation of state infrastructural violence, in particular, forces us to reconsider the very notion of war. It suggests that potentially boundless and continuous landscapes of conflict, risk and unpredictable attack are currently emerging, as the everyday technics of urban life that are so usually taken for granted and ignored become key geopolitical sites through their use as mechanisms for the projection of organized, structural demodernization. Increasingly, such interventions are occurring from a distance, as the infrastructural connections themselves become the site either of violence (as with 9/11, the Madrid bombs, "cyber-terror", and state-backed "computer network attack') or demodernization (as with the systematic demodernizations in Kosovo, Iraq, and Palestine). Indeed, there are signs that, in globalizing urban societies, which rely utterly and continuously on complex, multi-layered and often ignored technical systems, war *becomes* a strategy of deliberate "decyborganization" and demodernization through orchestrated assaults on everyday, networked technics (Luke 2004).

Such notions of war being literally "unleashed" from the boundaries of time and space – what Paul James has termed "metawar" (2003) – pushes a two-pronged doctrine to the centre of (particularly US) geopolitical strategy. On the one hand, this centres on the defence of everyday "critical" infrastructures in the "homeland" through improving its "resilience" to attack and manipulation and re-inscribing national, and urban, borders that were previously becoming more and more porous (Kaplan 2003). On the other, it involves the development of capabilities to systematically degrade, or at least control, the infrastructural connectivity, modernity, and geopolitical potential of the purported enemy, again, increasingly from afar. Such a strategy is, in essence:

> war in the most general possible sense; war that reaches into the tiniest details of daily life, reengineering the most basic arrangements of travel and communications in a time when everyday life, in a mobile and interconnected society, is increasingly organized around these very arrangements.
>
> (Agre 2001: 5)

The problem with such strategies, of course, is that they implicitly push for a deepening militarization of all aspects of contemporary urban societies. Everything from the design of subways, through the topology of water networks, to the thickness of aeroplane doors and the software that makes electricity systems work, becomes a site of subtle militarization. War, in this broadest sense, suggests Phil Agre (2001), becomes a continuous, distanciated event, without geographical limits, that is relaid live, 24/7, on TV and the Internet. Here domains of "war" blur into those of "peace". Instead, replacing such binaried landscapes are continuous time-spaces dominated by discourses of "security", which saturate, and militarize, the tiniest details of everyday urban life. Certainly, many US political and military elites are currently perpetuating such discourses of endless, boundless war as part of the construction of post 9/11 states of emergency and the so-called "war on terror" (Agamben 2002).

Our second conclusion is that the very real risk here is that the "securitization" of network-based urban societies against this new notion of war becomes such an overpowering obsession that it is used to legitimize a re-engineering of the everyday systems that are purportedly now so exposed to the endless, sourceless, boundless threat. There is already considerable evidence to support Agamben's view that "security thus imposes itself as the basic principle of state activity" (2002: 1). He even argues that the imperative of "security" is beginning to overwhelm other, historic functions of nation states that were built up over the nineteenth and twentieth centuries (such as social welfare, education, health, economic regulation, planning). "What used to be one among several decisive measures of public administration until the first half of the twentieth century", suggests Agamben, "now becomes the sole criterion of political legitimation" (Agamben 2002: 1).

Our final conclusion is that it is imperative that theorists and analysts of the geographies and geopolitics of contemporary warfare address the intersections of infrastructural warfare and forced demodernization with much more theoretical and empirical vigour than has thus been the case. As King and Martin suggest, "work in international relations in political science and related social science disciplines almost always ignores all but the most direct public health implications of military conflict" (2001: 2). The realities of "war on public health", and the geopolitics of state-backed efforts to ensure that entire societies endure the immiseration of what Agamben (1998) has called "bare life," tend to be overwhelmingly ignored in social, political and media analyses of war. This is because both media and analytical attention tend to turn, capriciously, to the formal, mediatized, violence, and the most obvious "collateral" casualties of the "next" war. The wider neglect of networked infrastructures in the social and political sciences compounds this systematic ignoring of state-backed infrastructural warfare (Graham and Marvin 2001).

ACKNOWLEDGEMENTS

This chapter is a shortened version of a paper "Switching cities off" that was published in *City*, 9(2), 2005. Thanks to the British Academy for support which made this research possible. Thanks also for comments from Phil Agre, Ash Amin, Zygmunt Bauman, David Campbell, and Bulent Diken on an earlier draft. The usual disclaimers apply.

NOTE

1   1389 was the year Serbia lost its medieval empire and was absorbed into the Ottoman empire – a battle that is a key event in the formation of Serbian national identity.

BIBLIOGRAPHY

Agamben, G. (1998) *Homo Sacer: Sovereign Power and Bare Life*. Stanford, CA: Stanford University Press
Agamben, G. (2002) "Security and terror," *Theory and Event*, 5(4): 1–2
Agre, P. (2001) "Imagining the next war: infrastructural warfare and the conditions of democracy," *Radical Urban Theory*, 14 September 2001, available at www.rut.com/911/Phil-Agre.html (February 2004)

Ashford M. (2000) "Closing plenary speech, IPPNW XIV World Congress, International Physicians for the Prevention of Nuclear War," Available at www.ippnw.org/Ashford Plenary.html (February 2004)

Barakat, S. (1998) "City war zones," *Urban Age*, Spring: 11–19

Bauman, Z. (2002) "Reconnaissance wars and planetary frontierland", *Theory, Culture and Society*, 19(4), 81–90

Blakeley R. (2003a) *Bomb Now, Die Later*. Bristol University: Department of Politics. Available at www.geocities.com/ruth_blakeley/bombnowdielater.htm, February 2004

Blakeley, R. (2003b) "Targeting water treatment facilities," Campaign Against Sanctions in Iraq, Discussion List, 24 January. Available at www.casi.org.uk/discuss/2003/msg 00256.html (February 2004)

Bolkcom, C. and Pike, J. (1993) *Attack Aircraft Proliferation: Issues for Concern*, Federation of American Scientists. Available at www.fas.org/spp/aircraft (February 2004)

De Cuellar, J. P. (1991) *Report S/22366 to the United Nations Security Council*. New York: UN Office of the Iraq Programme

Debrix, F. (2001) "Cyberterror and media-induced fears: the production of emergency culture," *Strategies*, 14(1): 149–167

Defense Intelligence Agency (1991) *Iraq Water Treatment Vulnerabilities*. Filename 511rept.91, Memo to Centcom, 18 January

Der Derian, J. (2001) *Virtuous War: Mapping the Military-Industrial-Media-Entertainment Complex*. Boulder, CO: Westview Press

Dillon, M. (2002) "Network society, network-centric warfare and the state of emergency," *Theory, Culture and Society*, 19(4): 71–79

Dochterman, K. (2002) "Shock and awe: media, state, and techno-science in the war against Iraq," *Aporia Journal*, available at http: //aporiajournal.tripod.com/issues.html (February 2004)

Enders, D. (2003) "Getting back on the grid," *Baghdad Bulletin*, 10 June 2003. Available at www.baghdadbulletin.com, February 2004

Felker, E. (1998) *Airpower, Chaos and Infrastructure: Lords of the Rings*. U.S. Air War College Air University, Maxwell Air Force Base, Alabama, Maxwell paper 14, July

Graham, S. (2001) "In a moment: On glocal mobilities and the terrorised city," *City*, 5(3): 411–415

Graham, S. (2003) "Lessons in urbicide", *New Left Review*, 19, Jan–Feb: 63–78

Graham, S. (ed.) (2004) *Cities, War and Terrorism: Towards an Urban Geopolitics*. Oxford: Blackwell

Graham, S. and Marvin, S. (2001) *Splintering Urbanism: Networked Infrastructure, Technological Mobilities and the Urban Condition*. London: Routledge

Hardt, M. and Negri, A. (2000) *Empire*. Harvard, MA: Harvard University Press

Harris, J. (2003) "Dreams of global hegemony and the technology of war," *Race and Class*, 45(2): 54–67

Hirst, P. (2001) *War and Power in the 21st Century*. Cambridge: Polity

Human Rights Watch, (2003) *Off Target: The Conduct of the War and Civilian Casualties in Iraq*. Washington, DC. Available at www.hrw.org (February 2004)

James, P. (2003) "The age of meta-war," *Arena Magazine*, Issue 64: 4–8

Kaika, M. and Swyngedouw, E. (2000) "Fetishising the modern city: The phantasmagoria of urban technological networks", *International Journal of Urban and Regional Research*, 24(1): 122–148

Kaplan, A. (2003) "Homeland insecurities: Reflections on language and space," *Radical History Review*, 85: 82–93

Keaney, T. and Cohen, E. (1993) *Gulf War Air Power Surveys* (GWAPS), Washington, DC: Johns Hopkins University and the US Air Force. Available at www.au.af.mil/au/awcgate/awc-hist.htm+gulf (February 2004)

King, G. and Martin, G. (2001) *The Human Costs of Military Conflict*. Overview paper for 29 September 2001 conference on *Military Conflict as a Public Health Problem*

Klein, N. (2001) "Poor services aid terrorists," *Guardian*, 26 October, 21

Kopp, C. (1996) *The Electromagnetic Bomb: A Weapon of Mass Electrical Destruction*, Military Library, Available at http://198.65.138.161/militarty/liv=braray/report/1996/apjemp.htm (February 2004)

Leslie, J. (1999) "Powerless," *Wired*, April: 119–183

Liang, Q. and Xiangsui, W. (1999) *Unrestricted Warfare*. Beijing: PLA Literature and Arts Publishing House

Little, R. (2002) "Controlling cascading failure: understanding the vulnerabilities of interconnected infrastructures," *Journal of Urban Technology*, 9(1): 109–123

Luke, T. (2003) *Postmodern Geopolitics in the 21st century: Lessons from the 9.11.01 Terrorist Attacks*, Center for Unconventional Security Affairs, Occasional Paper 2

Luke, T. (2004) "The co-existence of cyborgs, humachines and environments in postmodernity: getting over the end of nature." In S. Graham (ed.), *The Cybercities Reader*. London: Routledge, 106–110

Miller, D. (2004) "The domination effect," *Guardian*, 8 January: 24

Nagy, T. (2001) "The secret behind the sanctions: how the U.S. intentionally destroyed Iraq's water supply," *The Progressive*, 1–6 September. Available at www.progressive.org/0801issue/nagy0901.html (February 2004)

Pape, R. (1996) *Bombing to Win: Air Power and Coercion in War*. Ithaca, NY: Cornell University Press

Patterson, C. (2000) *Lights Out and Gridlock: The Impact of Urban Infrastructure Disruptions on Military Operations and Non-Combatants*. Washington, DC: Institute for Defense Analyses

Pawley, M. (1997) *Terminal Architecture*. London: Reaktion

Pax, S. (2003) "Baghdad Blogger ," 13 August 2003, *Guardian* 2: 4

Perry, D. (1995) "Introduction." In D. Perry (1995) *Building the Public City: The Politics, Governance and Finance of Public Infrastructure*. London: Sage, 1–20

Pineiro , R. (2004) *Cyberterror*. New York: Tor

Project for the New American Century (2000) *Rebuilding Americas Defenses*. Washington, DC

Rattray, G. (2001) *Strategic Warfare in Cyberspace*. Cambridge, MA: MIT Press

Rizer, K. (2001) "Bombing dual-use targets: legal, ethical, and doctrinal perspectives," *Air and Space Power Chronicles*, 5 January. Available at www.airpower.maxwell.af.mil/airchronicles/cc/Rizer.html (February 2004)

Rochlin, G. (1997) *Trapped in the Net: The Unanticipated Consequences of Computerization*. Princeton, NJ: Princeton University Press

Rowat, C. (2003) "Iraq: potential consequences of war," *Campaign Against Sanctions in Iraq Discussion List*, 8 November. Available at www.casi.org.uk/discuss/2002/msg02025.html (February 2004)

Skoric, I. (1999) "On not killing civilians," posted at amsterdam.nettime.org, 6 May

Smith, T. (2002) "The new law of war: Legitimizing hi-tech and infrastructural violence," *International Studies Quarterly*, 46: 355–374

Smith, C. (2003) "U.S. wrestles with new weapons," NewsMax.Com, 13 March. Available at www.newsmax.com/archives/articles/2003/3/134712.shtml (February 2004)

Star, S. (1999) "The ethnography of infrastructure," *American Behavioral Scientist*, 43(3): 377–391

Tahoub, T. (2003) "The war on al-Jazeera," *Guardian*, 4 October: 23

Thomas, W. (1995) *Scorched Earth: The Military's Assault on the Environment.* Philadelphia, PA: New Society

Thrift, N. and French, S. (2002) "The automatic production of space," *Transactions of the Institute of British Geographers*, 27(4) 309–335

United Nations Children's Fund (UNICEF) (1999) *Annex II of S/1999/356, Section 18.* Available at www.un.org/Depts/oip/reports (February 2004)

United States Air Force (1998) *Strategic Attack: Air Force Doctrine Document* 2–1.2, 20 May, Washington, DC

Verton, D. (2003) *Black Ice: The Invisible Threat of Cyberterrorism.* New York: Osborne McGraw Hill

Virilio, P. (2002) *Desert Screen: War at the Speed of Light.* London: Continuum

Warden, J. (1995) "The enemy as a system," *Airpower Journal*, 9 (1): 41–55

Zimmerman, R. (2001) "Social implications of infrastructure network interactions," *Journal of Urban Technology*, 8(3): 97–119

Index